THEORY OF CHARGE EXCHANGE

WILEY-INTERSCIENCE SERIES IN ATOMIC AND MOLECULAR COLLISIONAL PROCESSES

Advisory Editor

C. F. BARNETT

ION-MOLECULE REACTIONS *by E. W. McDaniel, V. Čermák, A. Dalgarno, E. E. Ferguson, L. Friedman*

THEORY OF CHARGE EXCHANGE *by Robert A. Mapleton*

DISSOCIATION IN HEAVY PARTICLE COLLISIONS *by G. W. McClure, J. M. Peek*

EXCITATION IN HEAVY PARTICLE COLLISIONS *by E. W. Thomas*

In preparation:

Ionization by Heavy Particle Collisions by J. W. Hooper, F. W. Bingham
Charge Exchange in Gases and Solids by C. F. Barnett, I. A. Sellin

NATIONAL STANDARD REFERENCE DATA SYSTEM
NSRDS

THEORY OF CHARGE EXCHANGE

ROBERT A. MAPLETON

Air Force Cambridge Research Laboratory
Bedford, Massachusetts

WILEY-INTERSCIENCE

A Division of John Wiley & Sons, Inc.
New York · London · Sydney · Toronto

Library of Congress Catalog Card Number: 75–168639

ISBN 0–471–56781–7

Printed in the United States of America.

10 9 8 7 6 5 4 3 2 1

SERIES PREFACE

The proliferation of scientific literature in the past two decades has exceeded the point where an individual scientist can either absorb or keep up with all of the publications even in narrow fields. In recognition of this situation, the Federal Council for Science and Technology, in 1963, established the National Standard Reference Data System with responsibility for its implementation assumed by the National Bureau of Standards of the Department of Commerce. In 1968, the existing authorities of the Department of Commerce to operate the NSRDS were supplemented by special legislation passed by the Congress of the United States. Within NBS, the standard reference data program is administered by the Office of Standard Reference Data which provides coordination and planning for data compilation efforts in the field of the physical sciences. As part of this effort, the Office of Standard Reference Data and the Division of Research of the United States Atomic Energy Commission have sponsored jointly the establishment of the Atomic and Molecular Processes Information Center at the Oak Ridge National Laboratory. The objectives of this Center are the collection, evaluation, and critical review of data in specialized topics of heavy particle atomic and molecular collisions. This monograph, *Theory of Charge Exchange*, is the second of a planned series and is written by R. A. Mapleton, a recognized authority in the field of atomic collision processes.

C. F. Barnett

PREFACE

The phenomenon of capture of electrons by α-particles was first observed by Henderson in 1923, and this interpretation of the experiment was validated by Rutherford a year later. The same year Fowler published the first theoretical analysis of the experiment (based on thermodynamic equilibrium) in an attempt to predict the observed variation with velocity of the ratio of concentrations, $[He^{++}]/[He^{+}]$. Thomas in 1927 developed the classical picture of electronic capture most fully; this method has been revived during the last five years. The following year Oppenheimer applied quantum mechanics to calculate approximately the first Born cross section for capture from hydrogenic atoms/ions by positively charged nuclei. Later Brinkman and Kramers evaluated these same Born cross sections exactly, but both Oppenheimer's and Brinkman and Kramers's studies were restricted to capture from and into atomic *s*-states.

This restriction and prevalent speculation about the importance of capture into other than *s*-atomic states led Saha and Basu in 1945 to use the same Born (OBK) approximation to calculate the cross section for capture from *s*-states into *p*-states for hydrogenic atoms and ions. Previously, Massey and Smith did an approximate calculation for electronic capture from helium by protons, a result that included the effect of the nucleus-nucleus (*n-n*) interaction in contrast to the previously cited Born (OBK) calculations, which did not include this interaction. It was in 1952 that Bates and Dalgarno approximately accommodated the effect of the *n-n* interaction in a first Born calculation for resonant charge transfer between protons and H(1*s*), and in the next year Jackson and Schiff succeeded in evaluating the same cross sections more accurately with the aid of Feynman integrals. A controversy developed about the correct method of including this *n-n* contribution in different approximating methods, and disagreement on this subject still persists.

These few papers form a brief history of the first two decades of accomplishment in the theoretical and calculational aspects of electronic capture from atoms; this literature now is dwarfed by contributions to this research published during the past two decades. Associated with this growth of literature, a jargon has emerged that denotes different levels of sophistication of the approximations used in both the classical and quantum mechanical disciplines. To the nonspecialist, or the newcomer eager to contribute to this field of research, the deciphering of this specialized language is a frustrating experience; determining the omissions between equations is an adequate initial challenge for most newcomers!

In 1967 C. F. Barnett, I. A. Sellin, and I decided that a need existed for a descriptive and explanatory treatment containing both the theoretical and experimental features of electronic capture. We had planned to write a single chapter on selected theoretical topics; the rest of this book was to be devoted to a collection and evaluation of measured values of cross sections for electronic capture from atoms and molecules. However, after evaluating the first draft, it was decided that this work constituted a book in itself. This book is neither a review nor a criticism of the approximating procedures contained in the literature; rather, it is a description and explanation of many of these approximations. I have attempted to present a painless introduction to the subject including explanations of the jargon that permeates the literature. Without apology I have omitted the methods of projection operators stemming from techniques developed by Feshbach; I have also omitted Fadeev's perturbation theory, which was reviewed by Bransden in 1965.

The first four chapters of this book (excluding one section) are devoted to quantum mechanical and semiclassical approximatimg methods. Both the wave and impact-parameter versions are discussed and explained. Particular attention is given to the coordinate systems used in these two versions. Chapter 5 covers classical methods that have been developed, and revived, during the last decade. Asymptotic variations of cross sections obtained by classical and quantum mechanical methods are compared. Twenty-six figures comparing calculated and experimentally determined cross sections are presented in Chapter 6 with associated discussions of what is calculated and what is measured. Of the three appendices, one is wholly devoted to the application of fractional parentage coefficients to problems of capture from many-electron atoms.

During the preparation of this book, I frequently sought the advice of, and was assisted by, C. F. Barnett of the Oak Ridge National Laboratory. For these favors I express my deep appreciation and thanks to him. To Mrs. Donna Cobble I express my thanks for her help with the second draft of the manuscript. I am particularly grateful to Mrs. Elizabeth Carter, who oversaw the completion of the second draft, the final revisions, and the monumental task of inserting corrections to the numerous equations and collecting the references. Without this collective assistance the preparation of this book would have been a difficult task. Finally, I express my appreciation and thanks to Professor F. J. Smith of the Queen's University of Belfast for reviewing the final manuscript.

ROBERT A. MAPLETON

Bedford, Massachusetts
May 1971

CONTENTS

THEORY OF CHARGE EXCHANGE

INTRODUCTION

The main purpose of this book is to provide a description and an explanation of the approximating methods that are used to calculate cross sections and probabilities for electronic capture. A secondary purpose is to compare different approximate predictions with each other and with experimentally determined values. These comparisons should enable us to determine which approximating procedures are most successful in predicting cross sections for different ranges of the initial relative velocity of the colliding aggregates; they also should indicate which methods show most promise for additional improvement. In connection with the main purpose, however, the descriptive and explanatory discussion is designed for the nonspecialist who wants to obtain knowledge of this subject quickly and painlessly. Without this type of aid, the nonspecialist is confronted with the laborious and frustrating task of deciphering the specialized language, the jargon, which is commonplace to the periodical literature on electronic capture.

This specialized language in the literature (as in most disciplines) originated chiefly from those pioneers who, without the aid of large-scale computers, had to rely upon their physical intuition to develop approximate equations to the very complicated exact mathematical equations. Different approximate equations are needed for different ranges of the energy of relative translation of the colliding aggregates, but usually these early research specialists had to be satisfied with different approximate solutions to the approximate equations. These different solutions were labeled with suggestive names, and the *jargon* was thus born. With the use of the present large-scale computers, however, exact numerical solutions to some of the approximate equations are feasible, and consequently some of the jargon is already vanishing. The literature will continue to be permeated by some of these descriptive terms, nevertheless, until the arrival of the computer that is capable of handling the scattering solutions to the Schrödinger equation. Until this speculated computer becomes available, the interpretation of these terms will be helpful to the nonspecialist and the newcomer to this field of research. In order to achieve these goals particular attention will be devoted to coordinate systems.

Different coordinate systems are advantageous to different approximating procedures; for example, there frequently is a benefit in using molecular orbitals for low velocities of relative translation in contrast to the gain obtained by using atomic orbitals at high velocities. In connection with the semiclassical method of impact parameters in which approximation the nuclei are often assumed to move rectilinearly, different sets of canonically conjugate systems of coordinates are related to one another by the unitary transformation that provides the correct transfer of translational motion to the active electron. (There is the implication of distinguishable electrons here, but this apparent contradiction is explained elsewhere in this book.) A particular advantage of two of these coordinate systems is that they provide the natural choice for the correct description of the initial and final states of the electronic capturing process. Although the coordinate systems used do not comprise the central theme of this book, we cannot understand the mathematical analysis of the scattering problems without knowing which variables are independent. Consequently, this feature of the explanations will be mentioned frequently in connection with differentiating and integrating operations. Since this book is concerned primarily with the techniques of calculation, it is appropriate here to mention the two books consulted most frequently: *The Theory of Atomic Collisions* by N. F. Mott and H. S. W. Massey,[1] and Chapter 14 by D. R. Bates in *Atomic and Molecular Processes.*[2] Other books on scattering theory and related topics are cited as needed. The periodical literature, however, comprises the chief source of references for this book.

The following notation is used in the mathematical description of the process;

$$\mathrm{P}^{f+}(M_f) + \mathrm{A}^{(i-1)+}(M_{i1}) \rightarrow \mathrm{P}^{(f-1)+}(M_{f1}) + \mathrm{A}^{i+}(M_i),$$
$$f = Z_f - n_p + 1, \quad i = Z_i - n_a + 1. \quad (1)$$

Equation 1 represents the capture of an electron by the incident ion (projectile), P^{f+}, from the atomic, or ionic, target, $\mathrm{A}^{(i-1)}$. (The symbol + occurring in the superscripts signifies that the resultant charge on the ion, or atom, is non-negative.) These two initial aggregates constitute the initial state of the system; the final state of the system consists of the projectile with its newly acquired electron in a bound state of the atom or ion and the other aggregate, the residual ion. Z_f^+ and Z_i^+ denote the respective nuclear charges of P and A; the number of electrons in P^{f+} and $\mathrm{A}^{i(-1)+}$ are $(n_p - 1)$ and n_a, respectively, and the number in $\mathrm{P}^{(f-1)+}$ and A^{i+} are n_p and $(n_a - 1)$, respectively. Usually the resultant initial charges of P and A are +1 and 0, respectively; however, the notation used is designed to include all cases of single electronic capture. It is pertinent to remark here that only processes typified by Eq. 1 are described mathematically in this book; although at places in the book comparisons are made between calculated cross sections for electronic capture from atoms and measured values for electronic capture by the same ion from the corresponding homonuclear diatomic molecules. The masses of the aggregates are

$$M_f = M_p + (n_p - 1)m, \qquad M_{f1} = M_f + m$$

and

$$M_i = M_a + (n_a - 1)m, \qquad M_{i1} = M_i + m,$$

in which m is the electronic mass and M_p and M_a denote the masses of the incident nucleus and atomic nucleus, respectively. The chemical scale O = 16 is used.

Other atomic parameters are: e, the electronic charge; $\hbar$, Planck's constant divided by 2π; M, the mass of the proton. From these definitions the derived quantities that constitute the atomic units (abbreviated a.u.) are: the unit of length (first Bohr orbit), $a_0 = \hbar^2/me^2$; the unit of velocity, $v_0 = e^2/\hbar$; the unit of energy, $2\epsilon_1 = me^4/\hbar^2$. An equation or physical quantity is expressed in terms of a.u. by equating e, m, and $\hbar$ each equal to unity.[3] Often it is convenient to use ε_1, the Rydberg, as the unit of energy, but this choice is always clearly stated.

The wave functions representing the initial and final states on the left- and right-hand sides of Eq. 1 are Φ_i and Φ_f, whereas the associated atomic states have wave functions ϕ_i and ϕ_f that generally represent the product of two atomic states. For the purpose of avoiding misinterpretation, note that the symbolic representation of Eq. 1 is precisely representative of the physical situation only if the corresponding aggregates are separated by large (infinite) distances. The symbol ψ, or Ψ, represents the wave function of the system, but specialized notation required for particular representations (expansions) of ψ (or Ψ) is introduced as needed. According to contemporary usage for the description of one-electron atomic states, the relevant quantum numbers are: n, principal; l, orbital angular momentum; m_l, magnetic quantum number or one of the $(2l + 1)$ components of l; s, spin; and m_s, one of the $(2s + 1)$ components of s. The additional labels needed to describe atoms containing more than one electron are defined as needed. If the atomic target is initially at rest in the observer's frame of reference, the laboratory system, the velocity of the incident aggregate is called the *impact velocity*, and the associated energy, the *impact energy*. Moreover, the impact velocity equals the initial velocity of relative translation. Only reactions occurring at non-relativistic velocities are analyzed in this book; as a practical rule this means that the impact energy of the incident aggregate must not exceed (roughly) one-tenth of its rest energy. The initial and final velocities of relative translation are denoted by v_i and v_f, respectively. Since some of the subsequent notation represents physical quantities that are inextricably connected with the motion of the center of mass of the system, this is a convenient place to define several systems of relative coordinates commonly used.

The motion of the center of mass is assumed to have been separated from the motion of relative translation, and only the coordinates used for the latter of the two motions are shown. In the subsequent relations the symbol μ, usually with identifying subscripts, denotes a reduced mass, and M_t always denotes the total mass. The first systems displayed are the prior and post coordinates that provide natural relative coordinates for the specification of the initial and final states and the scattering boundary conditions.[4,5] (As noted in Eq. 4, the Hamiltonians are diagonal in these *natural* coordinates.) For the process described by Eq. 1, the prior system labeled i and the post system labeled f are as follows.

i-System

$$\mathbf{x}_k = \mathbf{r}_k - \mathbf{r}_a, \qquad k = 1, 2, \ldots, n_a,$$
$$\mathbf{y}_j = \mathbf{r}'_j - \mathbf{r}_p, \qquad j = 2, 3, \ldots, n_p,$$

$$\mathbf{R}_i = \frac{M_p\mathbf{r}_p + m\sum_{j=2}^{n_p}\mathbf{r}'_j}{M_f} - \frac{M_a\mathbf{r}_a + m\sum_{k=1}^{n_a}\mathbf{r}_k}{M_{i1}}, \tag{2i}$$

$$M_f = M_p + (n_p - 1)m, \qquad M_{i1} = M_a + n_a m.$$

f-System

$$\mathbf{r}'_1 = \mathbf{r}_1, \qquad \mathbf{x}_k = \mathbf{r}_k - \mathbf{r}_a, \qquad k = 2, 3, \ldots, n_a,$$
$$\mathbf{y}_j = \mathbf{r}'_j - \mathbf{r}_p, \qquad j = 1, 2, \ldots, n_p,$$

$$\mathbf{R}_f = \frac{M_p\mathbf{r}_p + m\sum_{j=1}^{n_p}\mathbf{r}'_j}{M_{f1}} - \frac{M_a\mathbf{r}_a + m\sum_{k=2}^{n_a}\mathbf{r}_k}{M_i}, \tag{2f}$$

$$M_{f1} = M_p + n_p m, \qquad M_i = M_a + (n_a - 1)m.$$

The positions $\mathbf{r}_a$ and $\mathbf{r}_p$ of the respective nuclei A and P, and the electronic positions $\mathbf{r}_k$ and $\mathbf{r}'_j$, all are measured from a common origin fixed with respect to the observer's system. The $\mathbf{r}_k$ and $\mathbf{r}'_j$ refer to the electrons associated with the nuclei A and P, respectively. Although other relative coordinates can be used for the electrons in the prior and post systems, the relative vectors between the centers of masses of the respective initial and final aggregates $\mathbf{R}_i$ and $\mathbf{R}_f$ are always the same, and it is standard practice to denote the corresponding nonrelativistic Schrödinger equations by Eq. 3:[6,7]

$$\begin{aligned} H\Psi &= (H_i + V_i)\Psi = E\Psi, & H_i\Phi_i &= E\Phi_i, \\ H\Psi &= (H_f + V_f)\Psi = E\Psi, & H_f\Phi_f &= E\Phi_f. \end{aligned} \tag{3}$$

In Eq. 3 E is the sum of the internal energies and the energy of relative translation; V_i and V_f are the prior and post interactions, respectively. In the special case when the projectile of Eq. 1 is a positively charged nucleus, the special notation $\mathbf{x}_1 = \mathbf{x}_i$, $\mathbf{y}_1 = \mathbf{x}_f$ (cf. Eqs. 2i and 2f) is used in order to have a standard notation for this class of electronic capturing processes.

Since the prior and post coordinates are natural coordinates for the specification of scattering boundary conditions, they are also the natural system to use with the integral equation for scattering. This means that the inhomogeneous term and the Green's function (kernel) of the integral equation are expressed most simply in terms of the relative coordinates of Eqs. 2i and 2f, or equivalently, the equation is diagonal in these coordinates.[6,7] Due to lack of large-scale computational facilities, most of the calculational analysis has been restricted to systems containing only one electron. Thus systems of relative coordinates and corresponding Hamiltonians are given for this special case:

$$\mathbf{x}_i = \mathbf{r}_1 - \mathbf{r}_a, \qquad \mathbf{x}_f = \mathbf{r}_1 - \mathbf{r}_p,$$

$$\mathbf{R}_i = \mathbf{r}_p - \frac{M_a\mathbf{r}_a + m\mathbf{r}_1}{M_{a1}},$$

$$\mathbf{R}_f = \frac{M_p\mathbf{r}_p + m\mathbf{r}_1}{M_{p1}} - \mathbf{r}_a, \qquad \mathbf{R} = \mathbf{x}_i - \mathbf{x}_f,$$

$$M_t = M_a + M_p + m, \qquad \mu_i = \frac{M_pM_{a1}}{M_t}, \qquad \mu_1 = \frac{mM_a}{M_{a1}},$$

$$\mu_f = \frac{M_{p1}M_a}{M_t}, \qquad \mu_2 = \frac{mM_p}{M_{p1}},$$

$$H = -\frac{\hbar^2}{2\mu_i}\nabla_{\mathbf{R}_i}{}^2 - \frac{\hbar^2}{2\mu_1}\nabla_{\mathbf{x}_i}{}^2 + V(\mathbf{x}_i) + V_i(\mathbf{x}_i, \mathbf{R}_i) = H_i + V_i, \tag{4a}$$

$$H = -\frac{\hbar^2}{2\mu_f}\nabla_{\mathbf{R}_f}{}^2 - \frac{\hbar^2}{2\mu_2}\nabla_{\mathbf{x}_f}{}^2 + V(\mathbf{x}_f) + V_f(\mathbf{x}_f, R_f) = H_f + V_f, \tag{4b}$$

$$V = V(\mathbf{x}_i) + V_i(\mathbf{x}_i, \mathbf{R}_i) = V(\mathbf{x}_f) + V_f(\mathbf{x}_f, \mathbf{R}_f), \tag{4c}$$

$$\mathbf{x} = \mathbf{r}_1 - \frac{M_p\mathbf{r}_p + M_ar_a}{M_t - m}, \qquad \mathbf{R} = \mathbf{r}_p - r_a,$$

$$\mu = \frac{M_aM_p}{M_t - m}, \qquad \mu_1 = \frac{m(M_t - m)}{M_t},$$

$$H = -\frac{\hbar^2}{2\mu}\nabla_{\mathbf{R}}{}^2 - \frac{\hbar^2}{2\mu_1}\nabla_{\mathbf{x}}{}^2 + V, \tag{5}$$

$$\mathbf{s} = \mathbf{r}_1 - \frac{\mathbf{r}_p + \mathbf{r}_a}{2}, \qquad \mathbf{R} = \mathbf{r}_p - \mathbf{r}_a, \qquad \mu = \frac{M_a M_p}{M_t - m},$$

$$\mu_1 = \frac{4m\mu}{4\mu + m}, \qquad \mu_2 = \frac{M_a M_p}{M_p - M_a},$$

$$H = -\frac{\hbar^2}{2\mu}\nabla_{\mathbf{R}}^2 - \frac{\hbar^2}{2\mu_1}\nabla_{\mathbf{s}}^2 - \frac{\hbar^2}{2\mu_2}\nabla_{\mathbf{s}}\cdot\nabla_{\mathbf{R}} + V, \tag{6}$$

$$\mathbf{R} = \mathbf{r}_p - \mathbf{r}_a, \qquad \mathbf{x}_i = \mathbf{r}_1 - \mathbf{r}_a, \qquad \mu = \frac{M_a M_p}{M_t - M}, \qquad \mu_1 = \frac{mM_a}{M_{a1}},$$

$$H = -\frac{\hbar^2}{2\mu}\nabla_{\mathbf{R}}^2 - \frac{\hbar^2}{2\mu_1}\nabla_{\mathbf{x}_i}^2 - \frac{\hbar^2}{M_a}\nabla_{\mathbf{R}}\cdot\nabla_{\mathbf{x}_i} + V, \tag{7}$$

$$\mathbf{R} = \mathbf{r}_p - \mathbf{r}_a, \qquad \mathbf{x}_f = \mathbf{r}_1 - \mathbf{r}_p, \qquad \mu = \frac{M_a M_p}{M_t - m}, \qquad \mu_2 = \frac{mM_p}{M_{p1}},$$

$$H = -\frac{\hbar^2}{2\mu}\nabla_{\mathbf{R}}^2 - \frac{\hbar^2}{2\mu_2}\nabla_{\mathbf{x}_f}^2 - \frac{\hbar^2}{M_p}\nabla_{\mathbf{R}}\cdot\nabla_{\mathbf{x}_f} + V. \tag{8}$$

These are the forms assumed by the Schrödinger equation in the center-of-mass system if expressed in the given systems of relative coordinates. These equations all represent the motion of two nuclei and one electron, the electron being bound to one of the nuclei. It is appropriate at this point to describe the notation and units commonly used to express the conservation of energy in the wave mechanical version of electronic capture.

In this analysis each of the aggregates of the initial and final states is treated as a single particle. With the use of this simplification and the classical laws of conservation of energy and linear momentum, it is a straightforward task to relate the kinetic energies of the incident ion E_p, the atomic target E_a, and the energy of relative translation E_μ. This relation is conveniently described as

$$\frac{E_\mu}{E_i} = \frac{M_a - M_p}{M_t}\frac{E_p}{E_i} + \frac{M_p}{M_t} - 2\left[\frac{\mu}{M_t}\frac{E_p}{E_i}\left(1 - \frac{E_p}{E_i}\right)\right]^{1/2}\cos\theta_{ia},$$

$$E_i = \frac{M_p}{2}v_i^2 + \frac{M_a}{2}v_a^2 = \frac{\mu}{2}(\mathbf{v}_i - \mathbf{v}_a)^2 + \frac{M_t}{2}v^2, \qquad \mathbf{v} = \frac{M_p\mathbf{v}_i + M_a\mathbf{v}_a}{M_t}. \tag{9a}$$

In Eq. 9*a* θ_{ia} is the angle between the directions of $\mathbf{v}_i$ and $\mathbf{v}_a$, which correspond, respectively, to E_p and E_a.

These relations are specialized next to the situation commonly encountered in practice; that is, the initial velocity $\mathbf{v}_a$ and associated kinetic energy E_a are both zero. This means that $\mathbf{v}_i$ and E_i are the respective impact velocity and impact energy. Equation 9*a* now simplifies to Eq. 9*b* (in the notation of Eq. 1):

$$E_i = \frac{M_f}{2} v_i^2 = \frac{\mu_i}{2} v_i^2 + \frac{M_f}{2M_t} v_i^2, \qquad \mu_i = \frac{M_f M_{i1}}{M_t}. \tag{9b}$$

This relates E_i before the collision to the initial energy of relative translation and to the translational energy of the center of mass.[8] Since the center of mass moves uniformly, only the energy of relative translation is available for transfer of internal energy; this is expressed as[9]

$$E = \frac{\mu_i}{2} v_i^2 - \varepsilon_i = \frac{\mu_f}{2} v_f^2 - \varepsilon_f, \tag{10a}$$

or in terms of E_i as

$$\frac{\mu_i}{M_f} E_i - \varepsilon_i = \frac{\mu_f}{2} v_f^2 - \varepsilon_f. \tag{10b}$$

Next both sides of Eq. 10*b* are divided by ε_1 to get

$$\frac{\mu_i}{M_f} E_i - \varepsilon_i = \frac{\mu_f}{m} v_f^2 - \varepsilon_f, \tag{10c}$$

where it is understood that E_i, ε_i, and ε_f are in units of ε_1, and v_f is in units of $v_0 = e^2/\hbar$. If a.u. had been used, however, the unit of energy would have been $2\varepsilon_1$ and the factor $\frac{1}{2}$ would be needed in the first term on the right-hand side of Eq. 10*c*.

Other units that are frequently used with E_i are given approximately by

$$\frac{E_i(eV)}{\varepsilon_1} \simeq \frac{M}{m} E_i(25 \text{ keV}) = \frac{4M}{m} E_i(100 \text{ keV}), \qquad \varepsilon_1 = \frac{mv_0^2}{2} \tag{10d}$$

In the case when the incident ion is a proton, it is evident that

$$E_i(100 \text{ keV}) = \frac{m}{4M}\left[\frac{(M/2)v_i^2}{mv_0^2/2}\right] = \left(\frac{v_i}{2v_0}\right)^2,$$

which quantity depends upon only the velocities.[9] A feature of electronic capture that depends upon the smallness of the quantity $\Delta\varepsilon/E_i$, $\Delta\varepsilon = \varepsilon_i - \varepsilon_f$, is the angle of recoil of the residual ion.[10]

This information is readily derived from the classical laws of conservation of momentum and energy.[11] The relations among the angles and velocities are illustrated in Fig. 1. A particle of impact energy E_i is incident upon another particle, 2, initially at rest in the laboratory system. (This statement is somewhat redundant by virtue of the definition of E_i.) The bars represent corresponding quantities in the center-of-mass system. The two particles are infinitely separated initially and finally.

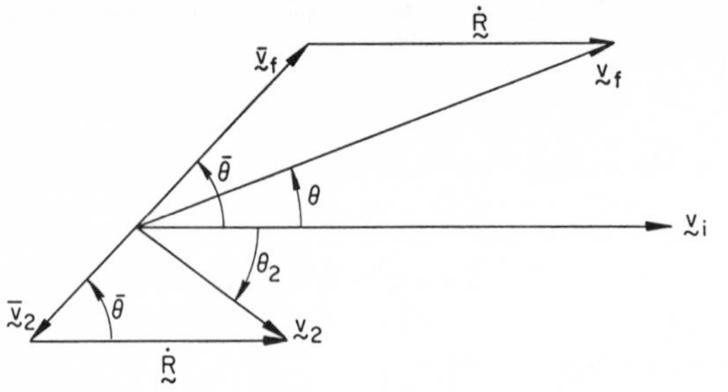

Fig. 1 Relations among velocities in the laboratory and center-of-mass systems.

E_i = impact energy of 1;
E_f = final energy of 1;
E_2 = translational energy of 2;
ε = energy transferred to internal motion of 2;
θ = angle of scattering;
θ_2 = angle of recoil;
$\dot{\mathbf{R}}$ = velocity of center of mass;
$\mathbf{v}_i$ = initial velocity of incident particle;
$\mathbf{v}_f$ = final velocity of incident particle;
$\mathbf{v}_2$ = velocity of recoil particle;
$\mathbf{v}$ = final relative velocity.

From conservation of linear momentum we obtain the relations

$$\dot{\mathbf{R}} = \frac{M_1}{M_t}\mathbf{v}_i, \qquad \bar{\mathbf{v}} = \bar{\mathbf{v}}_f - \bar{\mathbf{v}}_2 = \frac{M_t}{M_2}\bar{\mathbf{v}}_f,$$

and from conservation of energy we obtain

$$E_i = \frac{M_1}{M_t}E_i + \frac{M_2}{M_t}E_i = \frac{M_1}{M_t}E_i + \bar{E} + \epsilon, \qquad \bar{E} = \frac{\mu}{2}\bar{v}^2.$$

From the last two relations the solution for $\bar{v}_f$ is easily found to be

$$\bar{v}_f = \left[\frac{2M_2}{M_1M_t}\left(\frac{M_2}{M_t}E_i - \epsilon\right)\right]^{1/2}.$$

From the geometry of Fig. 1 and this relation, θ_2 can be calculated as follows:

$$\tan\theta_2 = \frac{\bar{v}_2\sin\bar{\theta}}{\dot{R} - \bar{v}_2\cos\bar{\theta}} = \frac{\sin\bar{\theta}}{(\dot{R}/\bar{v}_2) - \cos\bar{\theta}},$$

$$\frac{\dot{R}}{\bar{v}_2} = \frac{(M_2/M_t)(2E_i/M_1)^{1/2}}{\{(2M_2/M_1M_t)[(M_2/M_t)E_i - \varepsilon]\}^{1/2}}.$$

If $M_2E_i/M_t \gg \varepsilon$, an approximation to θ_2 is $\theta_2 \approx \pi/2 - \bar{\theta}/2$, and if $\bar{\theta}$ is small, $\theta_2 \approx \pi/2$. From Fig. 1 the solution for θ is given by

$$\tan\theta = \frac{\bar{v}_f \sin\bar{\theta}}{\dot{R} + \bar{v}_f \cos\bar{\theta}} = \frac{\sin\bar{\theta}}{\dot{R}/\bar{v}_f + \cos\bar{\theta}}, \qquad \frac{\dot{R}}{\bar{v}_f} \approx \frac{M_1}{M_2}.$$

Obviously θ is small if $\bar{\theta}$ is small. Although the last result is apparent, the first result requires the inequality between energies, and this is one of the points of the demonstration. Moreover, this deflection of the residual particles is observed experimentally.[12] If the masses of the particles are equal, another interesting possibility is $\bar{\theta} \approx \pi$, $\theta \approx \pi/2$, $\theta_2 \approx 0$. An approximate solution to a classical model for electronic capture has been obtained for these angular conditions, but we leave this problem to the section on classical methods and go on to discuss Wigner's rule for conservation of total spin.[13]

Although exact details of representations and coupling schemes for angular momenta must be specified in order to discuss this subject, Wigner's rule is mentioned here because it is implicit in the wave mechanical analysis of this book. Generally there are forces dependent upon spin between pairs of particles constituting the aggregates of Eq. 1, but cross sections originating from these forces usually are negligible relative to cross sections depending upon the corresponding electrostatic forces at nonrelativistic energies of relative translation. In order to use this fact the two aggregates constituting the initial and final states are first coupled using the rules for coupling two angular momenta in spin.[14] Wigner's rule, applied to these coupled aggregates, states that only those final states are strongly allowed if the corresponding final total spin is contained in the range of the initial total spin according to the rules for coupling two angular momenta.[14] This rule is applied in a later section where capture from complex atoms is discussed. We turn now to methods used at low velocities of relative translation.

REFERENCES

1. N. F. Mott and H. S. W. Massey, *Theory of Atomic Collisions* (3rd ed.), Oxford University Press, London (1965).
2. D. R. Bates, "Theoretical Treatment of Collisions between Atomic Systems," in *Atomic and Molecular Processes* (D. R. Bates, Ed.), Academic Press, New York (1962), Chapter 14.
3. H. A. Bethe and E. E. Salpeter, *Quantum Mechanics of One- and Two-Electron Atoms*, Academic Press, New York (1957), pp. 2–4.
4. Mott and Massey, *op. cit.*, pp. 421, 425.
5. A. Messiah, *Quantum Mechanics*, Vol. 1 (G. M. Temmer, Ed.), (1961); reprint ed., North-Holland, Amsterdam, and Interscience, New York (1963), pp. 361–366.

6. E. Gerjuoy, *Phys. Rev.*, **109**, 1806 (1958).
7. E. Gerjuoy, *Ann. Phys.* (*N. Y.*), **5**, 58 (1958).
8. Messiah, *op. cit.*, pp. 380–381.
9. J. D. Jackson and H. Schiff, *Phys. Rev.*, **89**, 359 (1953).
10. H. S. W. Massey and E. H. S. Burhop, *Electronic and Ionic Impact Phenomena*, Oxford University Press, London (1950), pp. 479–480; 482–483.
11. H. Goldstein, *Classical Mechanics*, Addison-Wesley, Reading, Mass. (1950), pp. 85–89.
12. Massey and Burhop, *op. cit.*, pp. 479–480.
13. *Ibid.*, pp. 426–427.
14. E. U. Condon and G. H. Shortley, *Theory of Atomic Spectra* (1935); reprint corr. ed., Cambridge University Press, New York (1953), Chapter 3.

CHAPTER 1

QUANTUM MECHANICAL METHODS FOR LOW VELOCITIES

This chapter is concerned with approximating procedures used to obtain cross sections at low velocities of relative translation. Low velocity usually signifies here a velocity less than $v_0 = e^2/\hbar = 2.188 \times 10^8$ cm/sec, although no exact value is defined. The impact energy of a proton is approximately 25 keV if its velocity is equal to v_0, and calculations using the first Born approximation, an approximation for high velocities, usually give unreliable predictions for lower velocities. Perhaps a more meaningful upper limit to low velocities is the velocity of the atomic electron in the outermost subshell of the atom from which capture occurs.[1] No definition of intermediate velocity is suggested in this book. The first topic treated in this chapter is the impact-parameter method, or version, as it is frequently called.

1.1 IMPACT-PARAMETER (IP) METHOD

Although the impact-parameter approximating technique is not restricted to use at low relative velocities, it is used so extensively in the mathematical analysis of electronic capture that an explanation of this method is given.

From classical mechanics we know that the impact parameter ρ and solid angle $d\Omega$ satisfy the equation

$$\rho\, d\rho\, d\phi = \rho(\Omega)\frac{d(\Omega)}{d\Omega}\, d\Omega, \tag{11}$$

and the relation between ρ(IP) and scattering angle is determined from the solutions to orbital equations for a given law of force.[2,3] Since the impact parameter is the perpendicular distance between a line through the center of force and the parallel line coincident with the initial position and direction of the incident particle, it is also clear that ρ is related to the classical angular momentum; this fact is useful in establishing relations between solutions using the impact-parameter and the wave versions.

It is assumed here that the relation between ρ and scattering angle is unique.[4] With the use of this assumption, Eq. 11 can be interpreted as a statement of conservation of particles: the number of particles passing normally through the area $\rho\, d\rho\, d\phi$ before scattering is equal to the number emerging from $d\Omega(\phi, \theta)$ after the scattering. The IP also occurs in the expression for the radial velocity,

$$\frac{dR}{dt} = v_i\left(1 - \frac{2U(R)}{\mu v_i^2} - \frac{\rho^2}{R^2}\right)^{1/2} = \left(v(R)^2 - \frac{\rho^2 v_i^2}{R^2}\right)^{1/2}, v(R)^2 = \frac{2}{\mu}[E - U(R)], \tag{12}$$

in which E is the energy of relative translation.[5] In the important case of electronic capture, the two aggregates are relatively massive (the mass of each aggregate being not less than the mass of the proton), and it is common practice to assume that the motion of these particles is classical. Furthermore, the usually small classical deflection of the two aggregates is commonly neglected, and in this approximation $\mathbf{v}_i$ and $\mathbf{v}(R)$ of Eq. 12 are equal.

In Fig. 2 this rectilinear motion is illustrated for the usual case in which the impact velocity is parallel to the Z-axis of the observer's frame of reference and equal to $\mathbf{v}_i$ in this frame; the center of mass C moves with the velocity shown, and in this center-of-mass system with C as the origin, the two velocities are

$$\mathbf{v}_p = \frac{M_{i1}}{M_t}\mathbf{v}_i \qquad \text{and} \qquad \mathbf{v}_a = -\frac{M_f}{M_t}\mathbf{v}_i.$$

These choices makes $\mathbf{v}_a$ vanish in the observer's frame of reference. As a result of these approximations the quantum-mechanical equations of motion, such as the Schrödinger equations (Eqs. 4–8), no longer contain the operator of kinetic energy of relative motion but contain only the electronic operator

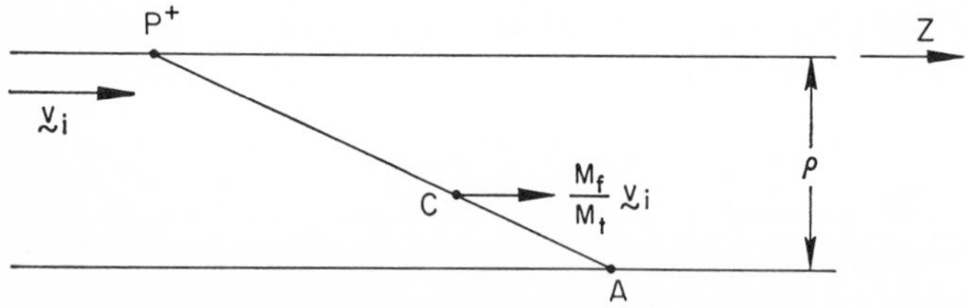

Fig. 2 Diagram of rectilinear motion assumed in the impact-parameter method. Motion of ion P⁺ and center-of mass C is relative to atomic target A at origin.

of kinetic energy and the interactions; moreover, the internuclear distance is a function of time and ρ, and the scalar operator E is replaced by $i\hbar(\partial/\partial t)$.

The Schrödinger equation, which is independent of time, has consequently been replaced by an equation that depends upon the time through the electrostatic interactions as well as the operator, $i\hbar(\partial/\partial t)$. Moreover, in order that the initial and final states satisfy this electronic equation, there must be allowance for motion of the electronic origin since the position of the electron is referred to several origins on the moving internuclear line. The electron, for example, is in motion with respect to an observer on C (see Fig. 2) if at rest with respect to an observer on A; hence this allowance for different origins on the internuclear line is described as "taking account of the translational motion of the separated atoms."[7] (This technique, incidentally, was originally correctly applied to a Born approximation in the pioneering paper by Brinkman and Kramers.[8]) In order to allow for these different electronic origins, attention is directed next to the required modification of the electronic wave function of the active electron. (The electron that is being captured in a theory of distinguishable electrons is called the active electron, and the remaining ones are labeled *passive electrons*.[9]) Symmetrization is discussed at another place in this chapter. This derivation is used since it illustrates the connection between a unitary transformation and Galilean transformation.

The point of departure is the expression

$$\delta\xi = -i\delta\alpha[\Theta, \xi], \tag{13a}$$

which relates the change in the dynamical variable ξ to a continuous parameter α and to a Hermitian operator Θ.[10] In the present application we put $\xi = p_x$, the x-component of linear momentum, and $\delta\alpha = -\delta p_x$ first, since our goal is the transformation that relates the linear momentum of the electron to a frame of reference in uniform motion relative to the atomic target. This choice of ξ and $\delta\alpha$ together with the commutative properties of p_x and x are used in Eq. 13*a* to get

$$\Theta_x = -\frac{x}{\hbar}, \qquad [\Theta_x, p_x] = -i. \tag{13b}$$

The infinitesimal operator of this transformation is

$$T(\delta\alpha) = 1 - i\Theta_x\delta\alpha = 1 + i\frac{x}{\hbar}\,\delta\alpha, \tag{13c}$$

and this last result is readily generalized to three dimensions.[11] Associated with this infinitesimal transformation is the finite transformation

$$T(m\mathbf{u}) = \exp\left(\frac{i}{\hbar}\,m\mathbf{u}\cdot\mathbf{r}\right), \tag{13d}$$

and this is the unitary transformation induced by the displacement in linear momentum

$$\mathbf{P} \to \mathbf{P} - m\mathbf{u}. \tag{13e}$$

(Here $\mathbf{u}$ is the referenced velocity between the moving frame of reference and the atomic target.)

The operator desired, however, represents $\mathbf{P} \to \mathbf{P} + m\mathbf{u}$, and this operator is

$$\begin{aligned} U_2 &= \exp\left[-\frac{i}{\hbar}\left(m\mathbf{u}\cdot\mathbf{x} + \frac{mu^2}{2}t\right)\right] \\ \mathbf{x} &= \mathbf{x}_i - \mathbf{u}t_i - a\boldsymbol{\rho}, \qquad \mathbf{u} = a\mathbf{v}_i, \\ t &= t_i, \qquad \mathbf{u}\cdot\boldsymbol{\rho} = 0, \qquad |a| \le 1, \\ \mathbf{P} &= \mathbf{P}_i - m\mathbf{u}. \end{aligned} \tag{14a}$$

[Here a is a parameter that defines the origin of $\mathbf{x}$ on the internuclear line (cf. Fig. 2).]

It is instructive here to transform to momentum space. Define

$$\Psi(\mathbf{x}_i, t_i) = (2\pi\hbar)^{-3/2}\int d\mathbf{P}' \exp\left(\frac{i\mathbf{P}'\cdot\mathbf{x}_i}{\hbar}\right)\Phi(\mathbf{P}', t_i), \tag{14b}$$

and neglect interactions for the present so that

$$i\hbar\frac{\partial\Phi(\mathbf{P}', t_i)}{\partial t_i} = \frac{(P')^2}{2m}\Phi(\mathbf{P}', t_i). \tag{14c}$$

Next, the corresponding transform of $U_2\psi(\mathbf{x}_i, t_i)$ is calculated using the definition in Eq. 14b:

$$\begin{aligned} F[U_2\psi(\mathbf{x}_i, t_i)] &= (2\pi\hbar)^{-3}\int d\mathbf{x}\exp\left(\frac{-i\mathbf{p}\cdot\mathbf{x}}{\hbar}\right)\exp\left[\frac{-i}{\hbar}\left(m\mathbf{u}\cdot\mathbf{x} + \frac{mu^2}{2}t\right)\right] \\ &\quad \times d\mathbf{P}'\exp\left[\frac{i}{\hbar}\mathbf{P}'\cdot(\mathbf{x} + \mathbf{u}t_i + a\boldsymbol{\rho})\right]\Phi(\mathbf{P}', t_i) \qquad (14d) \\ &= \exp\left\{\frac{i}{\hbar}\left[(\mathbf{P} + m\mathbf{u})\cdot\mathbf{u}t_i + a\mathbf{P}'\cdot\boldsymbol{\rho} - \frac{mu^2}{2}t\right]\right\}\Phi(\mathbf{P} + m\mathbf{u}, t_i). \end{aligned}$$

Equation 14c is used to establish that

$$\begin{aligned} i\hbar\frac{\partial\Phi'}{\partial t}(\mathbf{P}, t) &= \frac{P^2}{2m}\Phi'(\mathbf{P}, t), \\ \Phi'(\mathbf{P}, t) &= F[U_2\psi(\mathbf{x}_i, t_i)]. \end{aligned} \tag{14e}$$

The last result shows that the momentum $\mathbf{P}_i$ canonically conjugate to $\mathbf{x}_i$ has been replaced by $\mathbf{P}(\mathbf{P} = \mathbf{P}_i - m\mathbf{u})$ conjugate to $\mathbf{x}$. The factor $\exp[(i/\hbar)\mathbf{P}\cdot\boldsymbol{\rho}]$

containing the impact parameter is unimportant in this illustrative calculation, but this factor is canceled by another phase factor in an alternative derivation that utilizes the transformation to momentum space at the outset. The Galilean transformation is effected next, and U_2 emerges as the correct factor after transforming back to positional space.[12,13] The factor representing the translational energy of the electron (or a displacement in time) emerges from the alternative derivation, and this term is clearly necessary, as the subsequent applications will manifest.

The next point to note is the transformation of the operator, $\partial/\partial t_i$,

$$U_2^+\left(i\hbar\frac{\partial}{\partial t_i}\right)U_2\psi'(\mathbf{x}, t) = \left[-\frac{mu^2}{2} + i\hbar\left(\frac{\partial}{\partial t} - \mathbf{u}\cdot\nabla_{\mathbf{x}}\right)\right]\psi', \tag{14f}$$

$$\psi'(\mathbf{x}, t) = U_2^+\psi(\mathbf{x}_i, t_i).$$

Equation 14*a* was used to obtain the preceding result. It can easily be established, furthermore, that

$$U_2^+\left(H(\mathbf{x}_i, t_i)_i - i\hbar\frac{\partial}{\partial t_i}\right)U_2\psi'(\mathbf{x}, t) = \left[H(\mathbf{x}, t) - i\hbar\frac{\partial}{\partial t}\right]\psi'(\mathbf{x}, t), \tag{14g}$$

which manifests the invariance in form of the equations of motion although the separate operators are altered in form. Equation 14*g* is also correct for Hamiltonians containing Coulomb interactions, and this is evident since U_2 commutes with this type of operator. Although the values of the Coulomb interactions do not change, their dependence upon the coordinates does change, and thus they are not form-invariant. However, in the absence of these potentials the requirements of form-invariance and conservation of probability density (imaginary part of previously noted phase = 0) together uniquely determine the factor representing the translational motion of the electron.

The present results are inapplicable if the nuclear motion is not rectilinear but motion governed by a central field. In practice, however, allowance for departure from rectilinear motion is approximately accommodated by calculating the amplitude for electronic capture using the transformation U_2, and in the integral that represents this amplitude dt is replaced by the orbital relation in Eq. 12.[14] However, the volume of literature on this topic is increasing rapidly. In three recent papers corrections to the translational factor are derived for classical orbits in problems of ionization.[15–17] The equivalence of cross sections calculated by the IP method and the wave mechanical version can be established using the Jeffreys' or, equivalently, the WKB approximation for the nearly classical motion of relative translation. This is done in sections of this book treating specific approximating procedures. An example of the use of the IP version is given next.[18]

The process is represented by Eq. 1, where only one electron and two nuclei constitute the aggregates, the nuclei moving with constant relative velocity **v**. The wave function satisfies

$$\left(-\frac{\hbar^2}{2m}\nabla_s{}^2 - Z_i\frac{e^2}{x_i} - Z_f\frac{e^2}{x_f} - i\hbar\frac{\partial}{\partial t}\right)\psi = 0; \tag{15a}$$

the Hamiltonian of this equation is defined in Eq. 6 with $\nabla_R = 0$ and μ_1 approximated by m. (The reader may be confused by this reference to Eq. 6 since it refers to the center-of-mass system. Nevertheless, this electronic equation is correct and the relation to the impact energy is obtained directly from the impact velocity v occurring in Eq. 9*b*.) The relations among the three independent systems of coordinates—$(\mathbf{x}_i, t_i)$, $(\mathbf{x}_f, t_f)$, and $(\mathbf{s}, t)$—are given for convenience:

$$\begin{aligned} \mathbf{x}_i &= \mathbf{s} + \frac{\mathbf{v}t}{2} + \tfrac{1}{2}\boldsymbol{\rho} \\ t_i &= t \\ \mathbf{x}_f &= \mathbf{s} - \tfrac{1}{2}\mathbf{v}t - \tfrac{1}{2}\boldsymbol{\rho} \\ t_f &= t \end{aligned} \tag{15b}$$

Equation 15*a* is an equation in four dimensions, and although $t_i = t = t_f$ in Eq. 15*b*, $\partial/\partial t_i \neq \partial/\partial t \neq \partial/\partial t_f$ in positional space; these derivatives are equal in momentum space.

It is assumed that the wave function of the system can be represented by the following expansions:

$$\Psi = a_i(t_i)\phi_i(\mathbf{x}_i)\exp\left[-\frac{i}{\hbar}\left(\frac{m\mathbf{v}\cdot\mathbf{s}}{2} + \frac{mv^2t}{8} + \alpha_i t\right)\right]$$

$$+ b_f(t_f)\phi_f(\mathbf{x}_f)\exp\frac{i}{\hbar}\left[\frac{m\mathbf{v}\cdot\mathbf{s}}{2} - \frac{mv^2t}{8} - \alpha_f t\right] + \Gamma(\mathbf{s}, t). \tag{15c}$$

Here Γ is a function that can be represented by the set of eigenfunctions associated with nucleus A or the set associated with nucleus P, but neither the initial nor final state is included in either set; consequently, Γ is orthogonal to both the initial and final states.[18] The wave functions of the initial and final bound states satisfy

$$\left[-\frac{\hbar^2}{2m}\nabla^2_{\mathbf{x}_{i,f}} - Z_{i,f}\frac{e^2}{\mathbf{x}_{i,f}} - \alpha_{i,f}\right]\phi_{i,f}(\mathbf{x}_{i,f}) = 0. \tag{15d}$$

The initial state satisfies Eq. 15*a* if $\mathbf{R} \to \infty$ and $\mathbf{x}_i$ is finite, but the final state satisfies Eq. 15*a* if $\mathbf{x}_f$ instead of $\mathbf{x}_i$ remains finite.* There is a problem here

* The $\alpha_{i,f}$ occurring in Eq. 15*c* and subsequent equations involving the IP version are negative if they represent the energies (eigenenergies) of bound states. However, the eigenenergies appearing in equations of conservation of energy, such as $\epsilon_{i,f}$ in Eq. 10*a*, represent the absolute values of these energies and thus are positive.

since the perturbing potential is a Coulomb interaction, and this interaction produces effects at infinity. This problem is ignored here, however, since it is discussed in detail in the section on the continuum-distorted-wave approximation. Note that in the absence of bound atomic states the initial and final states represent a free electron. The wave function of a free electron with positive energy is a plane wave, or a superposition of them ($\phi_i = \phi_f = 1$, $\Gamma = 0$, in this example); thus, the Galilean transformation for this application insures a correct representation of the electron in the absence of interactions. If initially $a_p(-\infty) = \delta_{pi}$, then consistent with the postulated orthonormal set of functions, $|b_f(\infty)|^2$ represents the probability for electronic capture into state $\phi_f(\mathbf{x}_f) \exp(-i\alpha_f t/\hbar)$. Since $\rho\, d\rho\, d\phi$ is the classical differential cross section, the cross section for capture is

$$Q = \int_0^{2\pi} d\phi \int_0^{\infty} \rho\, d\rho |b_f(\infty)|^2, \tag{15e}$$

and $b_f(\infty)$ generally depends upon the impact velocity, azimuthal angle ϕ, and ρ. Methods of obtaining approximate solutions for b_f are discussed in other sections.

1.2 WAVE-MECHANICAL VERSION

The wave-mechanical version of the scattering equation is the unmodified Schrödinger equation and this is the version commonly used in treatises on nonrelativistic scattering theory. The main purpose of this short section is to deduce from this wave version some of the terms that were obtained from the Galilean transformations of the preceding section. For this purpose the Schrödinger equation is written in the relative coordinates of Eq. 5, and as in Section 1.1 the system consists of one electron and two nuclei.

Assume that the wave function can be represented by the expansion

$$\Psi = \sum_p F_p^{\,i}(\mathbf{R})\chi_p^{\,i}(\mathbf{x}, \mathbf{R}) + \sum_q F_q^{\,f}(\mathbf{R})\chi_q^{\,f}(\mathbf{x}, \mathbf{R}). \tag{16a}$$

This type of expansion is sometimes used in approximations for low velocities of relative motion, and in these applications the symbols $\chi_p^{\,i}$ and $\chi_q^{\,f}$ designate molecular wave functions that possess the desired asymptotic molecular property for large values of R.[19] As already mentioned in connection with Eq. 3, however, the relative coordinates, $\mathbf{R}_i$ and $\mathbf{R}_f$ of Eq. 4, are the appropriate ones for the correct specification of the asymptotic forms of the wave functions.

If the electron is bound initially to nucleus A, this initial state is

$$\Phi_i = \exp(i\mathbf{k}_p{}^i \cdot \mathbf{R}_i)\phi_p{}^i(\mathbf{x}_i), \tag{16b}$$

and similarly if the electron is captured by nucleus P, the final state is

$$\Phi_f = \exp(i\mathbf{k}_q{}^f \cdot \mathbf{R}_f)\phi_q{}^f(\mathbf{x}_f). \tag{16c}$$

The functions $F_p{}^i$ and $F_q{}^f$, however, are functions of $\mathbf{R}$ only, and therefore cannot tend to a plane wave in $\mathbf{R}_i$ or $\mathbf{R}_f$; furthermore, the molecular states tend to be discrete atomic states and thus contain no plane waves at large values of $\mathbf{R}$. If the dominant term of $F_p{}^i(\mathbf{R})$ is $\exp(i\mathbf{k}_p{}^i \cdot \mathbf{R})$ as $R \to \infty$, the additional factor, $\exp[-i(m/M_{a1})\mathbf{k}_p{}^i \cdot \mathbf{x}_i]$, is needed to achieve the correct asymptotic dependence upon $\mathbf{R}_i$ since

$$\mathbf{R}_i = \mathbf{R} - \left(\frac{m}{M_{al}}\right)\mathbf{x}_i.$$

Since $(m/M_{a1})\mathbf{x}_i \approx (m/M_a)[\mathbf{x} + M_p(M_p + M_a)^{-1}\mathbf{R}]$, however (cf. Eqs. 4 and 5), this desired dependence can be obtained, to a good approximation, by using the factor

$$\exp\left[-i\mathbf{k}_p{}^i \cdot \left(\frac{m}{M_a}\right)\mathbf{x}\right] = \exp\left[-i\hbar^{-1}\left(\frac{mM_p}{M_t}\right)\mathbf{v}_p{}^i \cdot \mathbf{x}\right],$$

and absorbing the other factor,

$$\exp[-i\mathbf{k}_p{}^i \cdot mM_pM_a{}^{-1}(M_a + M_p)^{-1}\mathbf{R}],$$

into $F_p{}^i(\mathbf{R})$. This artifice not only establishes the connection with the Galilean transformation but also elicits a connection between the wave-mechanical version using the representation given by Eq. 16*a* and the IP method. This exponential factor in $\mathbf{x}$ is exactly one of the factors obtained if the new electronic origin is C in Fig. 2 instead of the midpoint of the internuclear line used in connection with Eq. 15*c*. By a similar analysis we can readily deduce that the factor

$$\exp\left(\frac{i}{\hbar}\, mM_aM_t{}^{-1}\mathbf{v}_q{}^f \cdot \mathbf{x}\right)$$

must multiply the function $F_q{}^f(\mathbf{R})$ in order that the asymptotic form of this modified function is $\exp(i\mathbf{k}_q{}^f \cdot \mathbf{R}_f)$, which associated factor in $\mathbf{x}$ corresponds to the final state in Eq. 15*c*. [In this wave version the wave equation refers to the center-of-mass system. Thus the velocity of A relative to the point C in Fig. 2 is $-(M_p/M_t)\mathbf{v}_i$ $(M_f = M_p)$, and the velocity of P relative to C is $(M_a/M_t)\mathbf{v}_i$; these are precisely the velocities associated with the required factors in $\mathbf{x}$ provided that $\mathbf{v}_q{}^f \approx \mathbf{v}_p{}^i \approx v_i$, the usual case for massive nuclei.] These factors containing the electronic mass together allow for the transfer of linear momentum to the electron in the process of capture from the initial to the final atomic state.

Terms associated with this transfer of linear momentum are easily obtained from the electronic amplitude for capture calculated in the first Born approximation using the active electronic, or OBK (Oppenheimer-Brinkman-Kramers) interaction.[8,20] Although the analysis of this approximation is presented in another section, this feature of the transfer of linear momentum belongs here by reason of its close connection with the preceding material. It is important to remember that linear momentum is conserved, by definition, in the center-of-mass system, and the transfer of momentum in this system must not violate this conservation.[21] The coordinates of Eq. 4 are used in the integral, which represents this amplitude for electronic capture in a system consisting of one electron and two nuclei.[22] Dimensionless variables of integration are used:

$$f(\theta) = \frac{\mu_f z_i z_f a_0}{2\pi m} \int \frac{d\mathbf{x}_i\, d\mathbf{x}_f}{x_f} \phi_i(\mathbf{x}_i)\phi_f^*(\mathbf{x}_f) \exp\left[i(\mathbf{A}_i \cdot \mathbf{x}_i + \mathbf{A}_f \cdot \mathbf{x}_f)\right],$$

$$\mathbf{A}_i = \frac{M_i}{M_{i1}} \mathbf{K}_i - \mathbf{K}_f, \qquad \mathbf{A}_f = \frac{M_f}{M_{f1}} \mathbf{K}_f - \mathbf{K}_i, \tag{17a}$$

$$\hbar \mathbf{K}_i = \mu_i \mathbf{v}_i, \qquad \hbar \mathbf{K}_f = \mu_f \mathbf{v}_f.$$

[The asterisk ϕ_f^* (or bar $\bar{\phi}$) signifies the complex conjugate of ϕ_f.] The vectors $\mathbf{A}_i$ and $\mathbf{A}_f$ represent the change of linear momentum that is associated

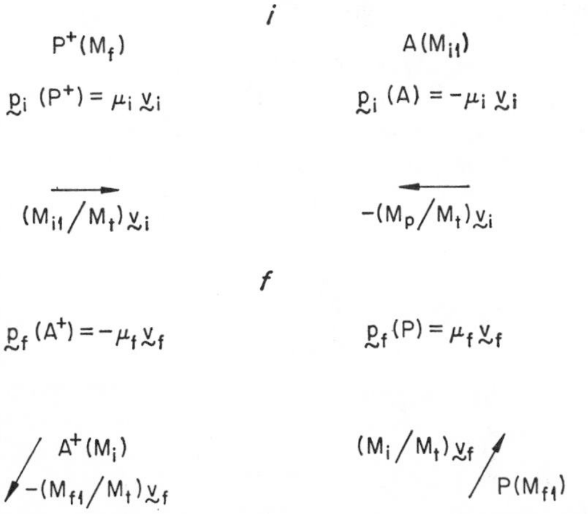

Fig. 3 Diagram of the constituent parts of the total linear momentum for the i- and f-systems. Note that total linear momentum of constituent parts is zero. P^+ is incident ion; A, atomic target; P, atom formed by capture; A^+, residual ion.

with $\mathbf{x}_i$ and $\mathbf{x}_f$, respectively. The linear momenta are illustrated in Fig. 3.

The velocities in Fig. 3 have the correct magnitudes and directions so that the corresponding total linear momentum, i or f, each vanishes in the center-of-mass system. An interesting relation from Eq. 17a is

$$\hbar(\mathbf{A}_f - \mathbf{A}_i) = 2\frac{M_i M_f}{M_t}(\mathbf{v}_f - \mathbf{v}_i) + m\left(\frac{M_i}{M_t}\mathbf{v}_f - \frac{M_f}{M_t}\mathbf{v}_i\right) \tag{17b}$$

(c.g.s. units are used here for clarity). The second term on the right-hand side of Eq. 17b represents the change of electronic momentum associated with the two velocities on the right of Fig. 3, and it is reemphasized that this change occurs in the center-of-mass system.

The retention of the electronic mass may appear to be an unnecessary refinement, but a study of the minimum value of $|\hbar M_t \mathbf{A}_f|$ dispels this notion. This minimum occurs in the forward direction (scattering angle $\theta = 0$, $\mathbf{v}_i \cdot \mathbf{v}_f = \mathbf{v}_i \mathbf{v}_f \cos\theta$), and for this example it is assumed that

$$\frac{2M_t}{M_f M_{i1}} \frac{\varepsilon_i - \varepsilon_f}{v_i^2} \ll 1,$$

which is usually fulfilled in processes of electronic capture. Conservation of energy, given by Eq. 10a, is used to get the dominant term,

$$|\hbar \mathbf{A}_f M_t| \sim \frac{M_t(\varepsilon_i - \varepsilon_f)}{v_i}, \tag{17c}$$

if m is omitted, but with the retention of m the corresponding value is

$$|\hbar M_t \mathbf{A}_f| \sim M_f\left[M_{i1} - \left(\frac{M_i M_{i1} M_f}{M_{f1}}\right)^{1/2}\right] v_i. \tag{17d}$$

Since Eqs. 17c and 17d are the dominant terms for large v_i, it is clear how sensitively the value of A_f depends upon the electronic mass. A final point of interest is the variation of

$$\hbar(\mathbf{A}_i + \mathbf{A}_f) = -m\left(\frac{M_f}{M_t}\mathbf{v}_i + \frac{M_i}{M_t}\mathbf{v}_f\right). \tag{17e}$$

If m is omitted, Eq. 17e vanishes for all scattering angles; but if m is retained, Eq. 17e can vanish only if $M_f = M_i$ and $\varepsilon_f = \varepsilon_i$, the condition for exact or symmetrical resonance.[23] Under this restrictive condition Eq. 17e vanishes at $\theta = \pi$, $\mathbf{v}_f = -\mathbf{v}_i$, a situation discussed in a subsequent section. Next we turn to the approximating method that has been fairly successful in predicting cross sections for exact resonance at low velocities of relative motion.

1.3 PERTURBED STATIONARY STATES METHOD

Apparently, Massey and Smith[24] first used the wave mechanical version of perturbed stationary states (PSS) in a calculation for exact resonance. A detailed study of this method was made later by Bates, Massey, and Stewart,[14] and some defects of the method were removed by Bates and McCarroll.[7,25] A description and explanation of this approximating procedure is in the treatise on atomic collisions by Mott and Massey.[26] Although the collection of these treatises was used to prepare the present version of this method, emphasis here is devoted to explaining the mathematical approximations and operations that are often confusing to nonspecialists.

In the PSS method the wave function of the scattering system is expanded in a series of atomic or molecular orbitals, but only the latter expansion is used in this section. For this description electronic capture is presumed to occur between two bare nuclei or cores, and in the case of many-electron cores, the associated electrons enter the process only by their effect on the charge and ionization potentials of the cores. The Hamiltonian and coordinates used are those of Eq. 5. The interaction V of Eq. 5, according to this assumption, is

$$V = V_i(\mathbf{x}_i) + V_f(\mathbf{x}_f) + U(R), \tag{18a}$$

with $\mathbf{x}_i$, $\mathbf{x}_f$, and $\mathbf{R}$ defined in Eqs. 4 and 5. In Eq. 18*a* $U(R)$ is the internuclear electrostatic interaction. The effective interaction between the separated systems, initially and finally, is assumed to decrease more strongly than the Coulomb potential, or else a qualifying remark is made at the appropriate place. It is assumed, furthermore, that Ψ is represented by Eq. 16*a*, reproduced here for convenience to the reader:

$$\Psi = \sum_p F_p{}^i(\mathbf{R})\chi_p{}^i(\mathbf{x}, \mathbf{R}) + \sum_q F_q{}^f(\mathbf{R})\chi_q{}^f(\mathbf{x}, \mathbf{R}). \tag{16a}$$

For fixed separation of the nuclei (or cores) the molecular wave functions χ are assumed to constitute an orthonormal set with respect to the independent variables $\mathbf{x}$, and the χ satisfy the equations

$$\left[-\frac{\hbar^2}{2\mu_1}\nabla_x{}^2 + V_i(\mathbf{x}_i) + V_f(\mathbf{x}_f) - \varepsilon(\mathbf{R})\right]\chi(\mathbf{R}, \mathbf{x}) = 0, \tag{18b}$$

which have the eigenfunctions $\chi_p{}^i$, $\chi_q{}^f$ and corresponding eigenenergies, $\epsilon_p{}^i$, $\epsilon_q{}^f$ as solutions.[26] Note $\mathbf{R}$ is a parameter and $\epsilon(\mathbf{R})$ is an abbreviation for

$$\varepsilon(\mathbf{R}) = E - U(\mathbf{R}), \tag{18c}$$

with E denoting the center-of-mass energy. The eigenfunctions of Eq. 18*b* approach the normalized atomic orbitals in the limits expressed by Eq. 18*d*:

$$\begin{aligned} &\lim_{R\to\infty} \chi_p{}^i(\mathbf{x}, \mathbf{R}) \to \phi_i(\mathbf{x}_i), \qquad \mathbf{x}_f = \mathbf{x}_i - \mathbf{R} \to \infty, \\ &\lim_{R\to\infty} \chi_q{}^f(\mathbf{x}, \mathbf{R}) \to \phi_f(\mathbf{x}_f), \qquad \mathbf{x}_i = \mathbf{x}_f + \mathbf{R} \to \infty. \end{aligned} \tag{18d}$$

It should be noted that the variable being kept finite is an important specification in the preceding limits, and this requirement will occur frequently in the subsequent description. It is also evident from Eq. 18*d* that Eq. 18*b* does approach the equation satisfied by the atomic orbitals that are defined by these limits. (We could also say that Eq. 18*d* defines the behavior of the χ-orbitals for large **R**.) The factors needed to adjust the asymptotic forms of Eq. 16*a* to the correct initial and final states already have been obtained following Eqs. 16*b* and 16*c*, and these requirements, together with the limits of Eq. 18*d*, necessitate care in the application of the operators of differentiation in the asymptotic regions. Bates and McCarroll[7] first modified Eq. 16*a* by the introduction of these factors, although Brinkman and Kramers[8] correctly used the same factors in their pioneering work on electronic capture. Since the PSS method is most simply applied to systems consisting of spherically symmetric initial and final atomic states, this is the example illustrated first; the wave-mechanical version is used for the case of nonresonance.

It is pertinent at this point to mention that complications occur in the application of the PSS method if the initial or final atomic states are spatially degenerate, but the discussion of these difficulties, and other difficulties associated with the rotation of the internuclear axis, is presented near the end of this section. The presentation for this first example is patterned from the treatment given by Bates and McCarroll.[7] If only the initial and final states are retained, the approximate wave function of Eq. 16*a* is

$$\begin{aligned} \Psi_2 &= F_i(\mathbf{R})\psi_i(\mathbf{x}; \mathbf{R}) + F_f(\mathbf{R})\psi_f(\mathbf{x}; R), \\ \psi_i &= \chi_i(\mathbf{x}; \mathbf{R}) \exp\left(-\frac{i}{\hbar} pm\mathbf{v}\cdot\mathbf{x}\right), \qquad p = \frac{M_p}{M}, \\ \psi_f &= \chi_f(\mathbf{x}; \mathbf{R}) \exp\left(\frac{i}{\hbar} qm\mathbf{v}\cdot\mathbf{x}\right), \qquad q = \frac{M_a}{M}, \qquad M = M_a + M_p. \end{aligned} \tag{19a}$$

The relative motion of the two nuclei is approximated by the constant velocity v, and although **v** is assumed to be small ($v \ll v_0$), terms containing **v** are retained for clarification. (If there is any deflection of the nuclei from rectilinear trajectories, however, as there is in the wave-mechanical solution to problems of electronic capture, **v** is not constant; it is thus important to realize that the exponential factors of ψ_i and ψ_f represent another approximation since v is constant in these factors. Moreover, this same approximation is invoked if the classical orbital equations are used to correct for nuclear deflection in the IP method.) Note that each of the functions, ψ_i and ψ_f, satisfy Eq. 5 if $\mu \gg \mu_1$ ($\mu_1 \approx m$) and v is zero since in this instance Eq. 5 is approximated by Eq. 18*b* for the molecular orbitals. We now investigate what approximate restrictions must be imposed upon X_i and X_f in order that each term of Ψ_2 satisfies Eq. 5 as $R \to \infty$.

The function Ψ_2 is inserted into Eq. 5, and the conditions of Eq. 18*d* are used. This laborious but straightforward calculation yields the equation

$$\exp\left(-\frac{i}{\hbar}pm\mathbf{v}\cdot\mathbf{x}\right)\left(-\frac{\hbar}{2\mu}(X_i\nabla_R{}^2F_i + F_i\nabla_R{}^2X_i + 2\nabla_R X_i\cdot\nabla_R F_i)\right.$$
$$-\frac{\hbar^2}{2m}\left\{-\frac{i}{\hbar}2mp\mathbf{v}\cdot\nabla_x X_i - \frac{m^2p^2v^2}{\hbar^2}X_i\right.$$
$$\left.\left.-\frac{2m}{\hbar^2}\Big[-E + \varepsilon_i(\mathbf{R}) + U(\mathbf{R})\Big]X_i\right\}F_i\right) + \exp\left(\frac{i}{\hbar}qm\mathbf{v}\cdot\mathbf{x}\right) \tag{19b}$$
$$\left(-\frac{\hbar^2}{2\mu}(X_f\nabla_R{}^2F_f + F_f\nabla_R{}^2X_f + 2\nabla_R X_f\cdot\nabla_R F_f) - \frac{\hbar^2}{2m}\left\{\frac{i}{\hbar}2mq\mathbf{v}\cdot\nabla_x X_f\right.\right.$$
$$\left.\left.-\frac{m^2q^2v^2}{\hbar^2}X_f - \frac{2m}{\hbar^2}\Big[\varepsilon_f(\mathbf{R}) + U(\mathbf{R}) - E\Big]X_f\right\}F_f\right) = 0.$$

Next Eqs. 4 and 5 are used to transform ∇_R and ∇_x to

$$\nabla_R = \nabla_{R(\mathbf{x}_i)} + p\nabla_{x_i}, \qquad \nabla_x = \nabla_{x_i} \tag{19c}$$

or

$$\nabla_R = \nabla_{R(\mathbf{x}_f)} - q\nabla_{x_f}, \qquad \nabla_x = \nabla_{x_f}. \tag{19d}$$

The subscripts $\mathbf{R}(\mathbf{x}_i)$, $\mathbf{R}(\mathbf{x}_f)$ signify that partial derivatives are calculated with $\mathbf{x}_i$ or $\mathbf{x}_f$ kept constant. The first part of Eq. 19*b* is rewritten using Eq. 19*c* as

$$(H - E)F_i(\mathbf{R})X_i(\mathbf{x};\mathbf{R})$$
$$= \exp\left(-\frac{i}{\hbar}m\mathbf{v}\cdot\mathbf{x}\right)\left\{X_i\left(-\frac{\hbar^2}{2\mu}\nabla_R{}^2 + \varepsilon_i - E + \frac{mp^2v^2}{2}\right)F_i(\mathbf{R})\right.$$
$$\left. + F_i\left(-\frac{\hbar^2}{2\mu}\nabla_R{}^2X_i + i\hbar p\mathbf{v}\cdot\nabla_{\mathbf{x}}X_i\right) - \frac{\hbar^2}{\mu}\nabla_{\mathbf{R}}F_i\cdot(\nabla_{\mathbf{R}}(X_i) + p\nabla_{x_i})X_i\right\}. \tag{19e}$$

Since the situation for $\mathbf{R} \to \infty$ is being investigated, $U(\mathbf{R}) = 0$ and $\varepsilon_i(R) = \varepsilon_i$; moreover, although the same functional symbol is used to denote F_i and X_i in all parts of Eq. 19*e*, it is understood that these functional forms are different if the independent variables ($\mathbf{x}_i$, $\mathbf{R}$) are used. The part of Eq. 19*b* that contains F_f and X_f is modified similarly using Eq. 19*d*, and this new form can be derived from Eq. 19*e* by replacing i with f, p with $-q$, and $\mathbf{x}_i$ with $\mathbf{x}_f$. These terms in X_f vanish, however, since X_f is a bound state and vanishes if $\mathbf{x}_i$ is finite and $\mathbf{R}$ is infinite since $\mathbf{x}_f$ is also infinite according to Eq. 18*d*. If the term $\nabla_R{}^2X_i$

is assumed negligible, and if the assumed asymptotic form of $F_i(\mathbf{R})$ is used,

$$\nabla_R F_i(\mathbf{R}) = \frac{i\mu\mathbf{v}}{\hbar} \exp\left(\frac{i\mu\mathbf{v}\cdot\mathbf{R}}{\hbar}\right) + \ldots, \tag{19f}$$

cancellation between the right-hand members of the second and third parentheses inside the braces occurs in Eq. 19*e*, and the remaining term $\nabla_{R(\mathbf{x}_i)}\chi_i$ must vanish in order to achieve the promised free equation for F_i at $R \to \infty$. This description clarifies the meaning of the corresponding discussion in the paper by Bates and McCarroll.[7] The free equation satisfied by F_f at $R = \infty$, $\mathbf{x}_f$ remaining finite, can be obtained similarly with the use of Eq. 19*d*. It is noteworthy that the use of Eqs. 19*d* and 19*c* is essential to the preceding demonstration since it would be confusing to extract the needed cancellation if only the $(\mathbf{R}, \mathbf{x})$-set were used.

In order to solve Eq. 19*b*, the equation is multiplied by ψ_i^* and ψ_f^*, successively, and with the use of orthonormal relations some of the integrals over $d\mathbf{x}$ can be effected to get Eqs. 20*a* and 20*b* (such integrals always signify that the integration is taken over all three-dimensional space of each spatial triad unless otherwise specified):

$$\begin{aligned}&\left\{\nabla^2 + k_i^2 - \frac{2\mu}{\hbar^2}[\varepsilon_i(R) + U(R)]\right\}F_i(\mathbf{R})\\ &\quad + \int \chi_i^*\left(\nabla_R^2\chi_i + 2\nabla_R\chi_i\cdot\nabla_R - \frac{i}{\hbar}2\mu p\mathbf{v}\cdot\nabla_x\chi_i\right)F_i(\mathbf{R})\,d\mathbf{x}\\ &\quad + \int \exp\left(\frac{i}{\hbar}mv\cdot x\right)\chi_i^*[F_f(\mathbf{R})\nabla_R^2\chi_f\\ &\quad + \left(\chi_f\nabla_R^2 + 2\nabla_R\chi_f\cdot\nabla_R + i\frac{2\mu q\mathbf{v}}{\hbar}\cdot\nabla_x\chi_f\right)F_f]\,dx = 0,\end{aligned} \tag{20a}$$

$$p + q = 1 \qquad k_i^2 = \frac{2\mu}{\hbar^2}\left[E - \frac{\hbar^2 p^2 v^2}{2m}\right].$$

$$\begin{aligned}&\left\{\nabla_R^2 + k_f^2 - \frac{2\mu}{\hbar^2}[\varepsilon_f(R) + U(\mathbf{R})]\right\}F_f(\mathbf{R})\\ &\quad + \int \chi_f^*\left(\nabla_R^2\chi_f + 2\nabla_R\chi_f\cdot\nabla_R + i\frac{2q\mu\mathbf{v}}{\hbar}\cdot\nabla_x\chi_f\right)F_f(\mathbf{R})\,d\mathbf{x}\\ &\quad + \int \exp\left(-\frac{i}{\hbar}mv\cdot x\right)\chi_f^*[F_i(\mathbf{R})\nabla_R^2\chi_i\\ &\quad + \left(\chi_i\nabla_R^2 + 2\nabla_R\chi_i\cdot\nabla_R - \frac{i}{\hbar}2p\mu\mathbf{v}\cdot\nabla_x\chi_i\right)F_i]\,d\mathbf{x} = 0,\end{aligned} \tag{20b}$$

$$k_f^2 = \frac{2\mu}{\hbar^2}\left(E - \frac{\hbar^2 p^2 v^2}{2m}\right).$$

These equations are solved using the method of distorted waves if the coupling is weak.[28] In Eq. 20*a* the term containing $F_f(\mathbf{R})$ is usually referred to as *back-coupling*, and analogously the name *forward-coupling* is given to the corresponding terms of Eq. 20*b* for the final state. In Eq. 20*b* the term containing F_f in the integral is omitted as unimportant in the first approximation. In the other integral both $\nabla_R{}^2\chi_i$ and $\nabla_R{}^2F_i$ are assumed relatively unimportant in slow collisions (small v) and are omitted. Equation 19*c* is now used to rewrite $\nabla_R\chi_i$, and the approximate equation for F_f becomes

$$\left\{\nabla_R{}^2 + k_f{}^2 - \frac{2\mu}{\hbar^2}[\varepsilon_f(R) + U(\mathbf{R})]\right\}F_f(R)$$

$$= \int d\mathbf{x} \exp\left(-\frac{i}{\hbar} m\mathbf{v}\cdot\mathbf{x}\right)\chi_f^*\left[\frac{i}{\hbar} 2p\mu\mathbf{v}\cdot(\nabla_x\chi_i) - 2\nabla_{R(\mathbf{x}_i)}\chi_i\cdot\nabla_R - 2p\nabla_{x_i}\chi_i\cdot\nabla_R\right]F_i(\mathbf{R}). \tag{20c}$$

If Eq. 19*f* together with the specification that $\nabla_{R(x_i)}\chi_i$ must vanish asymptotically are used in Eq. 20*c*, it is readily seen that the corresponding integrand vanishes as R tends to infinity. Equation 19*d* and the same procedure used to derive Eq. 20*c* are employed next to reduce Eq. 20*a* to

$$\left\{\nabla_R{}^2 + k_i{}^2 - \frac{2\mu}{\hbar^2}[\varepsilon_i(R) + U(\mathbf{R})]\right\}F_i(\mathbf{R})$$

$$= \int d\mathbf{x} \exp\left(\frac{i}{\hbar} m\mathbf{v}\cdot\mathbf{x}\right)\chi_i^*\left[-\frac{i2q\mu\mathbf{v}}{\hbar}\cdot(\nabla_x\chi_f) - 2\nabla_{R(\mathbf{x}_f)}\chi_f\cdot\nabla R + 2q(\nabla_{x_f}\chi_f)\cdot\nabla_R\right]F_f(\mathbf{R}). \tag{20d}$$

If $v = 0$, Eqs. 20*c* and 20*d* agree with Eqs. 126 and 127 of Mott and Massey,[29] and the present derivation clarifies the remarks following Eqs. 126 and 127 of that work.

The solution to Eq. 20*c* is next written for the case of small back-coupling, which corresponds to the neglect of the right-hand number of Eq. 20*d*. This solution is given to illustrate how a semiclassical approximation to the solution yields the same result as that obtained directly by the IP method for this case of weak coupling. It is reemphasized that the initial and final atomic systems are spherically symmetric; these atomic systems are not spatially degenerate; the velocity of relative motion is small, and the process is nonresonant. The procedure used in this illustrative example is adopted from Mott and Massey.[30]

The cross section for electronic capture is given by

$$Q = \frac{v_f}{v_i}\int_0^{2\pi} d\Phi \int_0^{\pi} d\Theta \sin \Theta |g(\Theta, \Phi)|^2, \tag{21a}$$

where the amplitude g obtained from Eqs. 20*c* and 20*d* is[31]

$$g(\Theta, \Phi) = \frac{1}{2\pi}\int \mathscr{F}_f(R', \pi - \theta)\chi_f^*(\mathbf{R}', \mathbf{x})\nabla_{R'(\mathbf{x}_i)}\chi_i(\mathbf{R}', \mathbf{x})\cdot \nabla_R F_i(R')\, d\mathbf{x}\, dR',$$

$$\cos \theta = \cos \Theta \cos \Theta' + (\sin \Theta \sin \Theta') \cos (\Phi - \Phi') \tag{21b}$$

$$k_{i,f}^2 \gg \frac{2\mu}{\hbar^2}[\varepsilon_{i,f}(R) + U(R)].$$

$\mathscr{F}_f$ is the solution to Eq. 20*c* if the right-hand member is omitted. The angles Θ', Φ' specify the direction of $\mathbf{R}'$, the variables of integration, and the inequality is a mathematical specification of negligible deflection of the two nuclei. This inequality also insures that the solution of the homogeneous part of Eq. 20*d* differs little from a plane wave, and thus the first and third terms of the integrand of Eq. 20*c* cancel to a very good approximation. Since the velocity of relative motion $\mathbf{v}$ is assumed to be small, the exponential in this same integrand may be approximated by unity, and the interaction of Eq. 21*b* results.

The approximate interaction occurring in the right-hand member of Eq. 20*c* and in the integrand of Eq. 21*b* is denoted by U_{if} in the subsequent semiclassical approximation to this approximate PSS solution. The right-hand member of Eq. 20*d* is henceforth neglected, and $\mathscr{F}_f$ and $U_{if}F_i$ of Eq. 21*b* are expanded in partial waves

$$\mathscr{F}_f = \sum_{l \geq 0} (2l + 1)i^l \exp (i\eta_{fl})F_{fl}(R')P_l[\cos (\pi - \theta)],$$

$$U_{if}F_i = U_{if} \sum_{l' \geq 0} (2l' + 1) \exp (i\eta_{il'})F_{il'}(R')P_{l'}(\cos \theta'). \tag{22a}$$

If the interaction U_{if} is spherically symmetric in $\mathbf{R}'$-space, the present procedure can be used; otherwise, a different and more elaborate analysis is needed.[7,32,38] (In two recent papers, Feynman's path-integral formulation of quantum mechanics is applied to this problem.[34,35])

The procedure employed is illustrative only, and is used to elicit several features; thus Eq. 22*a* can be considered an approximation. We return to the calculation and first change the variables of integration in Eq. 21*b* to $d\mathbf{R}\, d\mathbf{x}_i$, which is a complete transformation since the associated Jacobian is unity. Next the integration over solid angle $d\Omega(R')$ is affected in Eq. 21*b* to get

$$g(\Theta, \Phi) = \frac{i2\mu v}{\hbar} \sum_0^\infty (2l+1)i^l \exp[i(\eta_l - \pi l)] P_l(\cos\Theta)$$

$$\times \int_0^\infty dRR^2 F_{il}(R) U_{if}(\mathbf{R}) F_{fl}(R)$$

$$\eta_l = \eta_{fl} + \eta_{il}, \qquad F_{jl} = \frac{G_{jl}}{k_j R} \qquad j = i, f \tag{22b}$$

$$U_{if}(\mathbf{R}') = \int \chi_f^*(\mathbf{R}', \mathbf{x}_i) \frac{\partial}{\partial z_{x_i}} \chi_i(\mathbf{R}', \mathbf{x}_i)\, d\mathbf{x}_i.$$

[Although the same notation is here used for $\chi_{i,f}(\mathbf{R}, \mathbf{x}_i)$ as is used for $\chi_{i,f}(\mathbf{R}, \mathbf{x})$, the functional forms are, of course, different.] In the derivation of Eq. 22*b* the asymptotic relation of Eq. 19*f* is used for $\nabla_{R'} F_i(\mathbf{R}')$ with $\mathbf{v}$ assumed parallel to the direction of incidence, and the auxiliary radial functions G_{jl} are introduced for calculational convenience. (The requirements that the radial functions must fulfill are described elsewhere.[36]) Equation 21*a* is now integrated over scattering angles to get

$$Q = \frac{4\pi k_f}{k_i^3} \sum_0^\infty (2l+1)|\beta_l|^2,$$

$$\beta_l = \frac{2\mu v}{\hbar k_f} \int_0^\infty dR G_{fl}(R) U_{if}(R) G_{il}(R), \qquad \mu = \mu_i = \mu_f, \qquad \frac{v_f}{v_i} = \frac{k_f}{k_i}. \tag{22c}$$

Next the JWKB (WKB) approximation is used for the solutions to the radial equations,[37]

$$\frac{d^2 G_{jl}}{dR^2} + \left[k_j^2 - \frac{2\mu}{\hbar^2}\left(\epsilon_j(R) + U(R) + \frac{l(l+1)\hbar^2}{2\mu R^2}\right)\right] G_{jl} = 0 \qquad j = i, f. \tag{23a}$$

These solutions are

$$G_{jl} = \frac{1}{\mu_j^{1/2}} \sin\left(\frac{\pi}{4} + \int_{R_{mj}}^R \mu_j\, dR'\right), \qquad R > R_{mj},$$

$$G_{jl} = \frac{1}{\nu_j^{1/2}} \exp\left(\left(\int_{R_{mj}}^R \nu_j\, dR'\right)\right), \qquad R < R_{mj}, \qquad \nu_j^2 = -\mu_j^2 \geq 0, \tag{23b}$$

$$\mu_j^2 = k_j^2 - \frac{2\mu}{\hbar^2}\left[\varepsilon(R) + U(R) + \frac{(l+\frac{1}{2})^2\hbar^2}{2\mu R^2}\right].$$

Here R_{mj} is the largest value of R for which $\mu_j^2 = 0$. The integral, factor excluded, of Eq. 22*c* is

$$I = \int_0^\infty \frac{dR U_{if}(R)}{2(\mu_i \mu_f)^{1/2}} \left[\cos \int_{R>}^R (\mu_f - \mu_i)\, dR' + \sin \int_{R>}^R (\mu_f + \mu_i)\, dR'\right]. \tag{23c}$$

where $R>$ means the larger of the two R_{mj}; the relations

$$\sin A \sin B = \tfrac{1}{2}[\cos(\mathrm{A} - \mathrm{B}) - \cos(\mathrm{A} + \mathrm{B})]$$

and

$$\cos\left(\frac{\pi}{2} + \chi\right) = -\sin \chi$$

were used to obtain Eq. 23*c*. The term that contains $(\mu_i + \mu_f)$ is henceforth omitted since this argument of the sine function varies considerably for a small change of R, and this contribution to I is thus relatively negligible. We now use the assumption of nearly uniform rectilinear motion of the two nuclei and the definitions of Eq. 23*b* to obtain the approximations of Eq. 23*d*:

$$\frac{\mu v_j(R)}{\hbar} = \left\{k_j^2 - \frac{2\mu}{\hbar^2}[U(R) + \epsilon_j(R)]\right\}^{1/2},$$

$$\mu_j(R) = \frac{\mu v_j(R)}{\hbar}\left[1 - \frac{(l + \frac{1}{2})^2\hbar^2}{\mu^2 v_j^2(R)R^2}\right]^{1/2},$$

$$v_i(R) = \bar{v}(R) + \Delta v(R), \qquad v_f(R) = \bar{v}(R) - \Delta v(R),$$

$$\frac{\Delta v(R)}{\overline{v(R)}} \ll 1, \quad \overline{v(\infty)} = v \tag{23d}$$

$$\mu_i + \mu_f \approx \frac{2\mu\overline{v(R)}}{\hbar}\left\{1 - \left[\frac{(l + \frac{1}{2})\hbar}{\mu\overline{v(R)}R}\right]^2\right\}^{1/2},$$

$$\mu_i - \mu_f = \frac{\mu_i^2 - \mu_f^2}{\mu_i + \mu_f} \approx \frac{\Delta E(R)}{\hbar\overline{v(R)}\{1 - [(l + \frac{1}{2})\hbar/\mu\overline{v(R)}R]^2\}^{1/2}}$$

$$\Delta E(R) = \frac{\hbar^2}{2\mu}\left\{k_i^2 - k_f^2 - \frac{2\mu}{\hbar^2}[\varepsilon_i(R) - \varepsilon_f(R)]\right\}.$$

In this type of scattering the amplitude usually is sharply peaked in the forward direction, and an angular distribution that conspicuously demonstrates this feature is the delta function,

$$\frac{\delta(\theta - \theta')}{\sin \theta'} = \sum_{l=0}^{\infty} \frac{2l + 1}{2} P_l(\cos \theta) P_l(\cos \theta'). \tag{23e}$$

All angular momenta are here required to represent this function, and it is clear that the expression for Q of Eq. 22*c* likewise has significant contributions from the domain of large l for the assumed small deflections of the heavy nuclei. The sum over l consequently is approximated by an integral over the impact parameter that is obtained from the classical mechanical-quantum mechanical relation,

$$\mu v p = \hbar(l + \tfrac{1}{2}), \qquad (l + \tfrac{1}{2})\, dl = \left(\frac{\mu v}{\hbar}\right)^2 p\, dp, \tag{23f}$$

and thus the cross section becomes

$$Q = \frac{8\pi k_f}{k_i^3}\left(\frac{\mu v_f}{\hbar}\right)^2\left(\frac{2\mu v}{\hbar k_f}\right)^2 \int_0^\infty dpp \left|\frac{\hbar k_f}{2\mu v}\beta(p)\right|^2,$$

$$|f(p)|^2 = \frac{16}{k_i^3 k_f}\left(\frac{\mu v}{\hbar}\right)^4 \left|\frac{\hbar k_f}{2\mu v}\beta(p)\right|^2. \tag{23g}$$

Since the β_l with large l must contribute significantly to Eq. 22*c* if the deflection of the nuclei is small, it is evident that large values of the impact parameter are important in this approximation. The additional approximations,

$$(k_i^3 k_f)^{-1} \simeq \left(\frac{\hbar}{\mu v}\right)^4$$

and

$$\frac{1}{2(\mu_i \mu_f)^{1/2}} \approx \frac{\hbar R}{2\mu[R^2 \overline{v(R)}^2 - p^2 v^2]^{1/2}}, \tag{23h}$$

are used with Eqs. 23*c*, 23*d*, 23*f*, and 23*g* to get

$$|f(p)| = \left| 2\int_{R>}^{\infty} \frac{dRRU_{if}(R)}{[R^2\overline{v(R)}^2 - p^2v^2]^{1/2}} \times \cos\left(\int_{R>}^{R} \frac{dR'R'\Delta E(R')}{\hbar \overline{v(R')}[(R')^2 - p^2v^2/\overline{v(R')}^2]^{1/2}}\right)\right|. \tag{23i}$$

The lower limit to the integral in Eq. 23*i* has been altered from its value in Eq. 23*c* so that these classical approximations will not be violated. Two factors in the integrands of Eq. 23*i* are observed to correspond to the differential in time, *dt*, for the two-body central-force problem of classical mechanics;[38] this differential is given in Eq. 12. This refinement to the commonly assumed rectilinear motion in the IP treatment, moreover, does not affect the correction for the transfer or translational motion to the electron (omitted following Eq. 20*c*) in this order of the approximation.

In order to approximate Eq. 23*i* to the form that can be compared with the result of Bates and McCarroll,[7] R is related to the Z-axis, which is parallel to the direction of incidence, and the variation $\overline{v(R)}$ is omitted:

$$R'^2 = z^2 + p^2, \qquad dR'R' = dzz, \qquad \overline{v(R)} = v. \tag{23j}$$

These relations also are approximate in view of the relation between p and l in Eq. 23f. A change of variable is made, and by reason of the approximation $v(R) = v$ there is no radical in the denominators of Eq. 23i and the former limits, $R>$, can be replaced by $-\infty$. It is a straightforward exercise to transform this final result into the form

$$|f(p)| = \left| \int_{-\infty}^{\infty} dz U_{if}(p^2 + z^2)^{1/2} \exp\left[i \int_{-\infty}^{z} \frac{dz' \Delta\varepsilon(z'^2 + p^2)^{1/2}}{\hbar v} \right] \right|,$$
$$\Delta\varepsilon(z^2 + p^2)^{1/2} = \varepsilon_1(R) - \varepsilon_f(R). \tag{23k}$$

This agrees with Eqs. 14 and 15 of Bates and McCarroll[7] for small values of v. It must be understood, however, that Z of Bates and McCarroll is defined by $(\mathbf{v}\cdot\mathbf{R}) = vZ$ with the origin of time selected so that $Z - vt = 0$ when R is equal to ρ; moreover, Z is not the z-coordinate of either nucleus with respect to the spatial origin that is located at the midpoint of the internuclear line. Although this result can be obtained more directly by a less restricted method, the particular features of the following conditions have both been exemplified: (1) contributions from large values of the impact parameter and (2) the correction for the classical orbit of the heavy particles. The remaining application is charge transfer for exact (symmetrical) resonance.

In terms of the notation of Eq. 1, exact resonance means that $M_f = M_i$, $Z_f = Z_i$, $n_p = n_a$, and the internal energies of the initial and final atomic systems are equal. Since the IP method is commonly used for this application, this is the version that is explained here. Only electronic capture in alkali atom-alkali ion collisions are treated. The ground states of these atoms and ions are ^2S and ^1S, respectively, with this classification being defined by the LS (Russell-Saunders) coupling scheme.[39] In the case of atoms containing more than one electron the corresponding singly charged ion is treated as a core with a total charge of unity and no electronic spin, but the ionizing potential appropriate to that atom is used. Since the total spin and total charge of these atoms and ions are all the same, the next similarities to investigate are the corresponding Schrödinger equations.

The Hamiltonian that represents a single electron and two ions, each containing $(m - 1)$ electrons, is

$$H = H_0(1) + H_0(2) - \frac{\hbar^2}{2m} \nabla_{r\,(2m-1)}^2 + U_i + V_i,$$

$$U_i = e^2\left(\sum_{j=m}^{2m-2} \sum_{i=1}^{m-1} |\mathbf{r}_j - \mathbf{r}_i|^{-1} - m \sum_{i=1}^{m-1} |\mathbf{r}_i - \mathbf{r}_{n2}|^{-1} - m \sum_{j=m}^{2m-2} |\mathbf{r}_j - \mathbf{r}_{n1}|^{-1} \right),$$

$$V_i = -me^2(|\mathbf{r}_{2m-1} - \mathbf{r}_{n1}|^{-1} + |\mathbf{r}_{2m-1} - \mathbf{r}_{n2}|^{-1}) + e^2 \sum_{k=1}^{2m-2} |\mathbf{r}_{2m-1} - \mathbf{r}_k|^{-1}. \tag{24}$$

In Eq. 24, $H_0(1)$ and $H_0(2)$ denote the two ionic unperturbed Hamiltonians, and U_i is the interaction for the first $(2m - 2)$-electrons, whereas V_i represents the interaction of the $(2m - 1)$-electron with the other particles; the nucleus-nucleus interaction is purposely omitted since undeflected nuclear motion is assumed.

Next we introduce a set of relative coordinates and designate by $\mathbf{x}$ the distance of the (active electron) $(2m - 1)$-electron from the center of mass of the two ions; the other coordinates needed are the vector distance from the center of mass of one ion to the corresponding point on the other ion and the passive electronic nuclear distances,

$$\mathbf{x}_i = \mathbf{r}_i - \mathbf{r}_{n1}, \qquad i = 1, \ldots, m - 1, \qquad \mathbf{y}_j = \mathbf{r}_j - \mathbf{r}_{n2}, \qquad j = m, \ldots, 2m - 2,$$

of the ionic electrons from their respective nuclei. The motion of the center of mass is removed, and the terms $H_0(1)$, $H_0(2)$, and U_i are discarded, consistent with the assumptions of rectilinear motion of the two noninteracting ionic cores. Since the interaction V_i of the active electron is still cumbersome, this potential is simplified by approximating the distances from the active electron to the ionic electrons with the distances to the associated ionic nuclei:

$$\begin{aligned} \mathbf{r}_{2m-1} - \mathbf{r}_k &\approx \mathbf{r}_{2m-1} - \mathbf{r}_{n1}, \qquad k = 1, \ldots, m - 1, \\ \mathbf{r}_{2m-1} - \mathbf{r}_k &\approx \mathbf{r}_{2m-1} - \mathbf{r}_{n2}, \qquad k = m, \ldots, 2m - 2. \end{aligned}$$

With the use of these simplifications it is evident that the effective electronic Hamiltonian is approximated by

$$H = -\frac{\hbar^2}{2m}\nabla_x{}^2 - e^2(|\mathbf{x} + \tfrac{1}{2}\mathbf{R}|^{-1} + |\mathbf{x} - \tfrac{1}{2}\mathbf{R}|^{-1}). \tag{24a}$$

This result is the same as those of Eqs. 5 and 6 if ∇_R, $\nabla_R{}^2$ are omitted; μ_1 is approximated by m, and M_a is equated to M_p. In this approximation only the ionizing potentials distinguish between the $[H^+ - H(1s)]$-system and the other $[A^+(^1S) - A(^2S)]$ alkali ion-alkali atom systems; the relations between the impact velocity and the impact energy, of course, are different for each of the five processes. The application of the two-state approximation to this process is explained next.

The wave function of the two-state approximation satisfies

$$\left(H - i\hbar\frac{\partial}{\partial t}\right)\Psi_2 = 0, \qquad Z_i = Z_f = 1 \tag{24b}$$

with H defined by Eq. 24a. Consistent with Eq. 1 and Fig. 2, the incident ion P^+ moves with the constant velocity $\mathbf{v}$ parallel to the positive Z-axis, at a perpendicular distance ρ from the parallel line on which the atom A is located. The origin of coordinates is chosen so that $Z = 0$ at $t = 0$, and Eq. 15b gives the relations among the coordinate systems used in this problem. We should note, in particular, that $\mathbf{R} = \mathbf{Z} + \boldsymbol{\rho}$. Equation 19$a$, used in the wave mechanical version for nonresonance, is modified to suit the IP version for exact resonance (again effects on the phase due to the perturbing Coulomb interaction are ignored as in Section 1.1):

$$\Psi_2 = C^{(+)}(t)\psi^{(+)} \exp\left[-\frac{i}{\hbar}\int_{-\infty}^{t} dt'\eta^{(+)}(R)\right] + C^{(-)}(t)\psi^{(-1)}(t) \exp\left[-\frac{i}{\hbar}\int_{-\infty}^{t} dt'\eta^{(-)}(R)\right],$$

$$\psi^{(\pm)} = \tfrac{1}{2}\left\{[\chi^{(+)} + \chi^{(-)}] \exp\left(-\frac{i}{\hbar}\, m\mathbf{v}\cdot\mathbf{x}\right) \pm [\chi^{(+)} - \chi^{(-)}] \times \exp\left(\frac{i}{\hbar}\, m\mathbf{v}\cdot\mathbf{x}\right)\right\} \exp\left(-\frac{i}{8\hbar}\, mv^2 t\right),$$

$$H\chi^{(\pm)}(\mathbf{x}; \mathbf{R}) = \eta^{(\pm)}(R)\chi^{(\pm)}(\mathbf{x}; \mathbf{R}), \qquad \mathbf{x} = \mathbf{s}, \tag{25a}$$

$$\lim_{R\to\infty} \chi^{(\pm)} \to \frac{1}{\sqrt{2}}\,[\phi_i(\mathbf{x}_i) \pm \phi_f(\mathbf{x}_f)],$$

$$\lim_{R\to\infty} \eta^{(\pm)}(R) \to \varepsilon_i = \varepsilon_f.$$

Equation 25a contains the equations satisfied by the gerade (+) and ungerade (−) molecular wave functions $\chi^{(\pm)}$ for fixed values of the internuclear separation R. The same equation also contains the separated atom limits[14] satisfied by $\chi^{(\pm)}$ and $\eta^{(\pm)}$.

The limits of Eq. 25a corresponding to Eq. 18d of the nonresonant case are

$$\lim_{\substack{x_f\to\infty\\ t\to-\infty}} \chi^{(\pm)} \to \frac{\phi_i(\mathbf{x}_i)}{\sqrt{2}}, \qquad \lim_{\substack{x_i\to\infty\\ t\to\infty}} \chi^{(\pm)} \to \pm\frac{\phi_f(\mathbf{x}_f)}{\sqrt{2}}, \tag{25b}$$

consistent with the bound-state properties of the normalized atomic orbitals, $\phi_{i,f}$. In the present analysis the Z-axis of the $\mathbf{x}$-system is parallel to the velocity $\mathbf{v}$; however, if allowance is made for the effect of the rotation of the internuclear axis, the Z-axis is usually parallel to $\mathbf{R}$. The functions $\chi^{(\pm)}$ also satisfy the following relations:

$$\begin{aligned}\chi^{(\pm)}(-\mathbf{x}; -\mathbf{R}) &= \pm\,\chi^{(\pm)}(-\mathbf{x}; \mathbf{R}) = \chi^{(\pm)}(\mathbf{x}; \mathbf{R}),\\ \chi^{(\pm)}(\mathbf{x}; -\mathbf{R}) &= \pm\chi^{(\pm)}(\mathbf{x}; \mathbf{R}),\\ \chi^{(\pm)}(\mathbf{x}; -\mathbf{R}) &= \chi^{(\pm)}(\mathbf{x}; \rho, -z, \pi + \phi) = \chi^{(\pm)}(\mathbf{x}; \rho, -z, \phi)\\ &= \pm\chi^{(\pm)}(\mathbf{x}; \rho, z, \phi).\end{aligned} \tag{25c}$$

These symmetries are associated with the interchanging of the two identical nuclei $\mathbf{R} \to -\mathbf{R}$ and reflection of the electronic coordinate in the coordinate planes[40] $\mathbf{x} \to -\mathbf{x}$. [Only the molecular ion-states $\sum_g^+$ $(1s\sigma)$ and $\sum_u^+$ $(2p\sigma)$ are involved here.] The orthogonality integrals for the $\psi^{(\pm)}$ are simplified with the use of Eq. 25*a* to

$$\begin{aligned}&\int \overline{\psi^{(\pm)}}\psi^{(\mp)}\,d\mathbf{x}\\ &= \tfrac{1}{2}\int d\mathbf{x}\{\pm[\overline{\chi^{(+)}}\chi^{(-)} - \overline{\chi^{(-)}}\chi^{(+)}]\cos\gamma \mp i[\overline{\chi^{(+)}}\chi^{(+)} - \overline{\chi^{(-)}}\chi^{(-)}]\sin\gamma\}\\ &= 0; \qquad \gamma = \frac{m\mathbf{v}\cdot\mathbf{x}}{\hbar}.\end{aligned} \tag{25d}$$

The orthogonality expressed by Eq. 25*d* follows from the fact that the integrands are odd functions of $\mathbf{x}$, according to Eq. 25*c*. Equally easily simplified with the use of Eq. 25*a* are the two normalizing integrals whose values are

$$\begin{aligned}\int \overline{\psi^{(\pm)}}\psi^{(\pm)}\,d\mathbf{x} &= 1 \pm \sigma;\\ \sigma = \tfrac{1}{2}\int d\mathbf{x}[\overline{\chi^{(+)}}\chi^{(+)} &- \overline{\chi^{(-)}}\chi^{(-)}]\cos\gamma.\end{aligned} \tag{25e}$$

(Note from Eq. 25*b* that σ vanishes at $t = \pm\infty$.) Since $\psi^{(+)}$ is an even function of $\mathbf{x}$ and $\psi^{(-)}$ is an odd function of $\mathbf{x}$, and since the electronic operator in Eq. 24*b* does not alter this symmetry, two of the integrals of Eq. 25*f* vanish by reason of the fact that the associated integrands are odd functions of $\mathbf{x}$; in this connection also note that $(\mathbf{x}, t)$ constitute four *independent* variables:

$$\int \overline{\psi^{(\pm)}}\exp\left[\frac{i}{\hbar}\int_{-\infty}^{t}\eta^{(\pm)}\,dt'\right]\left(H - i\hbar\frac{\partial}{\partial t}\right)\psi^{(\pm)}\exp\left[-\frac{i}{\hbar}\int_{-\infty}^{t}\eta^{(\pm)}\,dt'\right]d\mathbf{x}. \tag{25f}$$

The two nonvanishing integrals of Eq. 25f have integrands denoted by $J(++)$ and $J(--)$, and their dependence on t is studied next:

$$\int J(++)\,d\mathbf{x} = if^+ + \tau^+, \qquad \frac{\partial}{\partial t_i} = \frac{\partial}{\partial t} - \frac{\mathbf{v}\cdot\nabla}{2},$$

$$\frac{\partial}{\partial t_f} = \frac{\partial}{\partial t} + \frac{\mathbf{v}\cdot\nabla}{2}, \qquad \nabla = \nabla_x, \qquad \gamma = \frac{m\mathbf{v}\cdot\mathbf{x}}{\hbar},$$

$$\tau^+ = \tfrac{1}{4}\int dx\Bigg(4\overline{\chi}^{(-)}\chi^{(-)}[\eta^{(-)} - \eta^{(+)}]\sin^2\left(\frac{\gamma}{2}\right) + \hbar\sin\gamma\Bigg\{[\overline{\chi^{(+)}} + \overline{\chi^{(-)}}]$$

$$\times \frac{\partial}{\partial t_f}[\overline{\chi^{(+)}} - \overline{\chi^{(-)}}] - [\overline{\chi^{(+)}} - \overline{\chi^{(-)}}]\frac{\partial}{\partial t_i}[\chi^{(+)} + \chi^{(-)}]\Bigg\}\Bigg),$$

$$\chi^{(\pm)}(\mathbf{r};\rho, -vt, \phi) = \pm\,\chi^{(\pm)}(\mathbf{r};\rho, vt, \phi), \tag{25g}$$

$$f^+ = \tfrac{1}{4}\int d\mathbf{x}\Bigg(2\overline{\chi^{(+)}}\chi^{(-)}[\eta^{(+)} - \eta^{(-)}]\sin\gamma - \hbar[\overline{\chi^{(+)}} + \overline{\chi^{(-)}}]$$

$$\times \frac{\partial}{\partial t_i}[\chi^{(+)} + \chi^{(-)}] - \hbar[\overline{\chi^{(+)}} - \overline{\chi^{(-)}}]\frac{\partial}{\partial t_f}[\chi^{(+)} - \chi^{(-)}]$$

$$-\hbar\cos\gamma\Bigg\{[\overline{\chi^{(+)}} - \overline{\chi^{(-)}}]\frac{\partial}{\partial t_i}[\chi^{(+)} + \chi^{(-)}]$$

$$+ [\overline{\chi^{(+)}} + \overline{\chi^{(-)}}]\frac{\partial}{\partial t_f}[\chi^{(+)} - \chi^{(-)}]\Bigg\}\Bigg).$$

If all superscripts, + and −, are interchanged in Eq. 25g, $J(--)$ is obtained. The symmetries of the $\chi^{(\pm)}$ associated with the cylindrical symmetry of Eq. 25c are reproduced in Eq. 25g since this property establishes that f^+ (or f^-) is antisymmetric in the variable t, and this fact simplifies the equations for the coefficients $\chi^{(\pm)}$ that are derived next.

Multiply Ψ_2 of Eq. 25a successively by $\int \overline{\psi^{(\pm)}} \exp\,[i\hbar^{-1}\int_{-\infty}^{t} dt'\eta^{(\pm)}]$ and integrate these results over $d\mathbf{x}$. With the use of the orthogonal conditions of Eq. 25d, the vanishing of the corresponding two integrals of Eq. 25f, and the definitions in Eq. 25g, these two results simplify to the two differential equations

$$i\hbar(1 \pm \sigma)\frac{dc^{(\pm)}}{dt} = [if^{(\pm)} + \tau^{(\pm)}]c^{(\pm)}. \tag{25h}$$

Since $f^{(\pm)}$ is antisymmetric in t and thus provides equal and opposite contributions to $c^{(\pm)}$ at $\pm t$, $f^{(\pm)}$ does not affect the solutions and can be omitted from the differential equations. The resulting approximate equations are those used by Bates and McCarroll.[7,25] The initial conditions are

$$\lim_{t\to-\infty} \psi^{(\pm)} = \frac{1}{\sqrt{2}} \phi_i(\mathbf{x}_i) \exp\left[-i\left(\frac{m\mathbf{v}\cdot\mathbf{x}}{2\hbar} + \frac{mv^2t}{8\hbar}\right)\right], \tag{25i}$$

and since we require that

$$\lim_{t\to-\infty} \Psi_2 \to \phi_i(\mathbf{x}_i) \exp\left[-i\left(\frac{m\mathbf{v}\cdot\mathbf{x}}{\hbar} + \frac{mv^2t}{8\hbar}\right)\right], \tag{25j}$$

the structure of Ψ_2 in Eq. 25*a* requires that

$$c^{(\pm)}(-\infty) = \frac{1}{\sqrt{2}}. \tag{25k}$$

Consequently, the solutions to Eq. 25*h* are

$$c^{(\pm)}(\infty) = \frac{1}{\sqrt{2}} \exp\,[i\beta^{(\pm)}], \qquad \beta^{(\pm)} = -\frac{1}{v}\int_{-\infty}^{\infty} \frac{\tau^{(\pm)}}{1 \pm \sigma}\, dz, \tag{25l}$$

in complete agreement with the results of Bates and McCarroll.[25]

We now use Eq. 25*a* to prove that

$$\lim_{t\to\infty} \frac{\psi^{(+)} - \psi^{(-)}}{\sqrt{2}} \to \phi_f(\mathbf{x}_f) \exp\left[i\left(\frac{m\mathbf{v}\cdot\mathbf{x}}{2\hbar} - \frac{mv^2t}{8\hbar}\right)\right], \tag{25m}$$

and then we use Eq. 25*b* to establish that the probability for electron transfer from A to P^+ is

$$P_{if} = \lim_{t\to\infty} \frac{1}{4}\left|\int (\overline{\psi^{(+)}} - \overline{\psi^{(-)}})\left(\psi^{+} \exp\left\{i\left[\beta^{(+)} - \frac{1}{v}\int_{-\infty}^{\infty} dz\eta^{(+)}\right]\right\}\right.\right.$$
$$\left.\left. + \psi^{(-)} \exp\left\{i\left[\beta^{(-)} - \frac{1}{v}\int_{-\infty}^{\infty} dz\eta^{(-)}\right]\right\}\right) d\mathbf{x}\right|^2 = \sin^2\alpha, \tag{25n}$$

$$\alpha = \frac{\beta^{(-)} - \beta^{(+)}}{2} + \frac{1}{2v}\int_{-\infty}^{\infty} [\eta^{(+)} - \eta^{(-)}]\, dz.$$

The cross section is given by

$$Q = 2\pi \int d\rho\rho \sin^2\alpha. \tag{25o}$$

An examination of the structure of α for low velocities reveals that $[\beta^{(-)} - \beta^{(+)}]$ is proportional to v; and with the use of Eq. 12 for rectilinear motion, it is readily found that a good approximation to α is

$$\alpha = \frac{1}{v}\int_{\rho}^{\infty} \frac{dRR[\eta^{(+)} - \eta^{(-)}]}{(R^2 - \rho^2)^{1/2}}. \tag{25p}$$

Although Bates and McCarroll[25] relate Eq. 25*p* to the wave mechanical formula of Massey and Smith,[24] Eq. 25*p* is needed here for later reference to a classical formula for the same process. In these approximations for low velocities the calculation of Q can be shortened by using the remarkably simple and effective approximating procedure of Firsov.[41] This method yields

$$\alpha(\rho^+) = \pi^{-1}, \qquad P_{if}(\rho^+) = \sin^2\left(\frac{1}{\pi}\right), \qquad Q = \frac{\pi}{2}(\rho^+)^2. \tag{25q}$$

In the PSS approximation described in the preceding paragraphs, the effect of the rotation of the internuclear line has been purposely omitted, and a short description of some of the modifications used to accommodate this rotational effect is now given.

We might expect the effect of rotation to increase as the impact parameter decreases since the corresponding angular velocity of the internuclear line increases for a given relative velocity of the assumed rectilinear motion of the two nuclei. The success of the IP version of the PSS approximation for symmetrical resonance is attributed to the relatively large contributions to the cross section from large values of the impact parameter, the region where the rotational effects can be neglected with small error.[42] (The approximation that neglects other molecular states, rotational effects, and deflection of the nuclei is sometimes referenced as an *adiabatic treatment*.[42,43]) Since these aspects of approximating procedures for collisions at low velocities are commonly misunderstood, the treatment by Bates and McCarroll is augmented.[25]

In the separated or united atom limits the angular part of the molecular eigenfunction is a spherical harmonic, $Y_l^m(\theta, \phi)$, of appropriate angles, θ and ϕ. Moreover, if the coordinate axes are rotated, these functions of the new angles are equal to a linear combination of the spherical harmonics belonging to the same irreducible representation.[44]

$$Y_{lm}(\theta', \phi') = \sum_{m'=-l}^{l} Y_{lm'}(\theta, \phi)(lm'|D(\alpha\beta\gamma)|lm). \tag{25r}$$

The angles α, β, γ occurring in this expansion are the Euler angles of the rotation. It is evident that rotation of the internuclear axis mixes the states of this example which are spatially degenerate. Consequently, molecular states that approach spatially degenerate atomic states at large values of R should all be included in the series that represents the wave function of the system; therefore, a two-state approximation in these cases cannot be expected to predict reliable cross sections.

In order to accommodate the rotational effects it is convenient to choose the Z'-axis parallel to $\mathbf{R}$ for the molecular eigenfunctions and to relate the coordinates of the active electron $\mathbf{s}' = (x', y', z')$ to the coordinates that are referred to a frame of reference whose Z-axis is parallel to the constant velocity $\mathbf{v}$. The transformation between these two systems that have the same origin is[45]

$$\begin{aligned} x' &= x \cos \Phi + y \sin \Phi, \\ y' &= (-x \sin \Phi + y \cos \Phi) \cos \Theta + z \sin \Theta, \\ z' &= (x \sin \Phi - y \cos \Phi) \sin \Theta + z \cos \Theta. \end{aligned} \tag{26a}$$

The angle between the coordinate axes z' and z is Θ, and Φ is the angle between the coordinate planes $x'z$ and xz. The transformation described by Eq. 26*a* refers equally well to the three sets $(\mathbf{s}', \mathbf{s})$, $(\mathbf{x}_i', \mathbf{x}_i)$, and $(\mathbf{x}_f', \mathbf{x}_f)$. As an example, the derivative, $\partial/\partial t_i = v(\partial/\partial Z_{x_i})$, of Eq. 25*g* is evaluated, and in this case $\mathbf{x}_i$ must not vary since $(\mathbf{x}_i, t_i)$ constitute the four independent variables; however, $\mathbf{x}_i'$ does vary with Θ and Φ through the relations of Eq. 26*a*:

$$\begin{aligned} \frac{\partial}{\partial t_i} &= \frac{v\partial}{\partial Z_{x_i}} = \mathbf{v}\cdot\nabla_{R_{xi}} = \mathbf{v}\cdot\hat{R}\frac{\partial}{\partial R_{x_i}} + \frac{\mathbf{v}\cdot\hat{\Theta}}{R}\frac{\partial}{\partial \Theta_{x_i}} + \frac{\mathbf{v}\cdot\hat{\Phi}}{R\sin\Theta}\frac{\partial}{\partial \Phi_{x_i}} \\ &= v\left(\cos\Theta\frac{\partial}{\partial R_{x_i}} - \frac{\sin\Theta}{R}\frac{\partial}{\partial \Theta_{x_i}}\right), \end{aligned} \tag{27a}$$

$$\frac{\partial}{\partial \Theta_{x_i}} = \frac{\partial'}{\partial \Theta} + \frac{\partial \mathbf{x}_i'}{\partial \Theta_x}\cdot\nabla_{x_i'}.$$

In Eq. 27*a*, $\hat{R}$, $\hat{\Theta}$, $\hat{\Phi}$ are the appropriate unit vectors with $\hat{\Phi}$ perpendicular to $\mathbf{v}$. Since the molecular wave functions are conveniently expressed in the variables $\mathbf{x}_i'$ and $\mathbf{R}$ in this example, the notion $\partial'/\partial\Theta$ is used to denote the derivative with respect to the explicit dependence upon Θ. (Equation 27*a*, of course, is an application of the chain rule for the differentiation of a compound function.) The derivative is easily found to be

$$\frac{\partial}{\partial t_i} = v\left[\frac{Z}{R}\frac{\partial}{\partial R_{x_i}} - \frac{\rho}{R^2}\left(\frac{\partial'}{\partial\Theta} - L_{x_i'}\right)\right]$$

$$L_{x_i'} = y_i'\frac{\partial}{\partial z_i'} - z_i'\frac{\partial}{\partial y_i'} = -\sin\phi_i'\frac{\partial}{\partial\theta_i'} - \cot\theta_i'\cos\theta_i'\frac{\partial}{\partial\phi_i'}, \tag{27b}$$

with z' the polar axis of the spherical coordinates (r_i', θ_i') of (x_i', y_i', z_i')

If the molecular wave functions are independent of the explicit dependence upon Θ, the result agrees with Eqs. 125 and 126 of Bates and McCarroll.[25] The wave function is now represented by

$$\Psi = \sum_p a_p(t)\psi_p{}^i(\mathbf{s}'; \mathbf{R}) \exp\left[-\frac{i}{\hbar}\int_{-\infty}^{t} \eta_p(\mathbf{R}')\, dt'\right] + \sum_q b_q(t)\psi_q{}^f(\mathbf{s}'; \mathbf{R}) \exp\left[-\frac{i}{\hbar}\int_{-\infty}^{t} \eta_q(\mathbf{R}')\, dt'\right]$$

$$\psi_p{}^i = \chi_p{}^i(\mathbf{R}) \exp\left[-\frac{i}{\hbar}\left(\frac{m\mathbf{v}\cdot\mathbf{s}}{2} + \frac{mv^2t}{8}\right)\right],$$

$$\psi_q{}^f = \chi_q{}^f(\mathbf{R}) \exp\left[\frac{i}{\hbar}\left(\frac{m\mathbf{v}\cdot\mathbf{s}}{2} - \frac{mv^2t}{8}\right)\right]. \tag{27c}$$

As in the previous case, Eq. 27*c* is inserted into Eq. 24*b* and with the use of Eq. 15*b* this result simplifies to

$$\frac{\partial a_i{}^i}{\partial Z}(t) + \sum_q \frac{\partial b_q{}^f}{\partial Z} \exp\left[-\frac{i}{\hbar}\int_{-\infty}^{t} (\eta_q{}^f - \eta_i{}^i)\, dt'\right] \int \overline{\chi_i{}^i} \exp\left(\frac{i}{\hbar} m\mathbf{v}\cdot\mathbf{s}'\right)\chi_q{}^f\, d\mathbf{s}'$$
$$= -\sum_{p\neq i} a_p{}^i(t) \exp\left[-\frac{i}{\hbar}\int_{-\infty}^{t} (\eta_p{}^i - \eta_i{}^i)\, dt'\right] \int \overline{\chi_i{}^i} \frac{\partial \chi_p{}^i}{\partial Z_{x_i}}\, d\mathbf{s}$$
$$- \sum_q b_q{}^f(t) \exp\left[-\frac{i}{\hbar}\int_{-\infty}^{t} (\eta_q{}^f - \eta_i{}^i)\, dt'\right] \times \exp\left(\frac{i}{2\hbar} m\mathbf{v}\cdot\mathbf{R}\right) \int \overline{\chi_i{}^i} \frac{\partial \chi_q{}^f}{\partial Z_{x_f}} \exp\left(\frac{i}{\hbar} m\mathbf{v}\cdot\mathbf{x}_f\right) d\mathbf{s},$$

$$\frac{\partial b_f{}^f}{\partial Z}(t) + \sum_p \frac{\partial a_p{}^i}{\partial Z}(t) \exp\left[-\frac{i}{\hbar}\int_{-\infty}^{t} (\eta_p{}^i - \eta_f{}^f)\, dt'\right] \int \overline{\chi_f{}^f} \exp\left[-\frac{i}{\hbar} m\mathbf{v}\cdot\mathbf{s}\right]\chi_p{}^i\, d\mathbf{s}$$
$$= -\sum_{q\neq f} b_q{}^f(t) \exp\left[-\frac{i}{\hbar}\int_{-\infty}^{t} (\eta_q{}^f - \eta_f{}^f)\, dt'\right] \int \overline{\chi_f{}^f} \frac{\partial \chi_f{}^f}{\partial Z_{x_f}}\, d\mathbf{s}$$
$$- \sum_p a_p{}^i(t) \exp\left[-\frac{i}{\hbar}\int_{-\infty}^{t} (\eta_p{}^i - \eta_f{}^f)\, dt'\right] \exp\left(\frac{i}{2\hbar} m\mathbf{v}\cdot\mathbf{R}\right) \times \int \chi_f{}^f \frac{\partial \chi_p{}^i}{\partial Z_{x_i}} \exp\left(-\frac{i}{h} m\mathbf{v}\cdot\mathbf{x}_i\right) d\mathbf{s}, \tag{27d}$$

$$\mathbf{v}\cdot\mathbf{R} - v^2t = \mathbf{v}\cdot\mathbf{x}_i - \mathbf{v}\cdot\mathbf{s} - \frac{\mathbf{v}\cdot\mathbf{R}}{2} = \mathbf{v}\cdot\mathbf{x}_f - \mathbf{v}\cdot\mathbf{s} + \frac{\mathbf{v}\cdot\mathbf{R}}{2}.$$

The integrals containing the derivatives $\partial/\partial Z_{x_i}$ and $\partial/\partial Z_{x_f}$ in Eq. 27*d* contain the rotational effects of the internuclear line through Eqs. 27*a* and 27*b*. This complication destroys the calculational simplicity of the two-state approximation which is incorrect in cases where rotational effects couple the states that are spatially degenerate at large values of R. Furthermore, the operator, $L_{x_i'}$, does not couple different atomic *s*-states. In order to emphasize additionally the importance of rotationally induced coupling, this effect in direct scattering processes is briefly described.

In two important papers Bates establishes the connection between the PSS and first Born approximations for direct processes.[46,47] In the first paper he uses the wave version and expands the wave function in a sum of products of molecular wave functions of the internuclear distance **R**. A rotating system of coordinates with the Z-axis parallel to **R** is used, and the atomic target consists of a nucleus and one electron, and the total Hamiltonian in the center-of-mass system is written as the sum of: (1) the operator of relative motion between the incident heavy particle (such as a proton) and the atomic target; (2) the operator for the electron in the field of the target; and (3) the interaction V arising from the presence of the incident system. The interacting potential is decomposed into a sum, $V = V_1 + V_2$; V_2 is combined with the electronic operator [preceding (2)] to form a quasi-molecular operator used to define a set of eigenfunctions, whereas V_1 is treated as a perturbing interaction. Since a Bornlike approximation is used, only the initial and final states are retained, but each of these states may be spatially degenerate.

The wave function is expanded in a sum of products of the previously discussed eigenfunctions and functions depending only upon the internuclear distance **R**, and an equation in this variable is obtained in the usual manner. The inhomogeneous term in the resulting three-dimensional Helmholtz equation consists of two parts. In the first approximation, called "**I**," $V_2 = 0$, so that the molecular wave functions become the wave functions of the atomic target; consequently, one of the inhomogeneous terms vanishes and the other term becomes a scalar function of R since the angle θ between the electronic position **s** and **R** is a variable of integration. This scalar quality emerges from the use of the rotating coordinate system, but, more important, this term vanishes unless the initial and final magnetic quantum numbers are the same. In the resulting amplitude for scattering the final wave vector $\mathbf{k}_n$ is parallel to **R** (infinite separation of initial and final aggregates), and the initial wave vector $\mathbf{k}_0$ is parallel to a Z-axis, which is nonrotating.

Next this Z-axis is taken parallel to $\mathbf{A} = \mathbf{k}_n - \mathbf{k}_0$, the momentum-change vector, and another expression for the amplitude is obtained. For a comparison the Born amplitude is calculated with the same atomic wave functions for the target, but these are referred to a nonrotating coordinate system in which **A** is again chosen parallel to the Z-axis. As a consequence, the previous scalar interaction, denoted by $U_{on}{}^m(R)$, is replaced by $V_{on}{}^m(\mathbf{R})$, and this is a function of a vector R since θ is *not* the angle between the new **s** and **R**. (Here, m denotes the magnetic quantum number, and o, n, which include m, refer to the initial and final atomic states, respectively.) It is also shown, using an expansion in spherical harmonics, that for $1s \rightarrow 1s$ transitions $V_{on}{}^m(\mathbf{R})$ can be replaced by $U_{on}{}^m(R)$, with the result that **I** and the first Born approximation are identical for this special case. For other transitions these two approximations are not the same, and **I** must be poor at moderate and high energies, since the electronic motion is unable to follow the rapid rotation of the internuclear line; or stated more precisely, coupling between different m-values of the irreducible representations (one or two azimuthal quantum numbers) is important, but only $m = 0$ occurs in the $1s$-$1s$ case. In the second approximation, hereafter referenced as **II**, $V_1 = 0$, and the expansion of the wave function would lead to the PSS approximation.

These molecular wave functions and the corresponding eigenenergies are obtained from perturbation theory using the first-order expansion in terms of the atomic wave functions of the target. The IP version of this approximation is analyzed in another important paper by Bates.[47] In this analysis the wave function is expanded in terms of molecular functions and unknown coefficients that depend upon time. First-order perturbation theory is again used for the molecular wave functions as in **II**. The coupled differential equations for the unknown coefficients are approximated by retaining only the initial and final atomic states, both of which may be spatially degenerate. Constant rectilinear motion of the colliding systems is assumed.

In the first case low velocities and a strong interaction are assumed. The equations are approximated well by the IP version of the PSS method. These equations are known as the PRA (perturbed rotating atom) approximation. If unperturbed eigenfunctions are used to facilitate the integration of these equations, this simplified version is denoted as the pRA approximation, and is the same as the IP version of **II**. (The lower-case p in pRA means that this is an approximation to the PRA approximation.)

In the second case high velocities and a weak interaction are postulated. The eigenenergies are replaced by their unperturbed values; although the resulting equations remain coupled, they are equivalent to the corresponding version of the first Born approximation. In summary, the difference between the pRA and the first Born approximation originates from the treatment of terms which couple states that differ (in the rotating frame of reference) only in the magnetic quantum numbers; in the pRA method this coupling is discarded, but in the first Born method the coupling is retained, although certain complex exponential multiplicative factors occurring in the coupling terms are equated to unity. A complementary view is that in the first Born approximation the perturbed eigenfunctions of the target are referred to a fixed frame of reference, whereas in the pRA approximation these functions are referred to the rotating frame of reference.[48] Since the equations for electronic capturing processes generally are more difficult to solve than the equations governing direct scattering processes, it is clear that spatial degeneracy greatly complicates the calculational procedure for electronic capture. An analysis of these effects for charge transfer and some suggested improvements for the associated approximate wave functions are discussed in a paper by Schneiderman and Russek.[49] A few additional remarks are made in conclusion to this section on the PSS method.

A pioneering calculation on charge transfer between protons and hydrogen atoms in the 1*s*-state was effected by Dalgarno and Yadav, who used the wave version of the two-state expansion in molecular wave functions.[50] Gurnee and Magee repeated this calculation using the IP version of a two-state expansion in terms of atomic orbitals without allowance for transfer of translational motion to the electron.[51] Ferguson calculated cross sections for the same process using the IP version of the two-state expansion in molecular orbitals including allowance for the transfer of translational motion to the electron.[52] Rapp and Francis also used the IP version to derive an expression resembling the PSS formula defined by Eqs. 25*n* and 25*o* to calculate cross sections for a number of singly charged ions capturing from their gaseous parents.[53] (In connection with the calculations of Gurnee and Magee and of Rapp and Francis consult Eqs. 40*k* and 41*q* of Section 3.1.) Bates and Williams used a four-state approximation that includes the atomic wave functions of the second quantum state at infinite separation of the colliding aggregates.[54] Comparisons of these and other predictions with the corresponding experimentally determined quantities are deferred to a subsequent section.

1.4 ACCIDENTAL RESONANCE AND NONRESONANCE

All processes of electronic capture between atomic systems that are not symmetrically resonant (exactly resonant) are nonresonant. If the internal energies $\epsilon_{i,f}$, of Eqs. 10*a* and 10*b* are exactly equal (or very closely equal), the initial and final colliding systems are accidentally resonant.[55] Symmetrical resonance, it will be recalled, also requires equality of the two operators, $(H_i + V_i)$ and $(H_f + V_f)$ of Eq. 3.

This equality means that $(H_f + V_f)$ is obtained from $(H_i + V_i)$ by substituting the $\mathbf{x}_f$-coordinates (post) for the corresponding $\mathbf{x}_i$-coordinates (prior) if the masses $M_{i,f}$ also are identical. Bates and Lynn analyzed and compared the cross sections for electronic capture for accidentally and exactly resonant systems.[56] Their study was devoted to low velocities of relative motion, and the IP version of the two-state approximation was used in that analysis. Although the two-state representation for this problem is easily inferred from Eqs. 19*a*, 25*a*, and 27*c*, this approximation is rewritten here for convenience:

$$\Psi_2 = a_i(t)X_i + a_f(t)X_f,$$

$$X_i = \left[p_i(\mathbf{R})\phi_i(\mathbf{x}_i)\exp\left(-\frac{i}{2\hbar}m\mathbf{v}\cdot\mathbf{s}\right) + p_f(\mathbf{R})\phi_f(\mathbf{x}_f)\exp\left(\frac{i}{2\hbar}m\mathbf{v}\cdot\mathbf{s}\right)\right]$$
$$\times \exp\left\{-\frac{i}{\hbar}\left[\int_{-\infty}^{t}\varepsilon_i(\mathbf{R}')\,dt' + \frac{mv^2t}{8}\right]\right\}, \tag{28a}$$

$$X_f = \left[q_i(\mathbf{R})\phi_i(\mathbf{x}_i)\exp\left(-\frac{i}{2\hbar}m\mathbf{v}\cdot\mathbf{s}\right) + q_f(\mathbf{R})\phi_f(\mathbf{x}_f)\exp\left(\frac{i}{2\hbar}m\mathbf{v}\cdot\mathbf{s}\right)\right]$$
$$\times \exp\left\{-\frac{i}{\hbar}\left[\int_{-\infty}^{t}\varepsilon_f(\mathbf{R}')\,dt' + \frac{mv^2t}{8}\right]\right\}.$$

The type of representation given in Eq. 28*a* admits the use of an atomic or molecular model simply by altering the functions $p_{i,f}$ and $q_{i,f}$. The coefficients $a_{i,f}$ are determined by the requirement that Ψ_2 satisfies Eq. 15*a*. Since this problem treats the case of two nuclei and a single electron, the electronic Hamiltonian is given by Eq. 6 with $\nabla_R = 0$.

The analysis is simplified by setting v equal to zero in the exponentials, $\exp[\pm i(2\hbar)^{-1}m\mathbf{v}\cdot\mathbf{s}]$ after the differentiating operations are performed. At these low velocities of relative motion it is easily established that

$$p_i^2 + 2p_ip_fS + p_f^2 = q_i^2 + 2q_iq_fS + q_f^2 = 1$$
$$S = \int d\mathbf{s}\overline{\phi_i(\mathbf{x}_i)}\phi_f(\mathbf{x}_f), \tag{28b}$$

which ensures the normalization of X_i and X_f. If Ψ_2 of Eq. 28*a* is inserted into Eq. 15*a*, and this equation is multiplied, successively, by X_i^* and X_f^*, two simultaneous first-order differential equations for $a_i(t)$ and $a_f(t)$ are derived by integrating the preceding two equations over $d\mathbf{s}$. Coefficients related to a_i and a_f are obtained next by rearranging Eq. 28*a* in the form

$$\Psi_2 = \left\{c_i(t)\phi_i(\mathbf{x}_i)\exp\left[-\frac{i}{\hbar}\left(\frac{m\mathbf{v}\cdot\mathbf{s}}{2} + E_it\right)\right]\right.$$
$$\left. + c_f(t_f)\phi_f(\mathbf{x}_f)\exp\left[\frac{i}{\hbar}\left(\frac{m\mathbf{v}\cdot\mathbf{s}}{2} - E_ft\right)\right]\right\}\exp\left(-\frac{imv^2t}{8\hbar}\right); \tag{28c}$$

$$c_i(-\infty) = 1, \qquad c_f(-\infty) = 0,$$

and with these incorporated initial conditions, the cross section for electronic capture from the bound state $\phi_i(\mathbf{x}_i)$ of energy E_i into the bound state $\phi_f(\mathbf{x}_f)$ of energy E_f is

$$Q_{if} = 2\pi \int_0^\infty |c_f(\infty)|^2 \rho \, d\rho. \tag{28d}$$

[The functions $p(\mathbf{R})$ and $q(\mathbf{R})$ of Eq. 28*a* have been absorbed into the coefficients $c_i(t)$ and $c_f(t)$ of Eq. 28*c*.] The initial and final atomic states are assumed to be spatially nondegenerate in order to avoid the coupling complications induced by the rotating internuclear axis.

In the molecular model of Bates and Lynn,[57] the variational principle is used to adjust $p_{i,f}$ and $q_{i,f}$ so that $X_{i,f}$ represent the relevant eigenfunctions and $\epsilon_{i,f}$ represents the corresponding eigenenergies; all are obtained by using a linear combination of atomic orbitals to represent the quasi-molecule formed by the electron and the two nuclei for fixed values of the internuclear distance, R. Out of this analysis emerge the two equations

$$\dot{a}_i = p_i q_f \dot{S} a_i - L(p, q) a_f \exp\left[\frac{i}{\hbar}\int_{-\infty}^{t} dt'(\varepsilon_i - \varepsilon_f)\right],$$

and

$$\dot{a}_f = p_f q_i \dot{S} a_f - L(q, p) a_i \exp\left[\frac{i}{\hbar}\int_{-\infty}^{t} dt'(\varepsilon_f - \varepsilon_i)\right], \tag{28e}$$

$$\dot{Q} = \frac{dQ}{dt}, \qquad L(p, q) = p_i \dot{q}_i + p_f \dot{q}_f + (p_i \dot{q}_f + p_f \dot{q}_i) S.$$

If the resonance is accidental, new coefficients defined by

$$b_i = a_i \exp\left(-\frac{1}{\hbar}\int_{-\infty}^{t} p_i p_f \dot{S} dt'\right), \qquad b_f = a_f \exp\left(-\frac{1}{\hbar}\int_{-\infty}^{t} q_i q_f \dot{S} dt'\right), \tag{28f}$$

are introduced, and the modified equations satisfied by $b_{i,f}$ relate $\dot{b}_{i,f}$ directly to $b_{f,i}$. According to Eq. 28*f*, $b_{i,f}(-\infty) = a_{i,f}(-\infty)$; these relations together with Eq. 28*b* and the other restrictions imposed on $p_{i,f}$, $q_{i,f}$ that relate them to $\epsilon_{i,f}$ together allow the initial conditions of Eq. 28*c* to be recast as $b_i(-\infty) = 1$, $b_f(-\infty) = 0$. After the change of variable $z = vt$ is made the equations satisfied by $b_{i,f}$ are

$$\frac{db_i}{dz} = -M_i(z) b_f \exp\left[-\frac{i}{\hbar v}\int_{-\infty}^{z} (\varepsilon_f - \varepsilon_i)\, dz\right], \tag{28g}$$

$$\frac{db_f}{dz} = -M_f(z) b_i \exp\left[\frac{i}{\hbar v}\int_{-\infty}^{z} (\varepsilon_f - \varepsilon_i)\, dz\right]. \tag{28h}$$

The coupling factors $M_{i,f}$ tend to finite values (not given) as v approaches zero, but the exponential factors oscillate increasingly with decreasing v; consequently, $b_i(z)$ differs very little from its initial value of unity at $z = -\infty$. From this variation it is clear that $b_f(z)$ is always much less than unity, and back-coupling is thus small; that is, $b_f(z)$ affects Eq. 28g negligibly. Since back-coupling is negligible, b_i can be equated to unity in Eq. 28h, and the resulting solution for $b_f(\infty)$ is

$$b_f(\infty) = -\int_{-\infty}^{\infty} M_f(z) \exp\left[\frac{i}{\hbar v}\int_{-\infty}^{z} dz'(\varepsilon_f - \varepsilon_i)\right] dz,$$
$$|b_f(\infty)| = |c_f(\infty)|. \tag{28i}$$

Equations 28d and 28i yield the cross section, Q_{if}. This approximation allows for distortion.[57] Distortion usually is discussed in connection with approximations using atomic functions, and as a correction to the atomic energies distortion can be interpreted as the difference between the effect of the interaction on the colliding systems when in the initial state and when in the final state; however, the correction vanishes for symmetrical resonance.[25] If

$$p_i = p_f = 2^{-1/2}(1 + S)^{-1/2} \qquad \text{and} \qquad q_i = -q_f = 2^{-1/2}(1 - S)^{-1/2},$$

the cross section for symmetrical resonance is easily obtained and is

$$Q_{if} = 2\pi \int_0^{\infty} d\rho\rho \sin^2 \eta, \qquad \eta = \frac{1}{v}\int_0^{\infty} (\varepsilon_f - \varepsilon_i)\, dz,$$

$$\varepsilon_i - \varepsilon_f = 2\,\frac{h_{fi} - Sh_{ff}}{1 - S^2} = 2\,\frac{h_{if} - Sh_{ii}}{1 - S^2},$$

$$h_{fi} = h_{if} = \int \overline{\phi_i(\mathbf{x}_i)} V^A(\mathbf{x}_i)\phi_f(\mathbf{x}_f)\, ds,$$

$$= \int \overline{\phi_f(\mathbf{x}_f)} V^P(\mathbf{x}_f)\phi_i(\mathbf{x}_i)\, ds, \tag{28j}$$

$$h_{ii} = h_{ff} = \int \overline{\phi_i(\mathbf{x}_i)} V^P(\mathbf{x}_f)\phi_i(\mathbf{x}_i)\, ds$$

$$= \int \overline{\phi_f(\mathbf{x}_f)} V^A(\mathbf{x}_i)\phi_f(\mathbf{x}_f)\, ds,$$

$$\varepsilon_i(\infty) = \varepsilon_f(\infty).$$

If $\varepsilon_{i,f}$ are replaced by the exact energies for the molecular wave functions, Eq. 28j agrees with the corresponding result derived by the method of perturbed stationary states.[56,58] The atomic model for this accidentally resonant process now is discussed briefly.

In this case $p_i = 1, p_f = 0, \varepsilon_i = E_i, q_i = 0, q_f = 1, \varepsilon_f = E_f$, and a glance at Eq. 28*a* confirms that Ψ_2 is spatially represented by the initial and final atomic eigenfunctions $\phi_{i,f}$ with their corresponding eigenenergies $E_{i,f}$. These coupled differential equations are easily rearranged into the forms

$$i\hbar \dot{a}_{i,f} = C_{i,f}a_{i,f} + D_{i,f}a_{f,i} \exp\left[\frac{i}{\hbar}(-, +)\beta t\right],$$
$$\beta = E_f - E_i, \qquad a_{i,f} = b_{i,f} \exp\left(-\frac{i}{\hbar}\int_0^t C_{i,f}\, dt'\right). \tag{28k}$$

The transformation between the $a_{i,f}$ and $b_{i,f}$, originated by Dirac,[59] is commonly referred to as "allowing for distortion."[57] It is an easy task to derive the coupled first-order differential equations corresponding to this model:

$$\frac{db_i}{dz} = -\frac{i}{v} D_i b_f e^{-i\zeta/v}$$

$$\frac{db_f}{dz} = -\frac{i}{v} D_f b_i e^{-i\zeta/v},$$

$$\zeta = \int_0^z (E_f - E_i + C_f - C_i)\, dz',$$

$$C_f = \frac{h_{ff} - Sh_{if}}{1 - S^2}, \qquad C_i = \frac{h_{ii} - Sh_{fi}}{1 - S^2},$$

$$D_f = \frac{h_{fi} - Sh_{ii}}{1 - S^2}, \qquad D_i = \frac{h_{if} - Sh_{ff}}{1 - S^2},$$

$$h_{ii} \neq h_{ff}, \qquad h_{if} \neq h_{fi}.$$

If b_i is equated to one in the differential equation for b_f, the resulting solution is labeled the *distortion-approximation*.

Attention is directed to the coupling factors $D_{i,f}/v$ of these differential equations, which increase without limit as v tends to zero. We thus see that the approximations used to solve the equations of the molecular model are unacceptable for the equations of the atomic model. Both back-coupling and forward-coupling are strong and cannot be omitted. We return to the molecular model and note from Eqs. 28*i* and 28*j* that the variation of the cross sections for accidentally and symmetrically resonant processes is radically different as the relative velocity tends to zero.

The other features reemphasized are the origins and definitions of the phrases, allowance for distortions, distortion-approximation, and back-coupling.

REFERENCES

1. N. F. Mott and H. S. W. Massey, *The Theory of Atomic Collisions* (3rd ed.), Oxford University Press, London (1965), p. 630.
2. H. Goldstein, *Classical Mechanics*, Addison-Wesley, Reading, Mass. (1950), pp. 85–89.
3. L. D. Landau and E. M. Lifshitz, *Mechanics*, Addison-Wesley, Reading, Mass. (1960), p. 49.
4. Mott and Massey, *op. cit.*, p. 98.
5. Landau and Lifshitz, *op. cit.*, p. 31.
6. *Ibid.*, p. 44.
7. D. R. Bates and R. McCarroll, *Proc. Roy. Soc.* (*London*)*A*, **245**, 175 (1958).
8. H. C. Brinkman and H. A. Kramers, *Proc. Acad. Sci. Amsterdam*, **33**, 973 (1930).
9. B. H. Bransden, A. Dalgarno, and N. M. King, *Proc. Phys. Soc.* (*London*)*A*, **67**, 1075 (1954).
10. A. Messiah, *Quantum Mechanics*, Vol. 2 (J. Potter, trans.) (1962); reprint ed., North-Holland, Amsterdam, and Interscience, New York (1963), pp. 650–652.
11. A. Messiah, *Quantum Mechanics*, Vol. 1 (G. M. Temmer, trans.) (1961); reprint ed., North-Holland, Amsterdam, and Interscience, New York (1963), p. 206.
12. R. A. Mapleton, "Classical and Quantal Calculations on Electron Capture," Air Force Cambridge Research Laboratories Report AFCRL 67–0351, Physical Science Research Papers, No. 328 (1967), pp. 42–45.
13. W. Pauli, "... Die allgemeinen prinzipien der Wellenmechanik ..." [reprint of *Handbuck der Physik*, 2 aufl., Band 24, 1 Teil, Springer-Verlag, Berlin (1933), Ann Arbor, Mich. (1947), p. 100.
14. D. R. Bates, H. S. W. Massey, and A. L. Stewart, *Proc. Roy. Soc.* (*London*)*A*, **216**, 437 (1953).
15. W. R. Thorson and H. Levy, II, *Phys. Rev.*, **181**, 230 (1969).
16. H. Levy, II, and W. R. Thorson, *Phys. Rev.*, **181**, 244 (1969).
17. H. Levy, II, and W. R. Thorson, *Phys. Rev.*, **181**, 252 (1969).
18. D. R. Bates, "Theoretical Treatment of Collisions between Atomic Systems," in *Atomic and Molecular Processes* (D. R. Bates, ed.), Academic Press, New York (1962), pp. 585–586.
19. Mott and Massey, *op. cit.*, p. 429.
20. J. R. Oppenheimer, *Phys. Rev.*, **31**, 349 (1928).
21. Landau and Lifshitz, *op. cit.*, pp. 18–41.
22. Mott and Massey, *op. cit.*, pp. 619–620.
23. *Ibid.*, pp. 654–655.
24. H. S. W. Massey and R. A. Smith, *Proc. Roy. Soc.* (*London*)*A*, **142**, 142 (1933).
25. D. R. Bates and R. McCarroll, *Phil. Mag. Suppl.*, **11**, 39 (1962).
26. Mott and Massey, *op. cit.*, pp. 428–436.
27. *Ibid.*, pp. 428–429.
28. *Ibid.*, p. 435.
29. *Ibid.*, p. 434.
30. *Ibid.*, pp. 77, 350, 354–356, 434–435.
31. *Ibid.*, p. 77.
32. D. R. Bates, *Proc. Roy. Soc.* (*London*)*A*, **240**, 437 (1957).
33. D. R. Bates and A. R. Holt, *Proc. Roy. Soc.* (*London*)*A*, **292**, 168 (1966).
34. P. Pechukas, *Phys. Rev.*, **181**, 166 (1969).

35. P. Pechukas, *Phys. Rev.*, **181**, 174 (1969).
36. Mott and Massey, *op. cit.*, Chapter 2.
37. *Ibid.*, p. 9.
38. Goldstein, *op. cit.*, p. 62.
39. E. U. Condon and G. H. Shortley, *Theory of Atomic Spectra* (1935); reprint corr. ed., Cambridge University Press, New York (1953), Chapter 7.
40. G. Herzberg, *Molecular Spectra and Molecular Structure*. Vol. I, *Spectra of Diatomic Molecules*, D. Van Nostrand, Princeton, N.J. (1950), pp. 28, 318.
41. O. B. Firsov, *Zh. Eksperim. i Theor. Fiz.*, **21**, 1001 (1951); (AEC-tr-1529).
42. D. R. Bates, *Quantum Theory*, Academic Press, New York (1961), pp. 290–293.
43. L. Ingber, *Phys. Rev.*, **139**, A35 (1965).
44. A. R. Edmonds, *Angular Momentum in Quantum Mechanics*, Princeton University Press, Princeton, N.J. (1957), p. 54.
45. Goldstein, *op. cit.*, pp. 107–109.
46. D. R. Bates, *Proc. Roy. Soc.* (*London*)*A*, **240**, 437 (1957).
47. D. R. Bates, *Proc. Roy. Soc.* (*London*)*A*, **243**, 15 (1957).
48. D. R. Bates, *Proc. Roy. Soc.* (*London*)*A*, **245**, 299 (1958).
49. S. B. Schneiderman and A. Russek, *Phys. Rev.*, **181**, 311 (1969).
50. A. Dalgarno and H. N. Yadav, *Proc. Phys. Soc.* (*London*)*A*, **66**, 173 (1953).
51. E. F. Gurnee and J. L. Magee, *J. Chem. Phys.*, **26**, 1237 (1957).
52. A. F. Ferguson, *Proc. Roy. Soc.* (*London*)*A*, **264**, 540 (1961).
53. D. Rapp and W. E. Francis, *J. Chem. Phys.*, **37**, 2631 (1962).
54. D. R. Bates and D. A. Williams, *Proc. Phys. Soc.* (*London*)*A*, **83**, 425 (1964)
55. Bates, *Atomic and Molecular Processes*, pp. 605–607.
56. D. R. Bates and N. Lynn, *Proc. Roy. Soc.* (*London*)*A*, **253**, 141 (1959).
57. Bates, *Atomic and Molecular Processes*, pp. 588–590.
58. Mott and Massey, *op. cit.*, p. 432.
59. P. A. M. Dirac, *The Principles of Quantum Mechanics* (4th ed.), Oxford University Press, London (1958), pp. 173–174.

CHAPTER 2

ELECTRON CAPTURE: SUNDRY TOPICS

2.1 CLASSICAL METHOD FOR SYMMETRICAL RESONANCE

In this chapter we discuss an approximate classical method for obtaining the cross section for electronic capture in alkali ion-alkali atom collisions.[1] A secondary purpose of defining those cross sections is to provide comparisons with the corresponding semiclassical and approximate quantum-mechanical cross sections.

In Fig. 4 the two ions N_1 and N_2 are imagined to be in uniform rectilinear motion along the parallel lines AB and CD. This assumption means that the nucleus-nucleus interaction is omitted and that the other electron-nucleus interactions produce negligible deflections. The electron e depicted in the plane of the paper is at distance x from the midpoint O of the internuclear line N_1N_2, and the cylindrical coordinates of e referred to O are ρ, Z, and ϕ

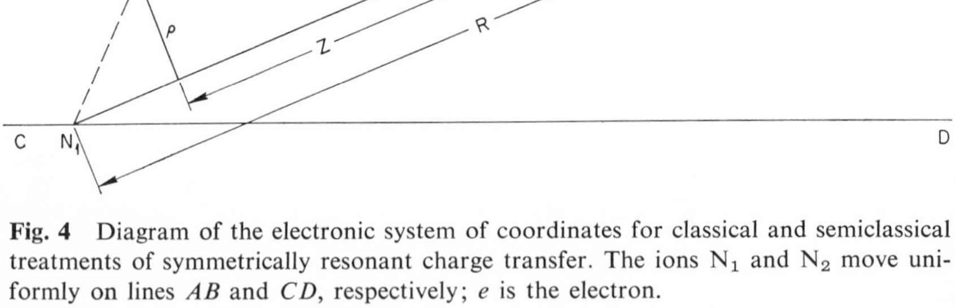

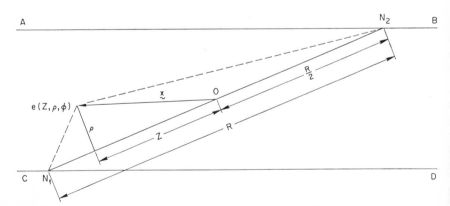

Fig. 4 Diagram of the electronic system of coordinates for classical and semiclassical treatments of symmetrically resonant charge transfer. The ions N_1 and N_2 move uniformly on lines AB and CD, respectively; e is the electron.

with the plane $\phi = 0$ being coincident with the plane of the paper. In the frame of reference with N_1 fixed, N_2 moves with constant velocity v along the line AB, and the problem is to calculate the classical cross section for the transfer of the electron from one ion to the other in a given potential field. The approximate potential field and the associated classical version of the Hamiltonian are deduced from Eq. 24*a*, and these functions are given (in atomic units) in Eq. 29*a*:

$$H(\text{cl}) = \tfrac{1}{2}v^2(\text{cl}) + V_i,$$

$$V_i \approx -\left[\frac{4}{R} + 8\,\frac{(2Z^2 - \rho^2)}{R^3}\right], \qquad \rho \ll R, \quad Z \ll R, \tag{29a}$$

where "cl" denotes classical. This approximation to V_i obtained from the binomial expansion includes only terms up to the second power in the cylindrical variables ρ and Z. The total energy of this classical electron in the presence of only *one ion* is

$$E' = -\frac{1}{2n^2}, \qquad n^2 = \frac{\varepsilon_0}{I}, \tag{29b}$$

where n is defined in terms of the ionization potential I.

In order to allow for the modification of E' by the presence of the other ion, however, we look at V_i. This interaction defined in Eq. 24*a* is approximately

$$V_i = -\frac{1}{r} - \frac{1}{|R - r|} \approx -\frac{1}{r} - \frac{1}{R} \tag{29c}$$

if R is much larger than the electron-ion distance r. Since only central forces and conservative fields are allowed here, the virial theorem of classical mechanics is applicable, and a corrected energy E for the electron is readily obtained:[2]

$$\bar{E}' = \overline{\left(E + \frac{1}{R}\right)} = E + \frac{1}{R} = \overline{(T + V)} = -\overline{\left(\frac{1}{2r}\right)} = -\frac{1}{2n^2}. \tag{29d}$$

The barred symbols denote the average values defined in the virial theorem. (The same result also emerges as the first-order correction to the energy according to quantum mechanics provided that the unperturbed wave function is the 1*s*-wave function of atomic hydrogen.[3]) This energy is used with the classical equation for the conservation of energy to obtain

$$\tfrac{1}{2}v^2(\text{cl}) - \left(\frac{3}{R} + \frac{8Z'^2}{R^3} - \frac{1}{2n^2}\right) = 0 \qquad Z'^2 = 2Z^2 - \rho^2, \tag{29e}$$

and this relation determines the regions of allowed and disallowed classical motion for electronic flow between the two ions.

The onset of classical electronic flow occurs at

$$R = R_0 = 6n^2, \qquad Z'^2 = 0, \tag{29f}$$

and although Z' vanishes for various pairs of values of (Z, p), the pair (0, 0) is of particular interest since this is a point of symmetry with respect to the two ions. According to the approximate relations given by Eqs. 29*e* and 29*f*, classical electronic flow can occur for all $R < R_0$ provided that the distance of closest approach $R(\text{c.a.}) < R_0$, and for a fixed value of R, the minimum value of $v^2(\text{cl})$ of Eq. 29*e* occurs at $Z = 0$. If we recall the restrictions imposed on ρ, Z, and R by the binomial expansion used in Eq. 29*a*, the classical probability that the electron is attached to a given ion (on a given side of the plane $Z = 0$) is 1/2. If one of the nuclei is imagined fixed, then as the nucleus-nucleus line of length R_0 assumes all orientations about this fixed nuclear point the area generated by the electronic midpoint of this line segment is $4\pi(R_0/2)^2$. The approximate classical cross section for electronic capture thus is given by

$$Q(\text{cl}) = \tfrac{1}{2} \times 4\pi\left(\frac{R_0}{2}\right)^2 = 18\pi n^4. \tag{29g}$$

This represents the probability times the area associated with the onset of classical electronic flow across the plane $Z = 0$ of Eq. 29*f* for the case $\rho = 0$. Next, mechanical similarity[4] is used to relate the classical quantities for different values of n.

It is readily established with the aid of Eq. 29*e* that lengths L, electronic velocities v, nuclear velocities V, and times dt satisfy the following relations for $n = n$ and $n = 1$:

$$\begin{aligned} L(n) &= n^2 L(1), \qquad v(n) = \frac{1}{n}\, v(1), \\ dt(n) &= n^3\, dt(1), \qquad V(n) = \frac{1}{n}\, V(1), \\ \frac{dR(n)}{dt(n)} &= \frac{V(n)[R^2(n) - \rho^2(n)]^{1/2}}{R(n)}, \\ Q(n) &= n^4 Q(1). \end{aligned} \tag{29h}$$

Equation 29*h* is obtained from Eq. 29*e* for $n = 1$ and $n = n$ by imposing the condition on length alone, and as a consequence we get the condition on the electronic velocity $v(n)$, which incidentally is the same as the one obtained between root-mean-square velocities calculated from quantum mechanics for atomic hydrogen.[5] Although the relation between impact velocities $V(1)$ and $V(n)$ is assumed to obey the same scaling law, it is not as easily interpreted since the classical cross section is independent of impact velocity. This independence is an obvious defect in the classical result, and although Q(cl) predicts the correct order of magnitude of the measured values at low impact velocities,[1] Q(cl) fails to increase as V decreases;[6] however, this defect is partially remedied by replacing the classical method by a semiclassical method which is explained in the next section.

2.2 SEMICLASSICAL METHOD FOR ALKALI ATOM-ALKALI ION COLLISIONS

In the semiclassical method the restriction of the classically forbidden region of Eq. 29*e* is removed; that is, the electronic motion can penetrate beyond the classical turning points where $(E - V_i)$ of Eqs. 29*a* and 29*d* are negative. This means that charge transfer can occur for $R > R_0$. The JWKB (Jeffreys-Wentzel-Kramers-Brillouin) approximating technique is used here to calculate this probability of penetration of the potential barrier.[7] For this treatment to be applicable, the electronic velocity must greatly exceed the relative nuclear velocity so that the nuclear motion can be ignored in the JWKB part of the calculation.

The electronic Hamiltonian of Eq. 24*a* is written in cylindrical coordinates, and the approximate interaction V_i of Eq. 29*a* and energy E of Eq. 29*d* are used again. Since this process is cylindrically symmetrical, the solution is independent of ϕ, and the variation of the wave function with ρ is assumed to be negligible; thus the approximate Schrödinger equation is

$$-\frac{d^2\psi}{dZ^2} + V_i\psi = E\psi, \tag{30a}$$

If the variation with ρ is retained, however, the unmodified Eq. 30*a* is exactly soluble. The equation in ρ is that of the two-dimensional harmonic oscillator,[8] and the parabolic cylinder functions satisfy the equation in Z.[9] These solutions, desirable as they are, can hardly be justified because of the many other approximations required in conjunction with this analysis. In order to get the cross section, the probability for electronic transfer between the ions is calculated, and the Firsov approximation of Eq. 25*q* is used to derive the cross section from this probability.

First, the JWKB approximate solution to Eq. 30*a* is used to get the associated coefficient of transmission:[10]

$$T(\rho) = \exp\left(-2\int_{Z-}^{Z+} v_1\, dZ\right),$$

$$v_1^2 = 32R^{-3}(A^2 - Z^2), \tag{30b}$$

$$A^2 = \frac{R^3}{16}\left(\frac{1}{2n^2} - \frac{3}{R} + \frac{8\rho^2}{R^3}\right), \qquad Z\pm = \pm A.$$

The domain of integration in Eq. 30*b* is between the classical turning points defined by $v(\mathrm{cl}) = 0$ in Eq. 29*a*. This transmission through the parabolic potential barrier is along the line $\rho =$ constant; therefore the next quantity calculated is the corresponding area of transmission $A(T)$:

$$A(T) = 2\pi\int_0^r d\rho\rho\, T(\rho) = 2\pi\int_0^r d\rho\rho \exp(-2\pi A)$$

$$= F(\rho)\left(\frac{R}{2}\right)^{3/2} \exp\left[-\pi\left(\frac{R}{2}\right)^{3/2}\left(\frac{1}{2n^2} - \frac{3}{R}\right)\right], \tag{30c}$$

$$F(r) = 1 - \exp\left[-\pi\left(\frac{2}{R}\right)^{3/2} r^2\right], \qquad r < \frac{R}{2}.$$

The radius of convergence of the binomial expansion used in Eq. 29*a* restricts the value of r as recorded in Eq. 30*c*, and according to Eq. 29*e* the minimum value of R is $R_0 = 6n^2$. Thus, if r is equated to $R/2$, $F(r)$ is close to unity even for R equal to its minimum value; consequently, this approximation for $A(T)$ is used.

Next the probability sought is decomposed into a product of two probabilities:

$$dP = P_1\, dP_2, \qquad P_1 = \frac{A(T)}{4\pi(R_0/2)^2}, \tag{30d}$$

$$dP_2 = \frac{w}{d}\, dt, \qquad dt = \frac{2dRR}{V[R^2 - (\rho^+)^2]^{1/2}}.$$

The probability P_1 is expressed as the ratio of the area $A(T)$ and the area available at the onset of classical electronic flow, a quantity already used in Eq. 29*g*. The other factor, dP_2, is composed of an electronic velocity w, a distance d traveled by the electron between impingements of the barrier, and the differential of time deduced from Eq. 12 for uniform rectilinear motion of the ions. The factor of 2 occurs in the last expression since the process is symmetric about the distance of closest approach, ρ^+. It is clear that V must be much smaller than w to insure that the frequency of electronic impingement of the barrier changes slowly with time; otherwise, the present approximating procedure would be inadequate. The velocity selected for w is the root-mean-square value for a hydrogenic atom, and is $v_n = n^{-1}$ in atomic units. It is noteworthy that this same root-mean-square value is obtained from the microcanonical distribution of classical statistical mechanics.[11] Since this value of velocity is associated with the energy of an atomic state, appeal is next made to the classical orbital equation for an electron of the same total energy,

$$\tfrac{1}{2}v^2 - \frac{1}{r} = -\frac{1}{2n^2}, \qquad v_n = n^{-1}.$$

From this relation it is observed that the maximum value of r is $r(\max) = 2n^2$. We may visualize the orbit of this electron being distorted by the presence of the other ion so that the perturbed orbit extends from $Z = 0$ to $Z = 5n^2$ in Fig. 4, and from this notion alone d is equated to

$$d = 2n^2 + \frac{R_0}{2} = 5n^2.$$

Although this selection of w and d is somewhat arbitrary, the values are probably good estimates of the average values; moreover, the dependence of cross sections upon moderate variations of w/d is not very sensitive; by *moderate variations* we mean changes in this quotient by a factor of two or three. The desired probability can now be calculated from Eq. 30d:

$$P = \frac{1}{180\pi\sqrt{2}\, n^7 V}\int_{\rho+}^{\infty} \frac{dR R^{5/2}}{[R^2 - (\rho^+)^2]^{1/2}} \exp\left[-\pi\left(\frac{R}{2}\right)^{3/2}\left(\frac{1}{2n^2} - \frac{3}{R}\right)\right], \tag{30e}$$

$$\rho^+ = \lambda R_0 = 6n^2\lambda, \quad \lambda \geq 1.$$

The convenient dimensionless parameter λ is introduced here for later us.

In order to evaluate the preceding integral approximately, the change of variable $R = \rho^+/u$ is made, and the terms in the exponential and the other terms are expanded in a Taylor's series about $u = 1$ with powers of $(u - 1)$ greater than the first being discarded since most contribution to the integral originates from the domain near $u = 1$. This approximate calculation yields

$$P = \frac{\sqrt{2}\,\lambda^{9/4}\exp[-(3\sqrt{3}/2)\pi n^{1/2}(\lambda - 1)]}{5\pi 3^{1/4} n^{5/2} v(3\lambda - 1)^{1/2}}. \tag{30f}$$

With the use of Firsov's approximation (cf. Eq. 25q) the cross section, with its dependence upon V, is obtained:

$$Q(F) = \frac{\pi}{2}(\rho^+)^2 = \frac{\pi}{2}(6n^2\lambda)^2,$$
$$V = \frac{\sqrt{2}\,\lambda^{9/2}\exp[-(3\sqrt{3}/2)\pi n^{1/2}(\lambda - 1)]}{5\pi 3^{1/4} n^{5/2}\sin^2(1/\pi)(3\lambda - 1)^{1/2}}. \tag{30g}$$

Recall that $Q(F)$ is expressed in units of a_0 $(=\hbar^2/me^2)$, and the velocity V in units of v_0 $(=e^2/\hbar)$. The relation between ρ^+ and V can be approximated by replacing λ by unity except in the exponential of Eq. 30g;

$$\rho^+ \approx 6n^2 + \frac{2n^2}{(\rho^+)^{1/2}}[0.145 - 2.59\log_{10}(2n^2) - 2.07\log_{10} V]. \tag{30h}$$

Although this cross section does increase as V decreases, the variation is different from the one obtained using the Firsov approximation with the method of perturbed stationary states.[6] It is thus desirable and instructive to use the Firsov approximation to obtain a relation corresponding to Eq. 30h for the method of perturbed stationary states.

2.3 FIRSOV APPROXIMATION FOR EXACT RESONANCE*

According to Eqs. 25n, 25p, and 25q, the leading term of α in the Firsov approximation at low velocities is

$$\alpha = \frac{1}{2v}\int_{-\infty}^{\infty}[\eta^{(+)} - \eta^{(-)}]\,dz = \int_0^{\infty}[\eta^{(+)} - \eta^{(-)}]\,dt$$
$$= \frac{1}{v}\int_{\rho^+}^{\infty}\frac{dRR}{[R^2 - (\rho^+)^2]^{1/2}}[\eta^{(+)} - \eta^{(-)}] = \frac{1}{\pi}. \tag{31a}$$

(Quantities are expressed in a.u. in this section, and the impact-parameter version is used.) It is clear from the structure of the integrand in dR that $[\eta^{(+)} - \eta^{(-)}]$ must be expressed in the variable R in order to achieve a $(\rho^+ - v)$ relation.

* The discussion in this section is modeled on the description given by Bates and McCarroll.[6]

If most contribution to the cross section originates from large values of the internuclear distance, the quasi-molecular functions $\chi^{\pm}$ of Eq. 25*a* can be adequately approximated by a linear combination (LCAO) of real, spherically symmetrical atomic wave functions,

$$\chi^{\pm} = \frac{1}{\sqrt{2}} [\phi(\mathbf{x}_i) \pm \phi(\mathbf{x}_f)]. \tag{31b}$$

At these low velocities the exponential factors representing transfer of translational motion differ negligibly from unity; consequently, the two-state wave function of Eq. 25*a* with $c^{\pm}(t) = 1/\sqrt{2}$ becomes

$$\begin{aligned}\Psi(\mathbf{x}, t) = \tfrac{1}{2}\Bigg(\phi(\mathbf{x}_i)\Bigg\{\exp\left[-i\int_{-\infty}^{t} \eta^{(+)}\, dt'\right] + \exp\left[-i\int_{-\infty}^{t} \eta^{(-)}\, dt'\right]\Bigg\} \\ + \phi(\mathbf{x}_f)\Bigg\{\exp\left[-i\int_{-\infty}^{t} \eta^{(+)}\, dt'\right] - \exp\left[-i\int_{-8}^{t} \eta^{(-)}\, dt'\right]\Bigg\}\Bigg).\end{aligned} \tag{31c}$$

Next, the condition

$$\int_{-\infty}^{t} [\eta^{(+)} - \eta^{(-)}\, dt' = n\pi, \qquad n = 0, 1, \ldots, \tag{31d}$$

is imposed upon $[\eta^{(+)} - n^{(-)}]$, and Eq. 31*c* is multiplied successively by $\exp\,[i\int_{-\infty}^{t} dt'\,\eta^{(+)}]$ and $\exp\,[\mathrm{i}\int_{-\infty}^{t} dt'\,\eta^{(-)}]$ to get

$$\Psi(\mathbf{x},t) \exp\left[i\int_{-\infty}^{t} \eta^{(\pm)}\, dt'\right] = \tfrac{1}{2}[\phi(\mathbf{x}_i)(1 + e^{\pm i n\pi}) \pm \phi(\mathbf{x}_f)(1 - e^{\pm i n\pi})]. \tag{31e}$$

These wave functions represent the active electron in one of two possible stationary states at the times t which satisfy Eq. 31*d*. The electron is bound to nucleus A^+ if n is even, and it is bound to nucleus P^+ if n is odd.

Next, let I denote the *probability current* that is flowing across the plane of symmetry that is perpendicular to the internuclear line. We now imagine successive instants of time t_0 and t_1 that correspond to n being even and odd, respectively, and write

$$\int_{t_0}^{t_1} dt\, I = 1. \tag{31f}$$

By reason of the preceding interpretation this integral states that it is certain that the electron crosses the plane of symmetry in its transfer from nucleus A^+ to nucleus P^+ during the interval of time $(t_1 - t_0)$. According to the definition of I, we obtain[12]

$$I = \frac{i}{2}\int \left(\Psi' \frac{\partial}{\partial z'} \Psi'^* - \Psi'^* \frac{\partial \Psi'}{\partial z'}\right) \underset{z'=0}{dS}$$

$$= -\tfrac{1}{2}\int \left[\phi(\mathbf{x}_i) \frac{\partial}{\partial z_f'} \phi(\mathbf{x}_f) - \phi(\mathbf{x}_f) \frac{\partial}{\partial z_i'} \phi(\mathbf{x}_i)\right] \underset{z'=0}{dS}$$

$$\times \sin\left\{\int_{-\infty}^{t} [\eta^{(+)} - \eta^{(-)}]\, dt'\right\}, \quad (31g)$$

$$z' = \hat{R}\cdot\mathbf{x}, \qquad z_i' = \hat{R}\cdot\mathbf{x}_i, \qquad z_f' = \hat{R}\cdot\mathbf{x}_f,$$

$$\frac{\partial}{\partial z'} = \frac{\partial}{\partial z_i'} = \frac{\partial}{\partial z_f'}.$$

The element of surface dS is in the plane of symmetry, and the normal derivatives are equal since only one of the two transformations, $(\mathbf{x}, t) \to (\mathbf{x}_f, t_f)$ and $(\mathbf{x}, t) \to (\mathbf{x}_i, t_i)$, is used in each part of the integrand (see remarks following Eq. 15*b* and note Eqs. 19*c* and 19*d*.) Equation 31*f* is satisfied if

$$\eta^{(+)} - \eta^{(-)} = -\int \left[\phi(\mathbf{x}_i) \frac{\partial}{\partial z_f'} \phi(\mathbf{x}_f) - \phi(\mathbf{x}_f) \frac{\partial}{\partial z_i'} \phi(\mathbf{x}_i)\right] \underset{z'=0}{dS}, \quad (31h)$$

since this relation reduces Eq. 31*g* to the form

$$dtf(t) \sin\left[\int_{-\infty}^{t} f(t')\, dt'\right] = -d\left\{\cos\left[\int_{-\infty}^{t} f(t')\, dt'\right]\right\},$$

which is an exact differential.

We now use Eq. 31*e* to establish that Eq. 31*h* does satisfy Eq. 31*f*. From the defining relations of z', z_i', and z_f' in Eqs. 31*g* and 15*b*, it is seen that $z' = 0$ is equivalent to $z_i' = R/2$ and $z_f' = -R/2$; moreover, the tangential components of $\mathbf{x}_i$ and $\mathbf{x}_f$ are equal in the plane $z' = 0$, and this permits Eq. 31*h* to be put into the following equivalent form:

$$\eta^{(+)} - \eta^{(-)} = -\frac{\partial}{\partial z_f'}\int [\phi(\mathbf{x}_f)^2] \underset{z_f'=-R/2}{dS},$$

$$\phi(\mathbf{x}_i)]_{z_i'=R/2} = \phi(\mathbf{x}_f)]_{z_f'=-R/2}, \quad (31h')$$

$$\left.\frac{\partial\phi(\mathbf{x}_i)}{dz_i'}\right]_{z_i'=R/2} = -\left.\frac{\partial\phi}{\partial z_f'}(\mathbf{x}_f)\right]_{z_f'=-R/2}.$$

This is evaluated by expressing x_f^2 in terms of its perpendicular components

$$\eta^{(+)} - \eta^{(-)} = -\frac{\partial}{\partial z_f'} \cdot 2\pi \int_0^\infty [\phi(\mathbf{x}_f)^2]_{z_f' = -R/2}\, p\, dp = -2\pi \frac{\partial}{\partial z_f'} \int_{R/2}^\infty \phi(\mathbf{x}_f)^2\, x_f\, dx_f$$

$$= \pi R \phi\left(\frac{R}{2}\right)^2, \qquad x_f^2 = p^2 + z_f'^2. \tag{31i}$$

In the evaluation of the derivative in Eq. 31*i*, use is made of the rule for differentiating an integral with respect to its lower limit. The relation between derivatives in Eq. 31*h* is consistent with the relations between derivatives in Eq. 31*g*,

$$\frac{\partial}{\partial z'} = \frac{\partial}{\partial z_i'}, \qquad \frac{\partial}{\partial z'} = \frac{\partial}{\partial z_f'}.$$

For the assumed spherically symmetric functions the slopes in z_i'- and z_f'-spaces are antisymmetric with respect to $z_i' = 0$ and $z_f' = 0$, respectively. Also note that this change of sign does emerge in Eq. 31*i*. For this particular application the function ϕ is

$$\phi(\mathbf{x}) = N\left[\frac{(2\alpha)^{2\gamma+1}}{4\pi(2\gamma)!}\right]^{1/2} x^{\gamma-1} e^{-\alpha x}, \tag{31j}$$

$$\alpha = (2\varepsilon_i)^{1/2}, \qquad \gamma = \alpha^{-1},$$

where ε_i is the ionization potential of the active electron and N is a constant of the order of unity. (If ε_i represents the ionization potential of atomic hydrogen and $N = 1$, ϕ is the 1*s*-wave function for atomic hydrogen.) Equation 31*a* is rewritten using Eq. 31*j* in Eq. 31*i*:

$$\frac{2N^2\alpha^{2\gamma+1}}{v(2\gamma)!} \int_{\rho^+}^\infty \frac{R^{2\gamma} e^{-\alpha R}\, dR}{[R^2 - (\rho^+)^2]^{1/2}} = \frac{1}{\pi}. \tag{31k}$$

The change of variable $R = y + \rho^+$ is made, and the integrand, excluding the exponential, is expanded in a Taylor series about $(y/\rho^+) = 0$. If all powers of y/ρ^+ greater than the first are discarded, Eq. 31*k* reduces to

$$N^2 e^{-\alpha\rho^+} (\alpha\rho^+)^{2\gamma-1/2} \frac{(2\pi^3\alpha^2)^{1/2}}{(2\gamma)!\, v} \left(1 + \frac{\gamma - 1/8}{\alpha\rho^+}\right) = 1, \tag{31l}$$

and if $\alpha\rho^+ \gg \log \alpha\rho^+$, $\alpha\rho^+ \gg (\gamma - 1/8)$, then Q of Eq. 25*q* can be written in the form of Eq. 31*m*:

$$Q = (a - b \log v)^2 \tag{31m}$$

in contrast to the variation displayed in Eq. 30h. The foregoing requirements on $\alpha\rho^+$ are actually implicit since Eq. 31b is supposed to be a good representation of the molecular functions.

2.4 PSEUDOCROSSING OF POTENTIAL ENERGY SURFACES

The concept of pseudocrossing of potential energy surfaces is utilized to construct another approximation within the framework of the two-state approximation. This method is not restricted to low velocities of relative motion of the colliding systems, although the approximations commonly used are adequate only at low velocities. By reason of its attractive simplicity the method of curve-crossing has been applied and described extensively.[13–15] The explanation here however, is based on selected features of two articles that critically analyze the approximations and their implied restrictions.[16,17] (Mention is also made of other recent research that has contributed to some refinement of the original treatment.[18–21])

Consider the process of Eq. 32 representing electronic transfer from A to B:

$$(\mathrm{A} + e)_n + \mathrm{B} \rightarrow \mathrm{A} + (\mathrm{B} + e)_m. \tag{32}$$

In the absence of interaction the left- and right-hand members of this equation can be imagined to form two quasi-molecules, and Fig. 5 depicts the associated potential energy surfaces in the absence of interaction (the curves containing the dashed lines) between the two systems; in this hypothetical circumstance the curves cross at an internuclear distance $R = R_c$. If there is an interaction, these curves usually do not cross provided that the matrix element of the associated interaction does not vanish.[22] (This discussion is limited to diatomic quasi-molecules.) In the region of the pseudocrossing R_c the difference in the two potential energies is smallest, and a transition is possible, or in the present application electronic capture may occur. In order to analyze this possibility we turn next to the previously noted paper by Bates.[16]

In that paper the Landau-Zener formula is derived from the two-state approximation. The IP version is used, and the system of coordinates of Eq. 6 is employed for the active electron; also in the notation of that section, A and B of Eq. 32 correspond to A^+ and P^+, respectively, if A and P are neutral atoms. The nuclear motion is determined from the classical mechanical Eq. 12. Let $\psi_n(\mathbf{s}; \mathbf{R})$ and $E_n(\mathbf{R})$ represent the exact electronic eigenfunction and associated eigenenergy for the quasi-molecule formed as the two aggregates of the system on the left of Eq. 32 approach each other

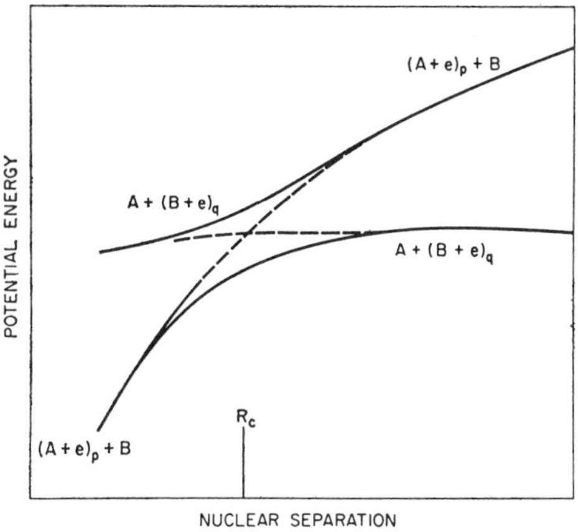

Fig. 5 Pseudocrossing of potential energy surfaces. The full curves represent $E_p(R)$ and $E_q(R)$; the broken curves represent $h_{pp}(R)$ and $h_{qq}(R)$. In the region to the right of pseudocrossing $E_p(R)$ and $h_{pp}(R)$ lie above $E_q(R)$ and $h_{qq}(R)$.

adiabatically (infinitely slowly) from infinite separation, and let $\psi_m(\mathbf{s}; \mathbf{R})$ and $E_m(\mathbf{R})$ denote the corresponding quantities for the system on the right side of Eq. 32. It is assumed that a pseudocrossing of the curves E_m and E_n occurs at some internuclear distance R_c so that for $R \ll R_c$, ψ_n and ψ_m represent A + (B + $e)_m$ and (A + $e)_n$ + B, respectively. In order to have functions that represent the left and right members of Eq. 32 for any allowed values of R, two orthogonal linear combinations of ψ_m and ψ_n are sought such that for any value of R, ϕ_n represents (A + $e)_n$ + B, and ϕ_m represents A + (B + $e)_m$. This construction is explained next.[17]

These functions, of course, are the unperturbed functions in the absence of interaction,

$$
\begin{aligned}
&\phi_n = c_{11}\psi_n + c_{22}\psi_m, && \phi_m = c_{12}\psi_n + c_{22}\psi_m,\\
&\int \bar{\psi}_i\psi_j \, d\mathbf{s} = \delta_{ij}, && \int \bar{\phi}_i\phi_j \, d\mathbf{s} = \delta_{ij},\\
&i, j = m, n, && \delta_{ij} = 1, \quad i = j,\\
&\delta_{ij} = 0, && i \neq j.
\end{aligned}
\tag{33a}
$$

(Although the initial and final atomic states for electronic capture are non-orthogonal, the states ϕ_n and ϕ_m have been orthogonalized by the Schmidt process.[23]) The coefficients c_{ij} are real and are the elements of a real unitary matrix; consequently, there are the following relations among the corresponding matrix elements of c:

$$
\begin{aligned}
c_{11}^2 + c_{12}^2 = c_{22}^2 + c_{21}^2 &= 1,\\
c_{11}c_{12} + c_{22}c_{21} &= 0,\\
c_{11}c_{22} - c_{12}c_{21} &= 1.
\end{aligned}
\tag{33b}
$$

The last condition of Eq. 33*b* is imposed on the determinant of c to obtain the additional equation that is needed.[17] The reason that we can represent ϕ_n and ϕ_m as a linear combination of ψ_n and ψ_m rests on the assumption that *each* of the two sets, (ϕ_n, ϕ_m) and (ψ_n, ψ_m), constitutes a *complete* set. As a consequence of this assumption, the Hamiltonian of Eq. 6 ($\nabla_R{}^2 = \nabla_R = 0$) applied to the unperturbed functions yields

$$\begin{aligned} H\phi_n &= h_{nn}\phi_n + h_{mn}\phi_m, \\ H\phi_m &= h_{nm}\phi_n + h_{mm}\phi_m, \\ H\psi_{m,n} &= E_{m,n}(R)\psi_{m,n}, \\ h_{ij} &= \int \bar{\phi}_i H\phi_j \, d\mathbf{s} = h_{ji}. \end{aligned} \tag{33c}$$

The characteristic energies $E_{m,n}(R)$ of the functions $\psi_{m,n}$ show the dependence upon R explicitly for emphasis (see Fig. 5). The curves containing the dashed portions represent the diagonal matrix element of H, $h_{ii}(R)$, and at infinite values of R, h_{ij} vanishes; moreover, the h_{ii} approach the energies E_i of the associated atomic states that the ϕ_i approach. The c_{ij} also depend upon R through the energies E_p and the matrix elements h_{ij}.

Equation 33*a* is solved to yield

$$\begin{aligned} \psi_n &= c_{11}\phi_n + c_{12}\phi_m, \qquad \psi_m = c_{21}\phi_n + c_{22}\phi_m, \\ H\psi_n &= E\psi_n = c_{11}[E_n{}^0 + V]\phi_n + c_{12}[E_m{}^0 + V]\phi_m, \\ H &= H_0 + V, \qquad H_0\phi_i = E_i{}^0\phi_i. \end{aligned} \tag{33d}$$

Equation 33*c* is next used in Eq. 33*d* and the resulting equation is multiplied successively by ϕ_n and ϕ_m followed by integration over $d\mathbf{s}$; then with the use of Eq. 33*a* the following homogeneous equations for the determination of E are obtained:[24]

$$\begin{aligned} (h_{nn} - E)c_{11} + h_{nm}c_{12} &= 0 \\ (h_{mm} - E)c_{12} + h_{mn}c_{11} &= 0 \\ \left.\begin{matrix} E_1 \\ E_2 \end{matrix}\right\} = \frac{h_{nn} + h_{mm}}{2} &\mp \tfrac{1}{2}[(h_{mm} - h_{nn})^2 + 4V_{nm}{}^2]^{1/2}. \end{aligned} \tag{33e}$$

Equation 33*b* is used with the preceding equation to obtain the values for the c_{ij}:

$$\begin{aligned} c_{11} &= \frac{V_{mn}}{(V_{mn}{}^2 + D_1{}^2)^{1/2}}, \qquad c_{22} = \frac{D_2}{(D_2{}^2 + V_{mn}{}^2)^{1/2}} = c_{11} \\ c_{21} &= \frac{V_{mn}}{(D_2{}^2 + V_{mn})^{1/2}}, \qquad c_{12} = \frac{D_1}{(V_{mn}{}^2 + D_1{}^2)^{1/2}} = -c_{21}. \\ D_1 &= E_1 - h_{nn} = h_{mm} - E_2, \qquad D_1 \leq 0 \\ D_2 &= E_2 - h_{nn} = h_{mm} - E_1, \qquad D_2 \geq 0. \end{aligned} \tag{33f}$$

(The U_1 and U_2 of Heinrichs' paper correspond to the h_{nn}, and h_{mm} of Bates' paper provided that $h_{mm} > h_{nn}$ for infinite separation.[16,17]) Although the values of these coefficients will be discussed for special values of R, the values at the crossing point can be determined now since by definition

$$h_{nn}(R_c) = h_{mm}(R_c), \tag{33g}$$

and from this result and Eq. 33*f* we find that

$$\begin{aligned} &R = R_c \\ E_2 - E_1 &= 2V_{nm}, \qquad & E_2 + E_1 &= 2h_{nn} = 2h_{mm}, \\ E_2 &= V_{nm} + h_{nn}, & E_1 &= h_{nn} - V_{nm} \\ D_1 &= -V_{nm}, & D_2 &= V_{nm} \end{aligned} \tag{33h}$$

$$c_{11} = c_{22} = c_{21} = -c_{12} = \frac{1}{\sqrt{2}}.$$

The approximate solution of the equations that describe Eq. 32 is considered next. In this analysis the motion is restricted to energies of relative translation that greatly exceed the difference between the exact eigenenergies of the two systems in the vicinity of R_c. It is clear from Eq. 33*h* that this is equivalent to

$$E_r = \frac{\mu v^2}{2} \gg 2|h_{mn}(R_c)| = 2|V_{mn}(R_c)|. \tag{33i}$$

Consistent with the two-state model, there are only two states—an initial state and a final state—and since the nuclear motion is assumed classical, the wave function can be represented by

$$\begin{aligned} \Psi &= c_n(t)\Phi_n(t) + c_m(t)\Phi_m(t), \\ \Phi_n &= \phi_n(\mathbf{s}, \mathbf{R}) \exp\left\{-\frac{i}{\hbar}\left[\frac{m\mathbf{v}\cdot\mathbf{s}}{2} + \int^t \left(h_{nn} + \frac{mv^2}{8}\right) dt'\right]\right\}, \\ \Phi_m &= \phi_m(\mathbf{s}, \mathbf{R}) \exp\left\{\frac{i}{\hbar}\left[\frac{m\mathbf{v}\cdot\mathbf{s}}{2} - \int^t \left(h_{mm} + \frac{mv^2}{8}\right) dt'\right]\right\}, \\ E_r &= \frac{\mu v^2}{2}. \end{aligned} \tag{34a}$$

Equation 34*a* corresponds to Eqs. 17 and 19*b* of Heinrichs' paper if the terms in v are included with his $U_{1,2}$.[17] In Eq. 34*a* μ is the reduced mass of Eq. 6 and v is the relative velocity of the colliding systems. The electronic Hamiltonian H of Eq. 33*c* is part of the operator,

$$\left(H - i\hbar\frac{\partial}{\partial t}\right),$$

which is applied successively to $\Phi_n(t)$ and $\Phi_m(t)$ of Eq. 34*a* to obtain

$$\left(H - i\hbar\frac{\partial}{\partial t}\right)\Phi_n = \left[h_{mn}\phi_m - i\hbar\left(\frac{\partial}{\partial t_i}\phi_n\right)\right]\exp\left[-\frac{i}{\hbar}\left(\frac{m\mathbf{v}\cdot\mathbf{s}}{2} + \int^t\left(h_{nn} + \frac{mv^2}{8}\right)dt'\right)\right],$$

$$\frac{\partial}{\partial t_i} = \frac{\partial}{\partial t} - \frac{\mathbf{v}\cdot\boldsymbol{\nabla}_{x_i}}{2},$$

$$\left(H - i\hbar\frac{\partial}{\partial t}\right)\Phi_m = \left[h_{nm}\phi_n - i\hbar\left(\frac{\partial}{\partial t_f}\phi_m\right)\right]\exp\left[\frac{i}{\hbar}\left(\frac{m\mathbf{v}\cdot\mathbf{s}}{2} - \int^t\left(h_{mm} + \frac{mv^2}{8}\right)dt'\right)\right], \quad (34b)$$

$$\frac{\partial}{\partial t_f} = \frac{\partial}{\partial t} + \frac{\mathbf{v}\cdot\boldsymbol{\nabla}_{x_f}}{2}.$$

It is pertinent to remark here that the velocity v is kept constant in the calculation of the probability of electronic capture, but in the calculation of the associated cross section allowance is occasionally made for the variation of v in the classical orbit.

Before proceeding with the calculation, the assumptions of the present analysis are listed. The relative velocity is sufficiently small so that: (1) the factors of electronic translational motion, $\exp(\pm im\mathbf{v}\cdot\mathbf{s}/2\hbar)$, may be replaced by unity; (2) the second bracketed term in each of the preceding equations may be neglected; (3) the wave function Ψ of Eq. 34*a* adequately represents the system; (4) the energy of relative motion is such that Eq. 33*i* is satisfied. With the use of assumptions (1), (2), and (3), the differential equations of motion, Eqs. 34*b* and 34*a*, simplify to[16]

$$i\hbar\frac{dc_n}{dt} = c_m h_{nm}\exp\left[-\frac{i}{\hbar}\int^t(h_{mm} - h_{nn})\,dt'\right],$$

$$i\hbar\frac{dc_m}{dt} = c_n h_{mn}\exp\left[-\frac{i}{\hbar}\int^t(h_{nn} - h_{mm})\,dt'\right], \quad (34c)$$

$$c_n(-\infty) = 1, \qquad c_m(-\infty) = 0.$$

Equation 34*c* also contains the assumed initial conditions in accord with Eq. 32. [These equations for c_n and c_m are equivalent to Heinrichs' Eqs. 21 and 22. As noted following Eq. 33*a* the atomic states ϕ_n and ϕ_m are orthogonal. Nevertheless, the associated coefficients $c_n(t)$ and $c_m(t)$ of Eq. 34*a* yield the same probabilities at $t = \pm\infty$ as do the coefficients $c_n'(t)$ and $c_m'(t)$ associated with the correct nonorthogonal initial and final atomic states. This fact is proven in Section 2.5 in connection with Eqs. 39*o* and 39*q*.]

Since it is supposed that the probability for transitions is appreciable only in a narrow zone about the pseudocrossing at $R = R_c$, the definition of the point R_c of Eq. 33*g* suggests a Taylor's expansion about this point for $(h_{nn} - h_{mm})$. The substitution $z = vt$ is made, and only the leading terms of this expansion are retained; therefore we have

$$h_{nn} - h_{mm} = (z - z_c)\frac{\alpha}{v}, \qquad v = v_r,$$

$$\alpha = v\frac{d}{dz}(h_{nn} - h_{mm})_{z=z_c}, \qquad z_c = \frac{\mathbf{R}_c \cdot v}{v}, \tag{34d}$$

$$\beta = h_{nm}(R_c) = h_{mn}(R_c), \qquad z = \frac{\mathbf{R}\cdot\mathbf{v}}{v} = vt.$$

If R_c is the origin of time, the simplified differential Eq. 34*c* thus becomes

$$\frac{d^2c_n}{dt^2} + \left(-\frac{i\alpha t}{\hbar} - \frac{i}{h_{nm}}\frac{d}{dt}h_{nm}\right)\frac{dc_n}{dt} + \frac{h_{nm}{}^2}{\hbar^2}c_n = 0,$$

$$c_m = \frac{i\hbar}{h_{nm}}\left(\frac{dc_n}{dt}\right)\exp\left[\frac{i}{\hbar}\left(\frac{\alpha t^2}{2} + K\right)\right], \tag{34e}$$

and these equations can be solved analytically to yield the probability for electronic capture:[17,25]

$$M = |c_n(\infty)|^2, \qquad 1 - M = |c_m(\infty)|^2,$$

$$M = \exp(-w), \qquad h'_{ii} = \left[\frac{dh_{ii}}{dz}\right]_{R=R_c} \tag{34f}$$

$$w = \frac{2\pi\beta^2}{\hbar\alpha} = \frac{\eta}{v}.$$

The system is in the state ϕ_n initially (Eq. 34*c*), and the two aggregates approach R_c from large distances. There are two ways that the system can make a transition so that the two aggregates recede to infinity in the state ϕ_m. There is the probability $(1 - M)$ for a transition to the state ψ_m on the route to the classical turning point along with the probability M that the system remains in this state upon passing the pseudocrossing on the return journey to infinity $(R \rightarrow \infty)$. Alternatively, there is the probability M that the system remains in the state ψ_n on passing R_c toward decreasing R along with the probability $(1 - M)$ that a transition to the state ψ_m upon passing R_c traveling toward increasing R. (Note that the exact quasi-molecular states with the interaction are being used here. Later it is shown that the associated probabilities are given by Eq. 34*f*.) It is evident that either of these two alternatives yields $M(1 - M)$ for the probability of a transition during the journey in and out, and thus the probability and cross section are

$$P = 2M(1 - M), \qquad Q = 2\pi \int_0^\infty P\rho \, d\rho. \tag{34g}$$

This is the formula of Landau and Zener.[26,27] Figure 6 displays the variation of P as a function of (v/η) (cf. Eq. 34*f*). The Landau-Zener formula shows that a high probability for electronic capture may exist even if the relative velocity is so small that the rapid variation of the exponential factors ($dt' = dz/v$) in Eq. 34*c* would be expected to cancel contributions to the associated integrated equations. Such a result is of great importance for a circumstance that usually is assumed to be adiabatic; that is, the system continually adjusts to the slowly changing perturbing forces and no transitions occur.[28]

Before the defects of this formula are described a study is made of c_n, c_m, and the c_{ij} of Eqs. 33*a* and 33*f*.[17] First, a transformation to the dimensionless variable z is made to transform Eq. 34*e* into

$$\frac{d^2f}{dz^2} + \left[\frac{1}{2} - \frac{z^2}{4} - \frac{i\beta^2}{\hbar\alpha}\right] f = 0, \qquad z = e^{i\pi/4}\left(\frac{\alpha}{\hbar}\right)^{1/2} t,$$
$$c_n = f(z) \exp\left[\left(\frac{z}{2}\right)^2\right], \qquad h_{nm} = \beta, \qquad \frac{dh_{nm}}{dz} = 0, \qquad n = -\frac{i\beta^2}{(\hbar\alpha)}. \tag{34h}$$

Next, Eq. 33*a* is used with Eq. 34*a* to obtain Ψ in terms of the assumed exact states ψ_n and ψ_m:

$$\Psi = [c_{11}c_n(t) \exp(-i\delta_n) + c_{12}c_m(t) \exp(-i\delta_m)]\psi_n + [c_{21}c_n(t) \exp(-i\delta_n) + c_{22}c_m(t) \exp(-i\delta_m)]\psi_m,$$
$$\delta_n = \frac{1}{\hbar}\int^t \left(h_{nn} + \frac{mv^2}{8}\right) dt', \tag{34i}$$
$$\delta_m = \frac{1}{\hbar}\int^t \left(h_{mm} + \frac{mv^2}{8}\right) dt'.$$

Since $h_{nn} > h_{mm}$ or $h_{nn} < h_{mm}$ as $t \to \pm\infty$ it is not difficult to show that

$$\lim_{t\to\pm\infty} |c_{11}| = \lim_{t\to\pm\infty} |c_{22}| = 1,$$
$$\lim_{t\to\pm\infty} |c_{12}| = \lim_{t\to\pm\infty} |c_{21}| = 0,$$

and with the aid of Eqs. 34*c* and 34*i*, we find that

$$\lim_{t\to-\infty} |\Psi| \to |\psi_n(-\infty)| = |\phi_n|,$$
$$\lim_{t\to\infty} \int ds \bar{\psi}_{n,m}\Psi = c_{n,m}(\infty)e^{-i\theta_{n,m}}. \tag{34j}$$

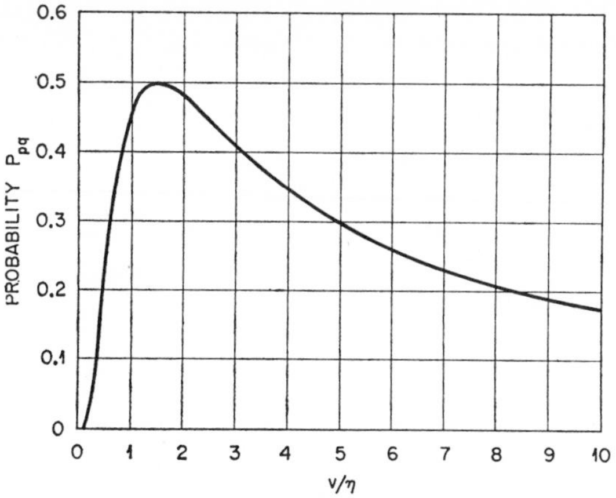

Fig. 6 Landau-Zener probability-velocity curve.

These results support the remarks following Eq. 34*f*. [In order to obtain the preceding results, the signs of α must be different for the two directions of motion on the asymptotic side ($t \to \pm\infty$) of R_c in the potential energy diagram.] The coefficients c_n and c_m can be determined exactly in terms of parabolic cylinder functions (the solutions to the differential equation in Eq. 34*h*). Heinrichs studied these solutions for $|z| \gtrless \beta(\alpha\hbar)^{-1/2}$ and refined Eq. 34*f* somewhat; we do not pursue this study here. However, the result for an effective width of the transition zone is given.

For this purpose the origin of time coincides with the crossing point, and T denotes the time for travel from R_0 to the edge of the effective zone in the assumed rectilinear motion. (The motion is assumed constant only in this region.) A characteristic value that enters into the expansions of the c_{ij} is used to define the half-width of the zone.[17] Since this characteristic parameter is $\gamma = \beta(a\hbar)^{-1/2}$, the width is

$$|z|_T = 2s\frac{\beta}{(\alpha\hbar)^{1/2}} = T\left(\frac{\alpha}{\hbar}\right)^{1/2}, \qquad R_T = z_T = vT, \qquad s \simeq 1. \tag{34k}$$

This formula also defines the value of the time T at the edge of the interacting width. (This characteristic parameter γ also enters into the asymptotic expansions of the parabolic cylinder functions. As another item of interest these functions or linear combinations of them are also the solutions of Eq. 30*a*, Section 2.2.) If we rewrite the right member of Eq. 34*k* in terms of R_T and the second member in terms of γ, we get

$$|z|_T = 2s\gamma^{1/2} = \frac{R_T}{v}\left(\frac{\alpha}{\hbar}\right)^{1/2} = \frac{\beta R_T}{v\hbar(\gamma)^{1/2}}, \qquad \gamma = \frac{\beta R_T}{2s\hbar v}. \tag{34l}$$

Since β can be represented by $2se^2f(R_T)$, with $f(R_T)$ an appropriate intermolecular potential function, we find that

$$\gamma \simeq R_T f(R_T)\frac{v_0}{v}, \qquad v_0 = \frac{e^2}{\hbar} \tag{34m}$$

and the condition $\gamma \gg 1$ means that the representative electronic orbital velocity is much greater than v, a condition commonly used to define the region of low relative velocities. Another point of interest is the relation obtained from Eq. 34k:

$$\frac{R_T\alpha}{v} = T\alpha = |z|_T(\alpha\hbar)^{1/2} = 2s\beta,$$

and with the use of Eq. 33i we obtain the inequality

$$\frac{R_T\alpha}{v} = |z|_T(\hbar\alpha)^{1/2} = 2s\beta \ll \tfrac{1}{2}\mu v^2, \tag{34n}$$

a relation that associates the half-width with the condition for the validity of a classical treatment of the motion of relative translation. From Eq. 34g it is also evident that P is proportional to γ if γ is small:

$$P = 2(2\pi\gamma) = \frac{4\pi\beta^2}{\alpha\hbar} = \frac{T^2\alpha}{4s^2\hbar} = \left(\frac{R_T}{v}\right)^2\left(\frac{\alpha}{4s\hbar}\right), \tag{34o}$$

but this (v^{-1})-dependence for fixed values of the impact parameter is not the same, for example, as the (v^{-2})-dependence obtained from various forms of the first Born approximations for s-s transitions if allowance is not made for the transfer of translational motion to the active electron.[29] [Note that α varies directly with v according to Eq. 34d (see Sections 1.1 and 1.2).] As Heinrichs[17] shows, nevertheless, the expansion used to obtain Eq. 34o is inconsistent with the condition for the asymptotic expansion of the parabolic cylinder functions whose leading terms yield the Landau-Zener (L-Z) formula through Eqs. 34f and 34g. Thus the expansion used to represent P in Eq. 34o is incomplete. However, the purpose of this chapter is not to criticize but rather to explain, and with this goal in mind we return to the analysis of Bates[16] who used a different criterion for estimating the transition width.

As Bates explains, transitions may occur at some distance from R_c, and since this zone is not precisely specified, he defines it to be that region inside of which a rapid variation of the exponential functions on the right-hand side of Eq. 34c does not result in pronounced cancellation in the corresponding integrated equations. (This argument is presented for convenience to the reader even though Heinrichs disagrees with the criterion.[17]) The width of the zone ΔZ can be estimated by equating $(\Delta Z/2)$ to the extent of the zone on each side of the crossing point and approximating $h_{nm}(R)$ by $h_{nm}(R_c)$ in this region. For the purpose of this estimation the exponential functions of Eq. 34*c* are evaluated using Eq. 34*d*. The sum of these two integrals times $(v\hbar)^{-1}$ is equated to $(s\pi)$ with s again a number on the order of unity:

$$\frac{1}{v\hbar}\left[\int_{Z_c-\Delta Z/2}^{Z_c} (h_{mm} - h_{nn})\, dz + \int_{Z_c}^{(Z_c+\Delta Z/2)} (h_{nn} - h_{mm})\, dz\right] = \frac{\alpha(\Delta Z)^2}{4v^2\hbar} = s\pi.$$

This restriction of the domain of integration is in accord with the previous assertion that the exponential functions do not oscillate many times in the zone ΔZ. From this analysis we can deduce that

$$\Delta Z = \left(\frac{4\pi s v\hbar}{\alpha}\right)^{1/2}, \tag{34p}$$

and the width of the zone is proportional to the square root of the velocity and is thus unbounded with velocity. [In this instance it is assumed that (α/v) is independent of velocity; (α/v) in this chapter equals the α of Bates.[16]]

If v is sufficiently large, the probability of a transition is small and c_n of Eq. 34*c* differs little from unity. We use this approximation and the preceding ones, in solving the second equation of 34*c* to obtain

$$c_m(Z) = \frac{h_{mn}(R_c)}{i\hbar v}\left\{\int_{-\infty}^{Z_c} dZ \exp\left[\frac{-i\alpha(Z_c - Z)^2}{2v\hbar}\right] + \int_{Z_c}^{\infty} dZ \exp\left[\frac{-i\alpha(Z - Z_c)^2}{2v\hbar}\right]\right\}.$$

If the width of the zone is unbounded, $Z \to \infty$, the change of variable $z = Z - Z_c$ is made, and the resulting integral is exactly integrable; we get

$$|c_m(\infty)| = \left|\left(\frac{2\pi}{\hbar v\alpha}\right)^{1/2} h_{mn}\right| = \left|\frac{h_{mn}}{\sqrt{2}\hbar}\right| \frac{\Delta Z}{v}. \tag{34q}$$

However, if contributions to the preceding integral stem only from finite values of Z, say $|Z| < Z_m$, the exponentials may be expanded, and the leading terms of this expansion (v large) yield the result

$$|c_m(\infty)| = \frac{2h_{mn}Z_m}{\hbar v}. \tag{34r}$$

In the first case the probability for electronic capture varies as v^{-1} for fixed impact parameter, but it should vary as v^{-2} for *s-s* transitions subject to the restrictions described after Eq. 34*o*; however, this (v^{-2})-dependence is achieved in the second case with a bounded width (independent of v), but this result is not derived from the L-Z formula.

In addition to the omission of the allowance for the transfer of translational motion to the electron, the other defects of this formula are discussed in detail in the Bates' paper. Nevertheless, since the chief purpose of this chapter is to clarify mathematical derivations, a paraphrase of the concluding remarks and a quotation from this paper are given. In spite of the defects of the Landau-Zener formula it is extremely useful for *s-s* transitions in that it provides a description of how curve-crossing influences collisions between atomic systems, and in some instances the predictions are very successful. Furthermore, the maximum value of P in Eq. 34*g* is one-half since M is positive and cannot exceed unity according to Eq. 34*f*. This fact limits the error that P or $(1 - P)$ may give.

If other than *s*-orbitals are used, the Landau-Zener formula is least satisfactory. This defect stems from the neglect of the effect of the rotating internuclear axis on the angular parts of the wave functions, a topic discussed in Section 1.3. Furthermore, as Bates puts it,[16]

> Since the course of a collision in general depends on interactions other than at the crossing, a simple treatment of wide applicability cannot be expected. If the Landau-Zener formula fails, it may often be possible to use Eq. 37 [Eq. 37 in Section 3.2 of Ref. 16] modified if necessary to allow for the effect of the change in the translational motion of the active electron; if this rather laborious alternative also fails, it may be difficult to make progress without solving the appropriate coupled equations numerically.

Most of us surmise that at some future date the appropriate equations will be solved numerically for all the processes of this chapter!

2.5 DETAILED BALANCING

Detailed balancing is inextricably connected with the symmetry of a system with respect to a change in the sign of the time. A system that possesses this symmetry can be imagined to run backward in time as well as forward. In order to express these statements mathematically it is first noted that the Hamiltonian is real and independent of spin for the applications of this chapter. It follows from this property of the Hamiltonian that to every solution $\Psi(t)$ of the corresponding Schrödinger equation there exists a time-reversed solution $\Psi_R(t)$. If $\Psi(t)$ and $\Psi_R(t)$ are constructed with the same normalization, using the opposite sign of time for $\Psi_R(t)$:

$$\Psi_R(t) = e^{i\delta}\Psi(-t)^*, \qquad |\varepsilon^{i\delta}| = 1. \tag{35a}$$

This follows from the associated Schrödinger equation, which is a linear first-order equation in time with a pure imaginary factor to this first derivative.[30] We note also that the assumed form of the Hamiltonian includes the electronic Hamiltonian used in the IP version of electronic capture, the problem of chief interest here.

Let P_{if} denote the probability of a transition that may represent the transfer of an electron from one atom to another one, a process designated by an initial state $\Psi(-\infty) = \Phi_i(-\infty)$ at $t = -\infty$ and a final state $\Phi_f(\infty)$ at $t = \infty$. If the sign of the time is changed, the directions of the velocities also are changed, and the associated time-reversed wave function is defined according to Eq. 35*a*. With the aid of the linear unitary operator $U(t, t_0)$ the evolution operator, and the antilinear operator of complex conjugation K, it can be demonstrated that the probability of a transition P_{ifR} calculated with the time-reversed wave function is equal to the probability P_{if} calculated with the ordinary wave functions.[31–35] No pretense to mathematical rigor is intended, and the reader is referred to the preceding references for such details.

$$\begin{aligned} P_{ifR} &= |[\langle\Phi_i(-\infty)K^+]U(\infty, -\infty)|K\Phi_f(\infty)\rangle|^2 \\ &= |(\Phi_i(-\infty), [KU(\infty, -\infty)K]\Phi_f(\infty))^*|^2 \\ &= |(\Phi_i(-\infty)^*, U(-\infty, \infty)\Phi_f(\infty)^*)|^2 \\ &= |(\Phi_f(\infty), U(\infty, -\infty)\Phi_i(-\infty))|^2 = P_{if}; \\ K^+ &= K, \qquad K|n\rangle = |n\rangle, \langle n|K = \langle n|, \\ (\langle vK)|n\rangle &= \langle v|n\rangle^*, \qquad \langle n|(Ku\rangle) = \langle n|u\rangle^*, \qquad \langle m|(KBK)|n\rangle = \langle m|B|n\rangle^*, \\ & U^+(t, t_0) = U(t_0, t), \\ U^+ &= U^{-1}, \qquad U(t, t_0) = K^+U^+(t, t_0)K = K^+U(t_0, t)K. \end{aligned} \tag{35b}$$

In Eq. 35*b* the operator B is linear; the properties of K and U are included for convenient reference; moreover, the manipulations used in Eq. 35*b* require that the integrals

$$(\phi, \psi) = \int \phi^*\psi \, d\mathbf{x}_1 \cdots d\mathbf{x}_n$$

over the appropriate spatial domains are defined.

Alternatively, we can derive Eq. 35*b* somewhat differently:

$$\begin{aligned} \Phi_i(-\infty)_R &= KU(\infty, -\infty)\Phi_i(-\infty), \\ \Phi_f(\infty)_R &= KU(-\infty, \infty)\Phi_f(\infty), \\ P_{ifR} &= |(\Phi_i(-\infty)_R, U(-\infty, \infty)\Phi_f(\infty)_R)|^2 \\ &= |(KU(\infty, -\infty)\Phi_i(-\infty), U(-\infty, \infty)KU(-\infty, \infty)\Phi_f(\infty))|^2 \\ &= |(U(\infty, -\infty)\Phi_i(-\infty), [K^+U(-\infty, \infty)K]U(-\infty, \infty)\Phi_f(\infty))^*| \\ &= |(U(\infty, -\infty)\Phi_i(-\infty)\ U(\infty, -\infty)U(-\infty, \infty), \Phi_f(\infty))^*|^2 \\ &= |(\Phi_f(\infty), U(\infty, -\infty)\Phi_i(-\infty))|^2 = P_{if}. \end{aligned} \tag{35c}$$

The second term in Eq. 35*c* shows that the transition is from $\Phi_f(\infty)_R$ to $\Phi_i(-\infty)_R$ in contrast to the forward transition from $\Phi_i(-\infty)$ to $\Phi_f(\infty)$. Exact solutions to the Schrödinger equation are assumed in these two derivations; however, exact solutions to the problems of electronic capture are seldom achieved, and we thus inquire whether or not detailed balancing can be satisfied using approximate solutions to the IP version of the problem.

This question has been answered affirmatively in papers by Green[36] and Browne.[37] In these analyses the deflection of the two nuclei is neglected with the nuclear paths confined to parallel lines separated by the distance ρ; the constant relative velocity is again denoted by v. (The magnitudes of the velocity and impact parameter ρ, however, are irrelevant to the proofs in these two papers.) The method of Green is explained here first. Although alternative expansions in atomic and molecular orbitals are given, only the proof using atomic orbitals (not necessarily orthogonal) is described since one of the methods is sufficient to clarify the proof.

A. Green's Proof

The systems of coordinates are given by Eqs. 5 and 15*b* with **s** replaced by **x** in Eq. 15*b*, and since a two-electron system is analyzed, there is a set of these coordinates for each electron:

$$\mathbf{x}_{ij} = \mathbf{x}_j + p\mathbf{R}, \qquad p = \frac{M_p}{M_t}, \qquad (M_p = M_f)$$

$$\mathbf{x}_{fj} = \mathbf{x}_j - q\mathbf{R}, \qquad q = \frac{M_a}{M_t}, \qquad (M_a = M_i) \tag{36a}$$

$$p + q = 1, \qquad j = 1, 2, \qquad \mathbf{R} = \boldsymbol{\rho} + \mathbf{v}t.$$

The center-of-mass of the nuclei is the origin of coordinates as depicted in Fig. 2; thus the electronic masses do not enter; the Z-axis of the electronic coordinates is parallel to **v**. The notation for the nuclear masses agrees with Eq. 5, and the correspondence with Eq. 1 is given parenthetically for convenience. The electronic coordinates $\mathbf{x}_{ij}$ and $\mathbf{x}_{fj}$ refer electron j to A and P, respectively. The electronic wave function is represented by

$$\Psi(C; t) = \sum_n a_n(C; t)\Phi_n(\mathbf{x}_1, \mathbf{x}_2; t) \tag{36b}$$

where the Φ_n constitute a finite set of linearly independent functions that are not necessarily orthogonal. These n states must contain the initial and final states in particular, as well as other approximate eigenstates of the electronic equation. The label C distinguishes between solutions associated with different initial conditions, and the coefficients $a_n(C; t)$ are determined by the requirement that

$$\left(\Phi_k, \left(H - i\hbar\frac{\partial}{\partial t}\right)\Psi(C; t)\right) = 0, \tag{36c}$$

is satisfied for each k of the sum in Eq. 36*b*. (H contains the electrostatic interactions for each pair of particles with the possible exclusion of the internuclear interaction.) The insertion of Eq. 36*b* into Eq. 36*c* yields the coupled equations whose solutions determine the set $a_k(C; t)$ for each C:

$$i\hbar \sum_n (\Phi_k, \Phi_n) \frac{d}{dt} a_n(C; t) = \sum_n a_n(C; t)\left(\Phi_k, \left(H - i\hbar \frac{\partial}{\partial t}\right)\Phi_n\right). \quad (36d)$$

Since the Hamiltonian contains the time in this IP treatment, its invariance under complex conjugation and time reversal is expressed by

$$H(x_1 y_1 - z_1; x_2 y_2 - z_2; -t)^* = H(\mathbf{x}_1, \mathbf{x}_2; t). \quad (36e)$$

It is evident from Eq. 36*a* that the electronic Hamiltonian satisfies Eq. 36*e*.

An important consequence of Eqs. 36*b*, 36*c*, and 36*e* is the independence of the scalar product upon time:

$$-i\hbar \frac{\partial}{\partial t}\left(\Psi(C; t), \Psi(D; t)\right) = \left(\Psi(C; t), \left(H - i\hbar \frac{\partial}{\partial t}\right)\Psi(D; t)\right)$$
$$-\left(\left(H - i\hbar \frac{\partial}{\partial t}\right)\Psi(C; t), \Psi(D; t)\right) = 0. \quad (36f)$$

H is added and subtracted so that each scalar product on the right vanishes since the same $\Phi_n(\mathbf{x}_1, \mathbf{x}_2, t)$ are used throughout; thus this independence of time has been demonstrated simply by adding and subtracting the *Hermitian* operator H inside the scalar products in Eq. 36*f*. If D is set equal to C, the usual conservation of probability ensues. In order to establish detailed balancing a condition must be imposed upon the Φ_{nR}:

$$\Phi_{nR} = \Phi_n(x_1 y_1 - z_1; x_2 y_2 - z_2; -t)^* = \varepsilon_n \Phi(\mathbf{x}_1; \mathbf{x}_2; t), \qquad \epsilon_n = \pm 1, \quad (36g)$$

and the Φ_n are supposed to become orthogonal initially and finally. (This condition on the Φ_n is satisfied, for example, by hydrogenlike wave functions if allowance is made for the translational motion of the electron.)

$$\lim_{t \to \pm\infty} (\Phi_n, \Phi_k) = \delta_{kn} \quad (36h)$$

It also follows from Eq. 36*d* that the inequality

$$\int_{-\infty}^{\infty} dt \left|\left(\Phi_k, \left(H - i\hbar \frac{\partial}{\partial t}\right)\Phi_n\right)\right| < \infty \quad (36i)$$

must be satisfied in order that solutions of $a_n(C; t)$ exist. This can be seen by formally integrating both sides of the absolute values of these coupled equations and using the Schwartz inequality on the right-hand sides.[23] Unless Eq. 36*i* is fulfilled, the sums are unbounded and the differential equations are meaningless. (The dependence of Eq. 36*i* upon the electrostatic interactions of H will be treated in a subsequent paragraph.)

Attention is next directed to the time-reversed functions

$$\Psi_R(C; \mathbf{x}_1, \mathbf{x}_2; t) = \Psi(C; x_1 y_1 - z_1; x_2 y_2 - z_2; -t)^*. \tag{36j}$$

With the use of Eq. 36*e* we may easily validate the modified Eq. 36*c*:

$$\left(\Phi_{kR}(C; t), \left(H - i\hbar\frac{\partial}{\partial t}\right)\Psi_R(C; t)\right) = 0, \tag{36k}$$

and with the aid of Eq. 36*g* we can confirm that $\Psi_R(C; t)$ may be represented by Eq. 36*b* if $a_n(C; t)$ are replaced with

$$a_{R,n}(C; t) = \varepsilon_n a_n(C; -t)^*. \tag{36l}$$

Moreover, the $a_{R,n}$ satisfy the same equation as the a_n, and consequently to every solution Ψ there is associated a solution Ψ_R related by Eqs. 36*j* and 36*l*. This is precisely what was found for the exact solutions of wave mechanics as recorded in Eq. 35*a*.

Particular solutions $\Psi^{\pm}(k; t)$ are next defined for each k according to initial conditions

$$\begin{aligned} &\lim_{t\to-\infty} \Psi^+(k; t) \to \Phi_k(\mathbf{x}_1, \mathbf{x}_2; t)]_{t\to-\infty}, \\ &\lim_{t-\infty} \Psi^-(k; t) \to \Phi_k(\mathbf{x}_1, \mathbf{x}_2; t)]_{t\to+\infty}, \end{aligned} \tag{36m}$$

with associated coefficients

$$\begin{aligned} &\lim_{t\to-\infty} a_n^+(k; t) \to \delta_{kn}, \\ &\lim_{t\to+\infty} a_n^-(k; t) \to \delta_{kn}. \end{aligned} \tag{36n}$$

Equations 36*g*, 36*j*, and 36*m* are used to identify $\Psi^{\pm}$ with

$$\begin{aligned} \lim_{t\to\infty} \Psi_R^+(k; t) &= \lim_{t\to\infty} [\Psi^*(k; -t)]^* = \lim_{t\to\infty} [\Phi_k(-z_1, -z_2; -t)]^* \\ &= \lim_{t\to\infty} \varepsilon_k \Phi_k(z_1, z_2; t) = \lim_{t\to\infty} \varepsilon_k \Psi^-(k; t); \end{aligned} \tag{36o}$$

$$\lim_{t\to-\infty} \Psi_R^-(k; t) = \lim_{t\to-\infty} \Phi_k(-t)^* = \lim_{t\to-\infty} \varepsilon_k \Phi_k(t) = \lim_{t\to-\infty} \varepsilon_k \Psi^+(k; t),$$

and $a_n^{\pm}(k; t)$ with

$$a_{R,n}^+(k; t) = \varepsilon_k a_n^-(k; t), \; a_{R,n}^-(k; t) = \varepsilon_k a_n^+(k; t). \tag{36p}$$

Recall that $a_n^+(k; \infty)$ is the amplitude that the system is in the state Φ_n at $t = \infty$ after the collision, provided the system was in the state Φ_k at $t = -\infty$ before the collision. We now use Eq. 36*f*, which expresses the independence of the scalar product upon time, and Eqs. 36*j* and 36*o* to get the following relations:

$$
\begin{aligned}
a_k^+(n;\infty) &= \lim_{t\to\infty} (\Psi^-(k;t), \Psi^+(n;t)) \\
&= \lim_{t\to\infty} (\Phi_k(\mathbf{x}_1, \mathbf{x}_2; \infty), \Psi^+(n;t)) = \lim_{t\to-\infty} (\Psi^-(k;t), \Psi^+(n;t)) \\
&= \lim_{t\to\infty} (\Psi^-(k;-t), \Psi^+(n;-t)) = \lim_{t\to\infty} (\Psi^+(n;-t)^*, \Psi^-(k;-t)^*) \\
&= \lim_{t\to\infty} (\Psi_R^+(n;t), \Psi_R^-(k;t)) = \lim_{t\to\infty} (\varepsilon_n \Psi^-(n;t), \varepsilon_k \Psi^+(k;t)) \\
&= \lim_{t\to\infty} \varepsilon_n \varepsilon_k (\Phi_n(\mathbf{x}_1, \mathbf{x}_2; t), \Psi^+(k;t)) = \varepsilon_n \varepsilon_k a_n^+(k;\infty).
\end{aligned} \tag{36q}
$$

(The independence of the scalar product upon time is used in the third line of Eq. 36*q*, and Eq. 36*b* is used in the last line.) The formula of detailed balancing for the IP version of capture,

$$
|a_k^+(n;\infty)|^2 = |a_n^+(k;\infty)|^2, \tag{36r}
$$

has thus been obtained, but we still need to validate Eq. 36*i*.

For the purpose of this demonstration the process

$$
\mathrm{He}^+(1s) + \mathrm{H}(1s) \to \mathrm{He}(1s^2) + \mathrm{H}^+ \tag{37a}
$$

is used. Reference is made to Eq. 36*a* for the coordinate systems. The wave functions Φ_n and Φ_k of Eq. 36*i* are selected to represent the initial and final states of Eq. 37*a*, respectively; they are

$$
\begin{aligned}
\Phi_n = \frac{1}{\sqrt{2}} \Big\{ &\phi_{1s}(\mathbf{x}_{i1}) \phi_{1s}(\mathbf{x}_{f2}) \exp\left[-\frac{im\mathbf{v}}{\hbar} \cdot (p\mathbf{x}_1 - q\mathbf{x}_2)\right] + \phi_{1s}(\mathbf{x}_{i2}) \phi_{1s}(\mathbf{x}_{f1}) \\
&\times \exp\left[-\frac{im\mathbf{v}}{\hbar} \cdot (p\mathbf{x}_2 - q\mathbf{x}_1)\right]\Big\} \exp\left[-\frac{i}{\hbar}\left(\frac{p^2mv^2}{2} + \frac{q^2mv^2}{2}\right)t\right] \\
&\times \exp\left\{-\frac{i}{\hbar}[(\varepsilon_a + \varepsilon_b)t_i - \phi_n(t)]\right\},
\end{aligned} \tag{37b}
$$

$$
\Phi_k = N_f \phi_{1s}(z\mathbf{x}_{f1}) \phi_{1s}(z\mathbf{x}_{f2}) \exp\left\{\frac{i}{\hbar}[mq\mathbf{v}\cdot(\mathbf{x}_1 + \mathbf{x}_2) - mq^2v^2t - \epsilon_f t_f]\right\}.
$$

The function $\exp[(i/\hbar)\phi_n(t)]$ occurring as a factor of Φ_n will be defined at an appropriate place. Although Φ_n is the exact electronic wave function, Φ_k represents an approximation to the final electronic state; also note that these spatial wave functions possess the symmetry appropriate to singlet electronic states (in the coordinates $\mathbf{x}_1$ and $\mathbf{x}_2$), the case assumed in this example. The transformations among the several sets of independent variables and the equations of the initial atomic states are recorded here for clarity:

$$(a)\quad \begin{cases} \mathbf{x}_{i1} = \mathbf{x}_1 + p\mathbf{R} \\ \mathbf{x}_{f2} = \mathbf{x}_2 - q\mathbf{R} \\ t_i = t \end{cases} \qquad \nabla_{x1} = \nabla_{x_{i1}}, \quad \nabla_{x2} = \nabla_{x_{f2}} \qquad \frac{\partial}{\partial t} = \frac{\partial}{\partial t_i} + p\mathbf{v}\cdot\nabla_{x_{i1}} - q\mathbf{v}\cdot\nabla_{x_{f2}},$$

$$(b)\quad \begin{cases} \mathbf{x}_{i2} = \mathbf{x}_2 + p\mathbf{R} \\ \mathbf{x}_{f1} = \mathbf{x}_1 - q\mathbf{R} \\ t_i = t \end{cases} \qquad \nabla_{x1} = \nabla_{x_{f1}}, \quad \nabla_{x2} = \nabla_{x_{i2}} \qquad \frac{\partial}{\partial t} = \frac{\partial}{\partial t_i} - q\mathbf{v}\cdot\nabla_{x_{f1}} + p\mathbf{v}\cdot\nabla_{x_{i2}};$$

$$\left(-\frac{\hbar^2}{2m}\nabla^2_{x_{im}} - \frac{e^2}{x_{in}} - \varepsilon_a\right)\phi_{1s}(\mathbf{x}_{in}) = 0,$$

$$\left(-\frac{\hbar^2}{2m}\nabla^2_{x_{fn}} - \frac{2e^2}{x_{fn}} - \varepsilon_b\right)\phi_{1s}(\mathbf{x}_{fn}) = 0, \qquad n = 1, 2. \tag{37c}$$

With the definitions given in Eqs. 37*b* and 37*c* it is found (omitting $\exp[(i/\hbar)\phi(t)]$ for the present) that

$$\left(H - i\hbar\frac{\partial}{\partial t}\right)\Phi_n = \left[-\frac{\hbar^2}{2m}(\nabla_{x1}{}^2 + \nabla_{x2}{}^2) - \frac{e^2}{x_{i1}} - \frac{2e^2}{x_{f2}} - \frac{2e^2}{x_{f1}} - \frac{e^2}{x_{i2}} + \frac{e^2}{|\mathbf{x}_1 - \mathbf{x}_2|} - i\hbar\frac{\partial}{\partial t}\right]\Phi_n$$

$$= \frac{e^2}{\sqrt{2}}\left\{\left(-\frac{2}{x_{f1}} - \frac{1}{x_{i2}} + \frac{1}{|\mathbf{x}_1 - \mathbf{x}_2|}\right)\phi_{1s}(\mathbf{x}_{i1})\phi_{1s}(\mathbf{x}_{f2}) \times \exp\left[\frac{-im\mathbf{v}}{\hbar}\cdot(p\mathbf{x}_{i1} - q\mathbf{x}_{f2}\right]\right.$$

$$\left. + \left(-\frac{2}{x_{f2}} - \frac{1}{x_{i1}} + \frac{1}{|\mathbf{x}_1 - \mathbf{x}_2|}\right)\phi_{1s}(\mathbf{x}_{i2})\phi_{1s}(\mathbf{x}_{f1}) \times \exp\left[\frac{-im\mathbf{v}}{\hbar}\cdot(p\mathbf{x}_{i2} - q\mathbf{x}_{f2})\right]\right\} \times \exp\left\{-\frac{i}{\hbar}\left[\frac{mv^2}{2}(p^2 + q^2) + \varepsilon_a + \varepsilon_b\right]t_i\right\}, \tag{37d}$$

$$\phi_{1s}(\mathbf{x}_{in}) = \frac{\exp(-|\mathbf{x}_{fn} + \mathbf{R}|/a_0)}{\sqrt{\pi}}, \qquad \phi_{1s}(\mathbf{x}_{fn}) = \sqrt{\frac{8}{\pi}}\exp\left(-\frac{2x_{fn}}{a_0}\right),$$

$$a_0 = \frac{\hbar^2}{me^2}, \qquad n = 1, 2.$$

Although the variables of integration are $\mathbf{x}_1$, $\mathbf{x}_2$ in Eq. 36*i*, it is convenient to use the $(\mathbf{x}_{f1}, \mathbf{x}_{f2})$ set here. The corresponding integral simplifies to

$$\frac{Ne^2}{2}\int d\mathbf{x}_{f1}\, d\mathbf{x}_{f2}\phi_{1s}(z\mathbf{x}_{f1})^*\phi_{1s}(z\mathbf{x}_{f2})^*$$

$$\times\left\{\phi_{1s}(|\mathbf{x}_{f1}+R|)\phi_{1s}(\mathbf{x}_{f2})\left[-\frac{2}{\mathbf{x}_{f1}}-\frac{1}{|\mathbf{x}_{f2}+\mathbf{R}|}+\frac{1}{|\mathbf{x}_{f1}-\mathbf{x}_{f2}|}\right.\right.$$

$$\left.+\frac{\partial\phi_n(t)}{e^2\partial t}\right]\exp\left(\frac{-im\mathbf{v}\cdot\mathbf{x}_{f1}}{\hbar}\right)+\phi_{1s}(|\mathbf{x}_{f2}+\mathbf{R}|)\phi_{1s}(\mathbf{x}_{f1})$$

$$\left.\times\left[\frac{-1}{|\mathbf{x}_{f1}+\mathbf{R}|}-\frac{2}{x_{f2}}+\frac{1}{|\mathbf{x}_{f1}-\mathbf{x}_{f2}|}+\frac{\partial\phi_n(t)}{e^2\partial t}\right]\exp\left(\frac{-im\mathbf{v}\cdot\mathbf{x}_{f2}}{\hbar}\right)\right\}$$

$$\times\exp\left(-\frac{i}{\hbar}\left\{(-\epsilon_f+\epsilon_a+\epsilon_b)t_f+\phi_n(t_f)\right.\right.$$

$$\left.\left.+\left[\frac{mq^2}{2}-p\left(\frac{p}{2}-1\right)mv^2\right]t_f\right\}\right). \tag{37e}$$

In the limits, $t = \pm\infty$, the exponential factors of the integrand, $\phi(|\mathbf{x}_{fn}+\mathbf{R}|)$, are expanded about $\mathbf{x}_{fn}+\mathbf{R}=0$ $(n = 1, 2)$ and can contribute to the integral only in the neighborhood of $\mathbf{x}_{fn} = -\mathbf{R}$. Consequently, if

$$\phi_n(t) = 2e^2\int_0^t \frac{dt'}{R}, \qquad \frac{\partial\phi_n}{\partial t} = \frac{2e^2}{R}, \tag{37f}$$

the leading terms of each of the two sums of the interactions are R^{-m} with $m = 2, 3, \ldots$. Since the integral

$$\int\frac{dt}{R^2} = \frac{1}{\rho v}\tan^{-1}\left(\frac{vt}{\rho}\right) \tag{37g}$$

is bounded at $t = \pm\infty$, the Schwartz inequality for integrals guarantees the convergence of Eq. 36*i*.

B. Browne's Proof: Differential-Equation Method[37]

In the two-state (or *n*-state) approximation, the wave functions represent exact or approximate eigenfunctions of the electronic equation

$$\left(H - i\hbar\frac{\partial}{\partial t}\right)\Psi = 0 \tag{38}$$

at $t = \pm\infty$, and, as always in the correct treatment using the IP method, the translational motion of the electron being captured is accommodated. We assume that the two moving atom wave functions have been orthogonalized and later show that the number of unknown parameters is consistent with the number of conditions imposed. It will also be established in the same analysis that this corresponding transformation of the coefficients $A'(t)$, $B'(t)$ does not alter the initial condition, $A(-\infty) = 1$, $B(-\infty) = 0$, or the probability for electronic capture $|B(\infty)|^2$.

The wave function of the system (in atomic units),

$$\Psi(t) = A'(t)\Psi'_1(t) + B'(t)\Psi'_2(t), \tag{39a}$$

is required to satisfy the equations and initial conditions

$$(\Psi'_m(t), G(t)\Psi(t)) = 0, \qquad G = H - i\frac{\partial}{\partial t}, \quad m = 1, 2, \tag{39b}$$

$$\begin{aligned} t \to -\infty, \qquad A'(-\infty) = 1, \qquad B'(-\infty) = 0, \\ A'(t) = a_1^+(1;t), \qquad B'(t) = a_2^+(1;t), \end{aligned} \tag{39c}$$

and these three equations yield two coupled, first-order differential equations for $A'(t)$ and $B'(t)$,

$$\begin{aligned} i\dot{A}'(t) &= G'(t)_{11}A'(t) + G'(t)_{12}B'(t), \qquad \dot{x} = \frac{dx}{dt}, \\ G'(t)_{mn} &= (\Psi'_m(t), G\Psi'_n(t)), \end{aligned} \tag{39d}$$

$$i\dot{B}'(t) = G'(t)_{21}A'(t) + G'(t)_{22}B'(t). \tag{39e}$$

In Eq. 39*c* A', B' are related to Green's $a_1^+(1;t)$, $a_2^+(1;t)$ (cf. Eqs. 36*d*, 36*l*, and 36*n*) in order to emphasize the initial conditions. As in most of this analysis, the scalar products (note the requirement following Eq. 35*b*) signify integration over the electronic coordinates referred to an origin on the internuclear line. Next we change the sign of t, form the complex conjugate of Eqs. 39*d* and 39*e*, and use new coefficients:

$$\begin{aligned} i\dot{C}(-t')^* &= G'(-t')^*_{11}C(-t')^* + G'(-t')^*_{12}D(-t')^*, \\ & \qquad\qquad t' = -t, \quad \dot{x} = \frac{dx}{dt'}, \end{aligned} \tag{39f}$$

$$\begin{aligned} i\dot{D}(-t')^* &= G'(-t')^*_{21}C(-t')^* + G'(-t)^*_{22}D(-t')^*, \\ \varepsilon_1\varepsilon_2 C(-t')^* &= a_1^-(2;t')^* = \varepsilon_1 a_{1R}^-(2;-t') = \varepsilon_1\varepsilon_2 a_1^+(2;-t'), \\ D(-t')^* &= a_2^-(2;t') = \epsilon_2 a_{2R}^-(2;-t') = a_2^+(2;-t'), \\ \varepsilon_m &= \pm 1, \quad m = 1, 2. \end{aligned} \tag{39g}$$

Again comparison with Green's notation is made, and his phases, $\varepsilon_{1,2}$, are also used (cf. Eq. 36*g*).

The preceding two differential equations are equivalent to Eqs. 39*d*, 39*e*, and 39*c* if Eqs. 39*h* through 39*j* are satisfied:

$$\Psi_m(-t')^* = \varepsilon_m \Psi_m(t'), \qquad \Psi'_m(-t')^* = \varepsilon_m \Psi'_m(t'),$$
$$G'(-t')^*_{mm} = G'(t')_{mm}, \qquad G'(t')_{11} = G'(t')_{22}, \qquad m = 1, 2, \tag{39h}$$

$$G'(-t')^*_{mn} = \varepsilon_m \varepsilon_n G'(t')_{mn}, \qquad G'(t')^*_{mn} = G'(t')_{nm},$$
$$m = 1, 2, \quad n = 2, 1, \tag{39i}$$

$$t' \to \infty, \mathrm{D}(-\infty)^* = 1, \qquad \mathrm{C}(-\infty)^* = 0. \tag{39j}$$

The fulfillment of Eqs. 39*h* and 39*i* makes the original Eqs. 39*d* and 39*e* invariant to the combined transformation of time reversal and complex conjugation. The initial condition of Eq. 39*j* makes the solutions of Eqs. 39*f* and 39*g* equivalent to those of Eqs. 39*d* and 39*e* with initial conditions the same as the initial conditions in Eq. 39*c*, which will be demonstrated shortly.

If the coefficients $\mathrm{A}'(t)$, $\mathrm{B}'(t)$, and $\mathrm{C}(-t')^*$, $\mathrm{D}(-t')^*$ are designated collectively by $\mathrm{X}_m(t)$ and $\mathrm{Y}_n(-t')^*$, respectively, it is clear that Eqs. 39*f* and 39*g* can be rearranged so that the $X_m(t)$ are effectively replaced by $\varepsilon_n \varepsilon_m Y_n(-t')^*$ This rearrangement is accomplished by multiplying the differential equation of Eq. 39*f* by $\varepsilon_1 \varepsilon_2$. These rearranged equations, with associated initial conditions, are rewritten as an aid to the subsequent analysis:

$$\begin{gathered} i\dot{\mathrm{D}}(-t')^* = G'(t')_{22}(t')\mathrm{D}(-t')^* + G'(t')_{21}[\varepsilon_1\varepsilon_2 \mathrm{C}(-t')^*], \\ t' \to \infty, \qquad \mathrm{D}(-\infty)^* = 1, \qquad \mathrm{C}(-\infty)^* = 0, \end{gathered} \tag{39k}$$

$$i[\varepsilon_1\varepsilon_2 \dot{\mathrm{C}}(-t')^*] = G'(t')_{12}\mathrm{D}(-t')^* + G'(t')_{11}[\varepsilon_1\varepsilon_2 \mathrm{C}(-t')^*]. \tag{39l}$$

In this analysis the differential equations of Eqs. 39*d* and 39*e*—with the initial conditions of Eq. 39*c*—are designated collectively by *D*, and Eqs. 39*k* and 39*l* are designated collectively by *R*. First, *D* in the range $(-\infty < t < 0)$ is compared with *R* in the range $(\infty > t' > 0)$. From Eqs. 39*h* and 39*i* it is clear that $G'(t)_{mn}(D) = \varepsilon_m \varepsilon_n G'(t')_{nm}(R)$, $m = 1, 2$, $n = 2, 1$; moreover, $t < 0$ in $\mathrm{X}_m(t)$ and $(-t') < 0$ in $Y_n(-t')^*$. Next, *D* in the range $(0 < t < \infty)$ is compared to *R* in the range $(0 > t' > -\infty)$. The relation between the matrix elements of G' is the same as in the previous ranges of t and t'; furthermore, $t > 0$ in $\mathrm{X}_m(t)$ and $(-t') > 0$ in $Y_n(-t')^*$. Therefore it is evident that the solution for $\mathrm{B}'(\infty)$ of *D* is equal to the solution for $[\varepsilon_1\varepsilon_2 \mathrm{C}(\infty)^*]$ of *R* since the defining differential equations and associated initial conditions are equivalent. Thus detailed balancing is satisfied as expressed by

$$\begin{aligned} \mathrm{P}_{12} &= |\mathrm{B}'(\infty)|^2 = |a_2^+(1;\infty)|^2 = |\varepsilon_1\varepsilon_2 \mathrm{C}'(\infty)^*|^2 \\ &= |\varepsilon_1\varepsilon_2 a_1^+(2;\infty)|^2 = \mathrm{P}_{21}. \end{aligned} \tag{39m}$$

Equation 39*a* may also be expressed without the superscript primes; this equation represents Ψ in terms of the original normalized, nonorthogonal wave functions. The equations that represent the transformation from the Ψ'_m-set to the Ψ''_m-set are

$$\Psi''_m(t) = \sum_{n=1}^{2} u_{mn}\Psi'_n(t), \qquad m = 1, 2. \tag{39o}$$

The coefficients of this matrix, the u_{mn}, are determined by the requirements that the $\Psi''_m(t)$ must satisfy Eq. 39p:

$$\begin{aligned} (\Psi''_m(t), \Psi''_n(t)) &= 0, \qquad m = 1, 2, \quad n = 2, 1, \\ (\Psi''_m(t), \Psi''_m(t)) &= 1, \qquad m = 1, 2, \end{aligned} \tag{39p}$$

and the $G'(-t')^*_{mn}$ must satisfy the conditions expressed by Eqs. 39h and 39i. Moreover, the transformation is additionally restricted by the requirements that $\mathrm{A}'(-\infty) = \mathrm{A}(-\infty) = 1$, $\mathrm{B}'(-\infty) = \mathrm{B}(-\infty) = 0$, $|\mathrm{B}'(\infty)|^2 = |\mathrm{B}(\infty)|^2$. The u_{mn} which preserve these relations and satisfy Eqs. 39h, 39i, and 39p are given by[38]

$$\begin{aligned}
u_{11} &= e^{i\delta_1}, \quad u_{12} = 0, \quad u_{21} = -Su_{22}, \quad u_{22} = e^{i\delta_2}(1 - |S|^2)^{-1/2}, \\
S &= S_{12} = (\Psi'_1(t), \Psi'_2(t)), \quad S^* = S_{21}, \\
G'(t)_{11} &= G'(t)_{22} = G(t)_{11} + \dot{\delta}_1, \\
G'(t)_{21} &= \frac{\exp[i(\delta_1 - \delta_2)]}{(1 - |S|^2)^{1/2}} [G(t)_{21} - S^*G(t)_{11}], \\
G'(t)_{12} &= \frac{\exp[-i(\delta_1 - \delta_2)]}{(1 - |S|^2)^{1/2}} [G(t)^*_{21} - SG(t)^*_{11}] = G'(t)^*_{21}, \\
G(t)^*_{mm} &= G(t)_{mm}, \qquad m = 1, 2, \\
G(t)^*_{21} &= G(t)_{12} + i\dot{S},
\end{aligned} \tag{39q}$$

$$\dot{X} = \frac{dX}{dt}, \quad \dot{\delta}_2 - \dot{\delta}_1 = f(t) = \frac{1}{1 - |S|^2}\Big[(1 - 2|S|^2)G(t)_{11} - G(t)_{22} + S^*G(t)_{12} + SG(t)^*_{12} + \frac{i}{2}(S^*\dot{S} - S\dot{S}^*)\Big],$$

$$\mathrm{A}'(t) = \frac{u_{22}\mathrm{A}(t) - u_{21}\mathrm{B}(t)}{D}, \qquad \mathrm{B}'(t) = \frac{u_{11}\mathrm{B}(t)}{D},$$

$$D = u_{11}u_{22}, \qquad S(\pm\infty) = 0.$$

In the preceding equation the phases $\delta_{1,2}$ are real functions of t, and the $G(t)_{mn}$ are defined in terms of the $\Psi'_m(t)$. Note also that $|\mathrm{A}'(\pm\infty)| = |\mathrm{A}(\pm\infty)|$, $|\mathrm{B}'(\pm\infty)| = |\mathrm{B}(\pm\infty)|$. With the use of the Hermitian property of H and the independence of (Ψ'_m, Ψ'_m) upon t (cf. Eq. 39b), it is straightforwardly demonstrated that $G(t)_{mm}$, $m = 1, 2$, are both real functions of t. Consequently, the reality of the $G(t)_{mm}$ and the structure of $f(t)$ in Eq. 39q together insure that $(\dot{\delta}_2 - \dot{\delta}_1)$ is a real function of t, as previously asserted. [The $(\mathbf{x}, t)$-set of independent variables is being used in this demonstration (cf., for example, Eqs. 14a, 15b, 15c.)] For the purpose of the present demonstration, δ_1 is equated to zero without loss of essential characteristics of the solution, and $\delta_2(t)$ is constructed to satisfy the properties of $\Psi''_m(t)$ required by Eq. 39h. The solution is

$$\delta_2(t) = \tfrac{1}{2}\int_{-t}^{t} dt' f(t') \tag{39r}$$

with $f(t)$ defined in Eq. 39q. According to Eq. 39h for $\Psi_m(t)$, and Eq. 39q for $f(t)$, it is evident that $f(t) = f(-t)$. It follows from Eq. 39r that $\delta_2(-t) = -\delta_2(t)$, and this structure satisfies the condition of the problem since $[e^{i\delta_2(-t)}]^* = e^{i\delta_2(t)}$. Thus this property of $\delta_2(t)$, the structure of u_{mn} in Eq. 39q, the property of $\Psi_m(t)$ in Eq. 39h, and the definition of $\Psi''_m(t)$ in Eq. 39o, collectively satisfy the time-reversed properties,

$$\Psi''_m(-t)^* = \varepsilon_m \Psi''(t).$$

This completes the verification of detailed balancing for the two-state case, and the n-state case is not given.

In the three-state case two of the states are associated either with the initial channel (direct scattering) or with the final channel into which capture occurs. These states are discriminated by subscripts 1, 2, and 3, and the equations corresponding to Eqs. 39d and 39e are replaced by three equations in $\mathrm{A}'(i; t)$ (initial), $\mathrm{B}'(i; t)$, and $\mathrm{C}'(i; t)$ (final). Conditions corresponding to Eq. 39h are imposed to determine the coefficients of transformation, u_{mn}. In order to establish detailed balancing, the additional condition $\varepsilon_1\varepsilon_2 G'(t)_{12} = \varepsilon_2\varepsilon_3 G'(t)_{23}$ is required; thus eight conditions must be fulfilled. Additional restrictions are required to insure that A′, B′, and C′ satisfy the same initial and final interpretations as A, B, and C, respectively. As a result of these imposed constraints, equations equivalent to the original ones are obtained by replacing $\mathrm{A}'(i; t)$, $\mathrm{B}'(i; t)$ and $\mathrm{C}'(i; t)$ with $\mathrm{C}'(f; -t)^*$, $\varepsilon_2\varepsilon_3\mathrm{B}'(f; -t')^*$, and $\varepsilon_1\varepsilon_3\mathrm{A}'(f; -t')^*$, respectively, with $t' = -t$ as before. With the initial conditions, $\mathrm{C}(f; -\infty)^* = 1$, $\mathrm{B}'(f; -\infty)^* = \mathrm{A}'(f; -\infty)^* = 0$, this analysis yields $|\mathrm{C}(i; \infty)|^2 = |\varepsilon_1\varepsilon_3\mathrm{A}(f; \infty)^*|^2$, as asserted, provided that solutions for the u_{mn} are physically realizable.

One other fact relative to this topic should be clarified. Suppose the two colliding aggregates have spins S_{p^+i}, S_{ai} initially and spins S_{pf}, S_{a^+f}, finally. The equation for detailed balancing now reads

$$(2S_{p^+i} + 1)(2S_{ai} + 1)\sigma_d = (2S_{pf} + 1)(2S_{a^+f} + 1)\sigma_r. \tag{39s}$$

In this equation the subscripts d and r designate direct and inverse, respectively.[32] This equation presupposes that the degeneracies are included in the definitions of σ_d and σ_r. Suppose the aggregates A^+ and A of Eq. 1 and Fig. 2 also have total orbital angular momentum L_i and L_f, respectively, and suppose that P^+ is a proton and P is a ^2S state of atomic hydrogen. Equation 39*s* is then modified to

$$(2S_i + 1)(2L_i + 1)\sigma_d = 2(2S_f + 1)(2L_f + 1)\sigma_r.$$

This relation is exemplified by applying it to the following direct processes:

$$H^+ + O(^3P) \rightarrow H(^2S) + O^+(^4S;\, ^2D;\, ^2P)$$

and the corresponding inverse process:[39]

$$H(^2S) + O^+(^4S;\, ^2D;\, ^2P) \rightarrow H^+ + O(^3P);$$

$$S_i = 1, \qquad S_f = \frac{3}{2},$$

$$L_i = 1, \qquad L_f = 0, \quad 9\sigma_d(^3P \rightarrow {}^4S) = 8\sigma_r(^4S \rightarrow {}^3P);$$

$$S_i = 1, \qquad S_f = \frac{1}{2},$$

$$L_i = 1, \qquad L_f = 2, \quad 9\sigma_d(^3P \rightarrow {}^2D) = 20\sigma_r(^2D \rightarrow {}^3P);$$

$$S_i = 1, \qquad S_f = \frac{1}{2},$$

$$L_i = 1, \qquad L_f = 1, \quad 9\sigma_d(^3P \rightarrow {}^2P) = 12\sigma_r(^2P \rightarrow {}^3P).$$

Although the cross sections for these processes were originally calculated using the wave version, the conditions of the capturing process are closely approximated by the IP conditions; consequently, Eq. 39*s*, modified to include the orbital angular momenta, is applicable to these reactions.

REFERENCES

1. D. R. Bates and R. A. Mapleton, *Proc. Phys. Soc. (London)*, **87**, 657 (1966).
2. H. Goldstein, *Classical Mechanics*, Addison-Wesley, Reading, Mass. (1950), pp. 69–71.
3. L. I. Schiff, *Quantum Mechanics*, McGraw-Hill, New York (1949), pp. 149–151.
4. L. D. Landau and E. M. Lifshitz, *Mechanics*, Addison-Wesley, Reading, Mass. (1960), pp. 22–24.
5. L. Pauling and E. B. Wilson, jr., *Introduction to Quantum Mechanics*, McGraw-Hill, New York (1935), p. 146.
6. D. R. Bates and R. McCarroll, *Phil. Mag. Suppl.*, **11**, 39 (1962).

7. D. Bohm, *Quantum Theory*, Prentice-Hall, New York (1951), Chapter 12.
8. Pauling and Wilson, *op. cit.*, p. 107.
9. W. Magnus and F. Oberhettinger, *Formulas and Theorems for the Special Functions of Mathematical Physics*, Chelsea, New York (1949), p. 94.
10. Bohm, *op. cit.*, pp. 276–277.
11. R. A. Mapleton, "Classical and Quantal Calculations on Electron Capture," Air Force Cambridge Research Laboratories Report AFCRL 67-0351, Physical Science Research Papers, No. 328, 1967, pp. 280–281.
12. L. D. Landau and E. M. Lifshitz, *Quantum Mechanics*, Addison-Wesley, Reading, Mass. (1958), p. 54.
13. *Ibid.*, pp. 304–313.
14. N. F. Mott and H. S. W. Massey, *The Theory of Atomic Collisions* (3d ed.), Oxford University Press, London (1965), pp. 351–354, 662–670, 804–808.
15. D. R. Bates, "Theoretical Treatment of Collisions between Atomic Systems," in *Atomic and Molecular Processes*, Academic Press, New York (1962), pp. 608–613.
16. D. R. Bates, *Proc. Roy. Soc.* (*London*)*A*, **257**, 22 (1960).
17. Jean Heinrichs, *Phys. Rev.*, **176**, 141 (1968).
18. C. A. Coulson and K. Zalewski, *Proc. Roy. Soc.* (*London*)*A*, **268**, 437 (1962).
19. A. M. Dykhne and A. V. Chaplik, *Zh. Eksperim. i Theor. Fiz.*, **43**, 889 (1962); [*Soviet Phys.-JETP*, **16**, 631 (1963)].
20. G. V. Dubrovskii, *Zh. Eksperim. i Theor. Fiz.*, **46**, 863 (1964); [*Soviet Phys.-JETP*, **19**, 591 (1964)].
21. V. K. Bykhovskiĭ, E. E. Nikitin, and M. Ya. Ovchinnikova, *Zh. Eksperim. i Theor. Fiz.*, **47**, 750 (1964); [*Soviet Phys.-JETP*, **20**, 500 (1965)].
22. D. R. Bates, *Quantum Theory*, Academic Press, New York (1961), p. 216.
23. R. Courant and D. Hilbert, *Methods of Mathematical Physics*, Vol. 1 (trans. ed.), Interscience, New York (1953), pp. 50–51.
24. Landau and Lifshitz, *Quantum Mechanics*, p. 364.
25. Mott and Massey, *op. cit.*, pp. 804–806.
26. *Ibid.*, 353, 805.
27. D. R. Bates, H. C. Johnston, and I. Stewart, *Proc. Phys. Soc.* (*London*), **84**, 517 (1964).
28. Bates, *Atomic and Molecular Processes*, p. 611.
29. *Ibid.*, p. 587.
30. M. L. Goldberger and K. M. Watson, *Collision Theory*, Wiley, New York (1964), pp. 51–57.
31. *Ibid.*, pp. 169–171.
32. Landau and Lifshitz, *Quantum Mechanics*, pp. 432–435.
33. Bates, *Quantum Theory*, pp. 259–261.
34. A. Messiah, *Quantum Mechanics*, Vol. 1 (G. M. Temmer, trans.) (1961); reprint ed., North-Holland, Amsterdam, and Interscience, New York (1963), pp. 310–314.
35. A. Messiah, *Quantum Mechanics*, Vol. 2 (J. Potter, trans.) (1962); reprint ed., North-Holland, Amsterdam, and Interscience, New York (1963), pp. 641–642, 664–675.
36. T. A. Green, *Proc. Phys. Soc.* (*London*), **86**, 1017 (1965).
37. J. C. Browne, *Proc. Phys. Soc.* (*London*), **86**, 419 (1965).
38. T. A. Green, H. E. Stanley, and You-Chien Chiang, *Hel. Phys. Acta*, **38**, 109 (1965).
39. R. A. Mapleton, *Phys. Rev.*, **130**, 1829 (1963).

CHAPTER 3

QUANTUM MECHANICAL METHODS FOR HIGH VELOCITIES

We now turn to quantum mechanical approximating procedures designed for higher impact velocities than the methods described in Sections 1.3 to 2.5. In these theoretical sections an *impact velocity* equal to the velocity of the first Bohr orbit, $v_0 = e^2/\hbar$, forms the boundary between high and low velocities, but this is a definition, and is not deduced from physical phenomena. Accordingly, some of the previous methods are successful for velocities greater than v_0, and some of the subsequent methods occasionally are successful for velocities less than v_0. Most attention will be devoted to the wave version of electronic capture, although the IP version is used in two subsections. It is particularly important that the reader is cognizant of the unknown limitations of those approximate treatments of the problem of electronic capture. At the time of this writing, for example, it was undecided whether any existing approximate solution agreed asymptotically with exact solution of the Schrödinger equation at high nonrelativistic impact velocities. This is a subject of continuing controversy and thus deserves special mention. Finally, only a select group of papers are treated since the purpose of this book is to explain and describe methods and not to review literature.

3.1 METHOD OF BATES

Bates' method was developed in part to eliminate the indeterminacy of the IP method upon arbitrary functions of the internuclear distance.[1] This arbitrariness in the results arises from representing the electronic wave function by expansions that do not explicitly contain the initial and final states, a fact also noted by others.[2–4] This defect (particularly Born expansions), which relates to nonorthogonality between initial and final states, is remedied by expanding the wave function as

$$\Psi = a_i(t)\psi_i + b_f(t)\psi_f + \gamma, \tag{40a}$$

in which decomposition ψ_i and ψ_f are the initial (prior) and final (post) states, respectively, and γ is orthogonal to ψ_i and ψ_f. Therefore γ can be expanded in the assumed complete orthogonal set of either the prior or post electronic Hamiltonians, omitting from these expansions the state of the same set occurring in the other two terms of Eq. 40*a* (see Eq. 2). As a consequence of these properties of the expansion used in Eq. 40*a*, Ψ is expressed in terms of an orthogonal set at large internuclear distances in which set ψ_f is explicitly separated from the initial state ψ_i. We turn next to the calculational details.[1,5–7]

The midpoint of the internuclear line is the common electronic origin, and thus the electronic Hamiltonian of Eq. 6 ($\nabla_R = \nabla_R^{\,2} = 0$) is appropriate for this problem. (This anaylsis is restricted to two singly charged cores and one electron; see comment following Eq. 40*b*.) The two nuclei are in uniform motion along lines that are parallel to the *Z*-axis, which is fixed in the observer's frame of reference. The functions of Eq. 40*a* and the relative coordinates are (Eqs. 4 and 6)

$$\psi_i = \phi_i(\mathbf{x}_i)\exp\left[-i\left(\frac{\mathbf{k}\cdot\mathbf{s}}{2} + \frac{mv^2}{8\hbar}t + \varepsilon_i t_i\right)\right],$$

$$\psi_f = \phi_f(\mathbf{x}_f)\exp\left[i\left(\frac{\mathbf{k}\cdot\mathbf{s}}{2} - \frac{mv^2 t}{8\hbar} - \varepsilon_f t_f\right)\right],$$

$$\mathbf{k} = \frac{m\mathbf{v}}{\hbar}, \qquad \mathbf{x}_i = \mathbf{s} + \tfrac{1}{2}\mathbf{R}, \qquad \mathbf{x}_f = \mathbf{s} - \tfrac{1}{2}\mathbf{R}, \qquad \mathbf{R} = \mathbf{v}t + \boldsymbol{\rho}, \tag{40b}$$

$$\gamma = \sum_{p\neq i} a_p(t)\phi_p(\mathbf{x}_i)\exp\left[-i\left(\frac{\mathbf{k}\cdot\mathbf{s}}{2} + \frac{mv^2}{8\hbar}t + \varepsilon_p t_i\right)\right]$$

$$= \sum_{q\neq f} b_q(t)\phi_q(\mathbf{x}_f)\exp\left[i\left(\frac{\mathbf{k}\cdot\mathbf{s}}{2} - \frac{mv^2 t}{8\hbar} - \varepsilon_q t_f\right)\right],$$

$$\nabla_s = \nabla_{x_i} = \nabla_{x_f}, \qquad \frac{\partial}{\partial t} = \frac{\partial}{\partial t_i} + \tfrac{1}{2}v\cdot\nabla_s = \frac{\partial}{\partial t_f} - \tfrac{1}{2}\mathbf{v}\cdot\nabla_s.$$

As noted twice previously the effects of the perturbing Coulomb interaction upon the phase of the wave function at $t = \pm\infty$ is ignored. The *Z*-axis of $\mathbf{s}$ is chosen parallel to $\mathbf{v}$. (The expansions representing γ are assumed to contain the appropriate states of the continuum, although these states are not explicitly shown.) For the purposes of this section we consider the transfer of an active electron from the atom A in atomic state ϕ_i to the incident ion P^+ whose final atomic state is $\phi_f(\mathbf{x}_f)$. Other electrons, if present, are ignored since they are assumed to be inactive. (Although the complicating feature of equivalent electrons in an outer subshell can be managed, this problem is deferred to a subsequent section.) The assumed solutions are substituted into the electronic equation to get the systems of differential equations

$$i\hbar\left\{\dot{a}_i(t) + \dot{b}_f(t)S_{if}\exp\left[-i(\varepsilon_f - \varepsilon_i)\frac{t}{\hbar}\right]\right\}$$
$$= b_f(t)K_{if}\exp\left[-i(\varepsilon_f - \varepsilon_i)\frac{t}{\hbar}\right] + \sum_p a_p(t)h_{ip}\exp\left[-i(\varepsilon_p - \varepsilon_i)\frac{t}{\hbar}\right],$$

$$i\hbar\left\{\dot{b}_f(t) + \dot{a}_i(t)S_{fi}\exp\left[-i(\varepsilon_i - \varepsilon_f)\frac{t}{\hbar}\right]\right\}$$
$$= a_i(t)K_{fi}\exp\left[-i(\varepsilon_i - \varepsilon_f)\frac{t}{\hbar}\right] + \sum_q b_q(t)h_{fq}\exp\left[-i(\varepsilon_q - \varepsilon_f)\frac{t}{\hbar}\right],$$

$$S_{if} = \int \phi_i(\mathbf{x}_i)^*\phi_f(\mathbf{x}_f)e^{i\mathbf{k}\cdot\mathbf{s}}\,d\mathbf{s}, \qquad S_{fi} = S_{if}^*, \tag{40c}$$

$$K_{if} = \int \phi_i(\mathbf{x}_i)^* V(\mathbf{x}_i)\phi_f(\mathbf{x}_f)e^{i\mathbf{k}\cdot\mathbf{s}}\,d\mathbf{s},$$

$$K_{fi} = \int \phi_f(\mathbf{x}_f)^* V(\mathbf{x}_f)\phi_i(\mathbf{x}_i)e^{-i\mathbf{k}\cdot\mathbf{s}}\,d\mathbf{s},$$

$$h_{ip} = \int \phi_i(\mathbf{x}_i)^* V(\mathbf{x}_f)\phi_p(\mathbf{x}_i)\,d\mathbf{s}, \qquad h_{fq} = \int \phi_f(\mathbf{x}_f)^* V(\mathbf{x}_i)\phi_q(\mathbf{x}_f)\,d\mathbf{s},$$

$$\int \phi_p(\mathbf{x}_i)^*\phi_m(\mathbf{x}_i)\,d\mathbf{s} = \int \phi_p^*(\mathbf{x}_f)^*\phi_m(\mathbf{x}_f)\,d\mathbf{s} = \delta_{pm}.$$

The interactions $V(\mathbf{x}_i)$ and $V(\mathbf{x}_f)$ are labeled by the relative coordinates $\mathbf{x}_i$ and $\mathbf{x}_f$ of the appropriate electron-nucleus distance; however, the prior and post interactions are $V_i = V(\mathbf{x}_f) + W(\mathbf{R})$ and $V_f = V(\mathbf{x}_i) + W(\mathbf{R})$, respectively. In the evaluation of the two normalizing integrals use is made of the fact that $(\mathbf{s}, t)$, $(\mathbf{x}_i, t_i)$, and $(\mathbf{x}_f, t_f)$ each constitutes a set of independent variables. The initial conditions are

$$a_p(-\infty) = \delta_{pi}, \qquad b_f(-\infty) = 0. \tag{40d}$$

With this choice of normalization the cross section for electronic capture into the atomic state $\phi_f(\mathbf{x}_f)$ is

$$Q_{i\to f} = \int_0^{2\pi} d\phi \int_0^\infty d\rho\rho|b_f(\infty)|^2. \tag{40e}$$

If we set $a_p(t) = 0$, $p \neq i$, and $b_q(t) = 0$, $q \neq f$, the two-state approximation results, and we study this case next.

If the second equation is multiplied by $S_{if} \exp[-i(\varepsilon_f - \varepsilon_i)t/\hbar]$ and subtracted from the first equation, and the first equation is multiplied by $S_{fi} \exp[-i(\varepsilon_i - \varepsilon_f)]t/\hbar$ and subtracted from the second equation, the two equivalent rearranged equations are

$$i\hbar\dot{a}_i(t) = (1 - |S_{if}|^2)^{-1}\left\{a_i(t)(h_{ii} - K_{fi}S_{if}) + b_f(t)(K_{if} - h_{ff}S_{if}) \exp\left[-i(\varepsilon_f - \varepsilon_i)\frac{t}{\hbar}\right]\right\}, \qquad (40f)$$

$$i\hbar\dot{b}_f(t) = (1 - |S_{if}|^2)^{-1}\left\{b_f(t)(h_{ff} - S_{fi}K_{if}) + a_i(t)(K_{fi} - h_{ii}S_{fi}) \exp\left[-i(\varepsilon_i - \varepsilon_f)\frac{t}{\hbar}\right]\right\}. \qquad (40g)$$

As previously mentioned in connection with Eq. 28*g*, the term containing $b_f(t)$ in Eq. 40*f* is commonly referred to as *back-coupling* from the final state to the initial state. Obviously, this qualifier is chiefly descriptive of the two-state approximations. The secular terms (those without the exponential factors) of Eqs. 40*f* and 40*g* are removed by a transformation similar to the one used to transform Eq. 28*k* of Section 1.4:

$$\begin{aligned} a_i(t) &= \mathrm{A}_i(t) \exp\left[-\frac{i}{\hbar}\int_{-\infty}^{t} \alpha_i(t')\, dt'\right], \\ \alpha_i &= \frac{h_{ii} - K_{fi}S_{if}}{1 - |S_{if}|^2} \\ b_f(t) &= \mathrm{B}_f(t) \exp\left[-\frac{i}{\hbar}\int_{-\infty}^{t} \beta_f(t')\, dt'\right], \\ \beta_f &= \frac{h_{ff} - S_{fi}K_{if}}{1 - |S_{if}|^2} \end{aligned} \qquad (40h)$$

$$i\hbar\dot{\mathrm{A}}_i(t) = \mathrm{B}_f(t)\frac{K_{if} - h_{ff}S_{if}}{1 - |S_{if}|^2} \exp\left\{-\frac{i}{\hbar}[(\varepsilon_f - \varepsilon_i)t - \delta_{if}]\right\} \qquad (40i)$$

$$\begin{aligned} \delta_{if} &= \int_{-\infty}^{t} dt'(\alpha_i - \beta_f) \\ i\hbar\dot{\mathrm{B}}_f(t) &= \mathrm{A}_i(t)\frac{K_{fi} - h_{ii}S_{fi}}{1 - |S_{if}|^2} \exp\left\{-\frac{i}{\hbar}[(\varepsilon_i - \varepsilon_f) + \delta_{if}]\right\}. \end{aligned} \qquad (40j)$$

The exponential, $\exp[\pm(i/\hbar)\delta_{if}]$, (see Eq. 28*f*) is labeled the *distortion*, and the retention of this factor is called *allowing for distortion*. The quantity δ_{if} can be interpreted as a correction to the difference of the unperturbed energies.[8,9]

An approximate solution to Eqs. 40h and 40j is readily obtained if $\mathrm{B}_f(t) \ll \mathrm{A}_i(t)$, in which case the approximate solution to Eq. 40j is

$$\mathrm{B}_f(t) = \int_{-\infty}^{t} dt' \frac{K_{fi} - h_{ii}S_{fi}}{1 - |S_{if}|^2} \exp\left\{-\frac{i}{\hbar}\left[(\varepsilon_i - \varepsilon_f)t' + \delta_{if}(t')\right]\right\},$$
$$\mathrm{A}_i(t) = 1. \tag{40k}$$

Since the variable of integration $\mathbf{s}$ is independent of t, it is clear from the definitions given in Eq. 40c that the integrand of Eq. 40k is unaltered by the addition of an arbitrary function of $\mathbf{R}$ to $V(\mathbf{x}_f)$, and this independence was one of the purposes for designing the expansion given by Eq. 40a. [It is instructive to note that the use of the approximation $b_f(t) = 0$, only in Eq. 40f, yields a solution to Eq. 40g that agrees in absolute value with Eq. 40k, $|b_f(t)| = |\mathrm{B}_f(t)|$.] Another point of interest is the integral h_{ii} in Eq. 40k. According to the definitions of h_{ii} in Eq. 40c and $\mathbf{x}_f$ in Eq. 40b, h_{ii} is the expectation value of $V(\mathbf{s} - \frac{1}{2}\mathbf{R})$ with $\mathbf{R}$ kept constant; this interpretation will be mentioned again in the following discussion of the wave version of electronic capture.

The motion of the center of mass and the associated coordinates are removed first, and the remaining independent variables are $(\mathbf{x}, \mathbf{R})$ so that the Hamiltonian of the resulting Schrödinger equation

$$(H - E)\Psi = 0 \tag{41a}$$

is given by Eq. 5 with H containing the interaction V defined by

$$V = V(\mathbf{x}_i) + V(\mathbf{x}_f) + W(\mathbf{R}). \tag{41b}$$

(In the typical problem $V(\mathbf{x}_{i,f})$ represent electron-nucleus electrostatic potentials, and $W(R)$ is the corresponding nucleus-nucleus interaction.) In this wave version Eq. 40a is expressed as

$$\Psi = \phi_i(\mathbf{x}_i)F_i(\mathbf{R}) \exp\left(-\frac{i}{\hbar} ml_1\mathbf{v}\cdot\mathbf{x}\right) + \phi_f(\mathbf{x}_f)G_f(\mathbf{R}) \exp\left(\frac{i}{\hbar} ml_2\mathbf{v}\cdot\mathbf{x}\right) + T(\mathbf{x}, \mathbf{R}),$$

$$\mathbf{x} = \mathbf{s} - \tfrac{1}{2}(l_1 - l_2)\mathbf{R}, \qquad l_1 = \frac{M_p}{M_t - m}, \quad l_2 = \frac{M_a}{M_t - m},$$

$$T = \sum_{p \neq i} \phi_p(\mathbf{x}_i)F_p(\mathbf{R}) \exp\left(-\frac{i}{\hbar} ml_1\mathbf{v}\cdot\mathbf{x}\right) \tag{41c}$$

$$= \sum_{q \neq f} \phi_q(\mathbf{x}_f)G_q(\mathbf{R}) \exp\left(\frac{i}{\hbar} ml_2\mathbf{v}\cdot\mathbf{x}\right), \qquad l_1 + l_2 = 1.$$

The function T is orthogonal to the other two terms of Ψ, as in the IP version. The exponential factors of Eq. 41c have been explained previously in connection with Eqs. 16a to 16c, and this structure of Ψ is used here to simplify subsequent comparisons between the wave and IP versions.

The magnitudes of the wave vectors of relative motion, $\mathbf{k}_i$ and $\mathbf{k}_f$, are restricted by conservation of energy, E:

$$E = \frac{\hbar^2 k_i^2}{2\mu} - \varepsilon_i + \frac{m l_1^2 v^2}{2} = \frac{\hbar^2 k_f^2}{2\mu} - \varepsilon_f + \frac{m l_2^2 v^2}{2}. \tag{41d}$$

If the expansion representing Ψ is inserted into Eq. 41a and the equation is multiplied successively by $\phi_p^* \exp[(i/\hbar)\, m l_1 \mathbf{v}\cdot\mathbf{x}]$ and $\phi_q^* \exp[(-i/\hbar) m l_2 \mathbf{v}\cdot\mathbf{x}]$ and these results are integrated over $\mathbf{x}$ and Eq. 41d is used, we obtain an infinite set of equations which the unknown function F_p and G_q must satisfy. In this section, however, the customary two-state approximation is again invoked and the resulting two equations (with $p = i$, $q = f$) are

$$(\nabla_R^2 + k_i^2) F_i(\mathbf{R}) = \frac{2\mu}{\hbar^2(1 - |S_{if}|^2)} \times \left\{ [h_{ii} - K_{fi} S_{if} + (1 - |S_{if}|^2) W(\mathbf{R})] F_i(\mathbf{R}) + (K_{if} - h_{ff} S_{if}) G_f(\mathbf{R}) \exp\left[-\frac{im}{2\hbar}(l_1 - l_2)\mathbf{v}\cdot\mathbf{R} \right] \right\} \tag{41e}$$

and

$$(\nabla_R^2 + k_f^2) G_f(\mathbf{R}) = \frac{2\mu}{\hbar^2(1 - |S_{if}|^2)} \times \left\{ [k_{ff} - S_{fi} K_{if} + (1 - |S_{if}|^2) W(\mathbf{R})] G_f(\mathbf{R}) + (K_{fi} - h_{ii} S_{fi}) F_i(\mathbf{R}) \exp\left[\frac{im}{2\hbar}(l_1 - l_2)\mathbf{v}\cdot\mathbf{R} \right] \right\}. \tag{41f}$$

In these two equations, the integrals denote the corresponding quantities in Eq. 40c with $\mathbf{s}$ replaced by $\mathbf{x}$.

As in the IP method, the qualification *back-coupling* is descriptive of the second term on the right-hand side of Eq. 41e, and what corresponds to the secular terms of Eqs. 40f and 40g can be interpreted as a *distortion* of the unperturbed values of k_i^2 and k_f^2. If both terms on the right-hand side of Eq. 41e are omitted, and this *plane-wave* approximation to $F_i(\mathbf{R})$ is used in Eq. 41f, omitting the other term on the right side, the resulting approximate solution for $G_f(\mathbf{R})$, corresponding to the correct boundary condition, is

$$G_f(\mathbf{R}) = -\frac{2\mu}{4\pi\hbar^2} \int d\mathbf{R}' \, \frac{\exp(i k_f |\mathbf{R} - \mathbf{R}'|)}{|\mathbf{R} - \mathbf{R}'|} B(\mathbf{R}') \times \exp\left[\frac{im}{2\hbar}(l_1 - l_2)\mathbf{v}\cdot\mathbf{R}' \right] \exp(i\mathbf{k}_i \cdot \mathbf{R}'). \tag{41g}$$

The associated amplitude for electronic capture is given[10] by

$$g_f(\theta, \phi) = \lim_{R\to\infty} RG_f(\mathbf{R})\, e^{-i\mathbf{k}_f\cdot\mathbf{R}}$$

$$= -\frac{\mu}{2\pi\hbar^2}\int \exp\left\{\frac{i[\mathbf{k}_i - \mathbf{k}_f + m(l_1 - l_2)\mathbf{v}]\cdot\mathbf{R}}{2\hbar}\right\}\mathrm{B}(\mathbf{R})\, d\mathbf{R}, \quad (41h)$$

$$\mathrm{B}(\mathbf{R}) = \frac{K_{fi} - S_{fi}h_{ii}}{1 - |S_{if}|^2}.$$

The cross section for electronic capture from the atomic state $\phi_i(\mathbf{x}_i)$ into the atomic state $\phi_f(\mathbf{x}_f)$ is, in this approximation,

$$Q = \int_0^{2\pi} d\phi \int_0^{\pi} d\theta \sin\theta\, |g_f(\theta, \phi)|^2. \quad (41i)$$

At this point, it is instructive to compare the exponential factors of the integrand with the corresponding factors of the integrand of the customary wave formula.[1] In the usual wave formula the coordinates $(\mathbf{R}_i, \mathbf{x}_i)$ and $(\mathbf{R}_f, \mathbf{x}_f)$ of Eq. 4 are used and these are related to the $(\mathbf{x}, \mathbf{R})$ and $(\mathbf{s}, \mathbf{R})$ sets by Eqs. 5 and 6. For this comparison we calculate $(\mathbf{K}_i\cdot\mathbf{R}_i - \mathbf{K}_f\cdot\mathbf{R}_f)$ with the aid of Eq. 41j:

$$\mathbf{R}_i \approx \mathbf{R} - \frac{m}{M_a}\mathbf{x}_i, \qquad \mathbf{R}_f \approx \mathbf{R} + \frac{m}{M_p}\mathbf{x}_f,$$

$$\mathbf{x}_i = \mathbf{x} + \frac{M_p}{M_t - m}\mathbf{R}, \qquad \mathbf{x}_f = \mathbf{x} - \frac{M_a}{M_t - m}\mathbf{R}, \quad (41j)$$

$$\mathbf{x} = \mathbf{s} - \frac{(l_1 - l_2)\mathbf{R}}{2}$$

$$\frac{m}{\mu}(l_1\mathbf{K}_i\cdot\mathbf{x}_i + l_2\mathbf{K}_f\cdot\mathbf{x}_f) \approx \frac{m\mathbf{v}}{\hbar}\cdot(l_1\mathbf{x}_i + l_2\mathbf{x}_f).$$

The approximate value is

$$\mathbf{K}_i\cdot\mathbf{R}_i - \mathbf{K}_f\cdot\mathbf{R}_f = (\mathbf{K}_i - \mathbf{K}_f)\cdot\mathbf{R} - \frac{\mu\mathbf{v}}{\hbar}\cdot\left(\frac{m\mathbf{x}_i}{M_a} + \frac{m\mathbf{x}_f}{M_p}\right)$$

$$= (\mathbf{K}_i - \mathbf{K}_f)\cdot\mathbf{R} - \frac{m\mathbf{v}\cdot\mathbf{x}}{\hbar} - \frac{(l_1 - l_2)m\mathbf{v}\cdot\mathbf{R}}{\hbar} \quad (41k)$$

$$= (\mathbf{K}_i - \mathbf{K}_f)\cdot\mathbf{R} - \frac{m\mathbf{v}\cdot\mathbf{s}}{\hbar} - \frac{(l_1 - l_2)m\mathbf{v}\cdot\mathbf{R}}{2\hbar},$$

$$E = \frac{\hbar^2 K_i^2}{2\mu_i} - \varepsilon_i = \frac{\hbar^2 K_f^2}{2\mu_f} - \varepsilon_f.$$

Since K_i/K_f and k_i/k_f differ only slightly according to the two different expressions for conservation of energy as displayed in Eqs. 41*d* and 41*k*, it is not surprising that the resulting exponential factors also differ slightly. The missing factor, $\exp(-im\mathbf{v}\cdot\mathbf{x}/\hbar)$, is contained in $B(\mathbf{R})$ of Eq. 41*h*. If high impact velocities are assumed, it is a simple task to derive the approximate formula,

$$k_i - k_f \approx \frac{1}{\hbar}\left[\frac{\varepsilon_i - \varepsilon_f}{v} - \frac{m(l_1 - l_2)v}{2}\right], \tag{41l}$$

and with the aid of the analyses of Frame[11] and Arthurs[12] we can then show that identical cross sections are obtained from Eq. 41*i* and Eq. 40*e*, with $b_f(\infty)$ being the solution of Eq. 40*k* without the correction for distortion ($\delta_{if} = 0$). [This analysis is applicable to (*s*-*s*)-transitions only. If $(1 - |S_{if}|^2)^{-1}$ is replaced by unity in Eq. 41*h* or 40*k* the resulting integral is the wave (or impact parameter) version of the first Born amplitude for the distorted wave interaction (see Section 3.5).] Since the wave version accurately accommodates the deflection of the nuclei, in a given approximation, it is natural to question the use of the IP version which always treats the nuclear deflection approximately.

This apparent disadvantage, however, is partially offset by the advantage that the electronic wave function Ψ' of Eq. 40*a* is integrable over **x**- (or **s**-) space:

$$\int |\Psi'(\mathbf{x}, t)|^2\, d\mathbf{x} = N. \tag{41m}$$

If N is equated to unity, Eq. 41*m* can be interpreted physically as a mathematical statement that the active electron is somewhere in **x**-space, conservation of electronic charge, and from the structure of Eqs. 40*a* and 40*b* we can deduce that each $|a_p(\infty)|^2$ and $|b_f(\infty)|^2$ is less than unity. Alternatively, it can be said that each of these quantities is less than unity since each represents a probability. Good approximate solutions should yield probabilities less than unity, and this bound on the coefficients often is used as a test of the quality of the approximating procedure. In addition to this advantage, the importance of close and distant encounters in electronic capture is provided by the IP method. We now return to Eqs. 40*f* and 40*g* and examine these equations for exact resonance.

In this circumstance it is recalled that $V_i = V_f$, $\varepsilon_i = \varepsilon_f$, $M_a = M_p$, and the systems preceding and following capture are identical. McCarroll showed that the resulting equations are exactly soluble.[13] In this case the following simplifications occur in Eqs. 40*c*, 40*f*, and 40*g* (the indistinguishability of the nuclei is ignored):

$$\phi_i = \phi_f, \quad \phi_i^* = \phi_i, \quad K_{if} = K_{fi} = K = K_1 + iK_2,$$
$$h_{ii} = h_{ff} = h, \quad S_{if} = S_{fi} = S,$$
$$i\hbar\dot{a}_i(t) = (1 - S^2)^{-1}[a_i(t)(h - KS) + b_f(t)(K - hS)], \tag{41n}$$
$$i\hbar\dot{b}_f(t) = (1 - S^2)^{-1}[b_f(t)(h - SK) + a_i(t)(K - hS)].$$

Since $V(\mathbf{x}_f)$ varies as $|\mathbf{s} - \frac{1}{2}\mathbf{R}|^{-1}$ and $V(\mathbf{x}_i)$ varies as $|\mathbf{s} + \frac{1}{2}\mathbf{R}|^{-1}$, it is made clear with the use of the transformation $\mathbf{s} \rightarrow > -\mathbf{s}$ that $S_{if} = S_{fi}$ is a real quantity. Moreover, from the definition of the absolute value of the difference of two vectors, $|\mathbf{A} - \mathbf{B}| = [(A_x - B_x)^2 + (A_y - B_y)^2 + (A_z - B_z)^2]^{1/2}$, and the two definitions, $\mathbf{R} = \boldsymbol{\rho} + \mathbf{v}t$, $\mathbf{v}\cdot\mathbf{s} = vs_z$, it is readily shown that $K_2(-t) = -K_2(t)$; that is, the imaginary part of K_{if} is antisymmetric with respect to time. Also, h_{ii} is a real quantity. The coupled equations of Eq. 41n are easily solved with the use of the substitution:

$$X(t) = a(t) + b(t),$$
$$Y(t) = a(t) - b(t), \tag{41o}$$

with the solution for $b(t)$ being

$$b(t) = -i\left[\sin\left(\int_{-\infty}^{t} \frac{K - Sh}{1 - S^2}\frac{dt'}{\hbar}\right)\right]\exp\left(-\frac{i}{\hbar}\int_{-\infty}^{t} \frac{h - SK}{1 - S^2}\,dt'\right). \tag{41p}$$

Since the imaginary part of $(h - SK)/(1 - S^2)$ is antisymmetric in the variable t, the real part of

$$-\frac{i}{\hbar}\int_{-\infty}^{\infty} \frac{h - SK}{1 - S^2}\,dt$$

vanishes; therefore the probability for capture is

$$|b(\infty)|^2 = \sin^2\left|\int_{-\infty}^{\infty} \frac{dt}{\hbar}\left(\frac{K - Sh}{1 - S^2}\right)\right|. \tag{41q}$$

At high impact velocities this probability is small, and it may be approximated by

$$\lim_{v \to \text{large}} |b(\infty)|^2 = \left|\int_{-\infty}^{\infty} \frac{dt}{\hbar}(K - Sh)\right|^2 \tag{41r}$$

since S is small for high values of v. This is the IP version of the same problem that was derived by Bassel and Gerjuoy,[14] whose method is explained in the next section. The interesting feature of Eq. 41p is that it represents the exact solution to the two-state approximation of Eq. 40a for exact resonance.

As explained at the beginning of this chapter, specialists in scattering theory formerly were unable to solve problems containing more than two states because of inadequate computing facilities. During the past ten years, however, this computational obstacle has been vanishing, and numerical solutions to expansions with six and eight states have been effected.[15] In order to exemplify this progress a brief description is given of the four-state approximations used by Lovell and McElroy.[16] They used the IP version of electronic capture to calculate cross sections for

$$H^+ + H(1s) \rightarrow H(1s; 2s) + H^+, \tag{41s}$$

and with reference to Eq. 40*c*, four amplitudes—$a_1(t)$, $a_2(t)$, $b_1(t)$ and $b_2(t)$—had to be determined numerically. The subscripts 1 and 2 refer to the 1*s*- and 2*s*-states, and the initial conditions are

$$a_1(-\infty) = 1, \qquad a_2(-\infty) = b_1(-\infty) = b_2(-\infty) = 0. \tag{41t}$$

Although this theory was developed for the four-state method, only three states were used in the calculation. These approximations are labeled $1sa|2sa|1sp$, $1sa|2sp|1sp$, $1sa|1sp|2sp$, and $1sa|2sa|2sp$. The letters *a* and *p* refer to initial and final nuclei, respectively, and the second state is the intermediate state. Thus for capture into H(1*s*), b_1 is sought, and the coefficients involved in the two different approximations are a_1, a_2, b_1 and a_1, b_2, b_1, respectively; whereas in the case of capture into H(2*s*), b_2 is sought, and these two approximations involve the respective sets a_1, b_1, b_2 and a_1, a_2, b_2. As a point of practical interest, Eq. 41*m* was closely satisfied at $t = \infty$ ($N = 1$), this feature being used as one test of the numerical solutions. We leave this impact-parameter analysis of capture now, and study a different approach to the problem.

3.2 METHOD OF BASSEL AND GERJUOY

According to the approximate solutions of Eq. 40*a*, explained in the preceding section, the cross section for electronic capture is independent of the nucleus-nucleus interaction. Moreover, we saw how to prove that approximate solutions to the wave version represented by Eq. 41*c* would also yield cross sections with this same property in the limit of undeflected nuclear motion. A correct theory should have this property since the nuclei act only as moving electron-nucleus interactions in this limit of uniform rectilinear nuclear motion. (This last sentence is not intended to imply that the method of solution is correct since a particular feature is correct.)

Approximate solutions to the formulation proposed by Bassel and Gerjuoy[14] satisfy this requirement as noted by McCarroll.[13] This method is based in part upon the integral equations of scattering, and these equations are written in terms of ket vectors or state vectors and the associated linear operators of these abstract spaces.[17–22] For the purpose of this description reference is made to Eq. 3, in which the Hamiltonian is decomposed into its constituent parts. (In these modes of the time-independent problem we are imagining the state of affairs at $t = -\infty$, the i-system, and at $t = \infty$, the f-system.) The notation of Eq. 3 is used, and we recall that E is the center-of-mass energy. The integral equations,

$$\Psi_i^{(+)} = \Phi_i + \frac{1}{E + i\varepsilon - H_i} V_i \Psi_i^{(+)} \tag{42a}$$

and

$$\Psi_f^{(-)} = \Phi_f + \frac{1}{E - i\varepsilon - H_f} V_f \Psi_f^{(-)} \tag{42b}$$

are solved, respectively, by

$$\Psi_i^{(+)} = \left(1 + \frac{1}{E + i\varepsilon - H} V_i\right)\Phi_i \tag{42c}$$

and

$$\Psi_f^{(-)} = \left(1 + \frac{1}{E - i\varepsilon - H} V_f\right)\Phi_f. \tag{42d}$$

Although references 17 through 22 explain these four equations, a brief qualitative explanation is included here. The parameter ε is a small positive number, and $+i\varepsilon$ and $-i\varepsilon$ signify that the solution is a diverging (outgoing) and converging (incoming) wave, respectively, for large separations of the two aggregates of the system after scattering. Thus Eqs. 42*a* to 42*b* can be represented symbolically by

$$\begin{aligned} \Psi_i^{(+)}&: \underset{R_i \to \infty}{P^+ + A} \to \underset{R_f \to \infty}{P + A^+}, \\ \Psi_f^{(-)}&: \underset{R_f \to \infty}{P + A^+} \to \underset{R_i \to \infty}{P^+ + A}. \end{aligned} \tag{42e}$$

The R_i and R_f of Eq. 42*e* are defined in Eqs. 2*i* and 2*f*. The initial state, Φ_i or Φ_f, and the sign of $\pm i\varepsilon$ together constitute the boundary condition of the scattering solutions. In the actual calculation, of course, representatives of these equations are formed, yielding the wave functions $\langle \mathbf{x} | \Psi_i^{(+)}, {}_f^{(-)} \rangle$ and Green's functions $\langle \mathbf{x} | 1/(E \pm i\varepsilon - H) | \mathbf{x}' \rangle$. [Other Green's functions are $\langle \mathbf{x} | 1/(E \pm i\varepsilon - H_i) | \mathbf{x}' \rangle$ and $\langle \mathbf{x} | 1/(E \pm i\varepsilon - H_f) | \mathbf{x}' \rangle$. In these representatives $\mathbf{x}$ and $\mathbf{x}'$ denote the set of independent variables appropriate for the number of particles in the center-of-mass system as in Eqs. 2*i* and 2*f*.] In practice the Green's functions corresponding to $(E \pm i\varepsilon - H)^{-1}$, the resolvent, are seldom known; consequently, solutions represented by Eqs. 42*c* and 42*d* can be calculated only by using an acceptable expansion for the associated resolvent. [If the resolvent Green's function corresponding to the Hamiltonian

H is known, all solutions to the associated eigenvalue-eigenfunction equation, $(H - \varepsilon)\psi = 0$, are known for real values of ε. This fact, in the author's opinion, is not mentioned sufficiently often.]

A central idea of the method of Bassel and Gerjuoy is to isolate the part of the wave function that contains only the elastic scattering. (Only distinguishable nuclei are assumed for the present, the case of identical nuclei being discussed in a subsequent section.) For this purpose we first seek an interaction that can cause only elastic scattering, and such a potential is the function $U_i(R_i)$ which is restricted in its functional form also by the requirement that

$$\lim_{R_i \to \infty} R_i U_i(R_i) \to 0 \tag{42f}$$

A function of R_i *only* fulfills this scattering requirement since the electrons are functions of the set of coordinates $\{\mathbf{x}_i\}$ (see Eq. 2*i*) which are independent of $\mathbf{R}_i$. The next step is to consult Eq. 42*c* and decompose V_i of H into $U_i + (V_i - U_i)$, and define the function $\chi_i^{(+)}$ by

$$\chi_i^{(+)} = \left(1 + \frac{1}{E + i\varepsilon - H_i - U_i} U_i\right)\Phi_i, \tag{42g}$$

which is the formal solution of the equation

$$\chi_i^{(+)} = \Phi_i + \frac{1}{E + i\varepsilon - H_i} U_i \chi_i^{(+)}. \tag{42h}$$

(Although these manipulations are straightforward, the mathematical justification of these operations is not trivial; the subject is treated in the theory of Hilbert and Banach spaces.[23]) Next Eq. 42*g* is subtracted from Eq. 42*c*, and the identity

$$\begin{gathered}\frac{1}{P - Q}(U_i + Q) - \frac{1}{P}U_i = \frac{1}{P}Q + \frac{1}{P - Q}Q\frac{1}{P}(U_i + Q) \\ P = E + i\epsilon - H_i - U_i, \qquad Q = V_i - U_i,\end{gathered} \tag{42i}$$

is used to get

$$\Psi_i^{(+)} - \chi_i^{(+)} = \left[\frac{1}{P}Q + \frac{1}{P - Q}Q\frac{1}{P}(U_i + Q)\right]\Phi_i. \tag{42j}$$

The terms on the right-hand side of Eq. 42*j* can be rearranged to yield

$$\begin{aligned}\frac{1}{P - Q}Q\left(1 + \frac{1}{P}U_i\right)\Phi_i + \left[\left(\frac{1}{P - Q}Q + 1\right)\frac{1}{P}Q - \frac{1}{P - Q}Q\right]\Phi_i \\ = \frac{1}{E + i\varepsilon - H}(V_i - U_i)\chi_i^{(+)},\end{aligned} \tag{42k}$$

since the factor of Φ_i in the second term on the left-hand side of Eq. 42*k* vanishes identically. Equation 42*k* is used in Eq. 42*j* to get the desired result,

$$\Psi_i^{(+)} = \chi_i^{(+)} + \frac{1}{E + i\varepsilon - H}(V_i - U_i)\chi_i^{(+)} \tag{42l}$$

In accordance with the meaning of $\chi_i^{(+)}$, the amplitude for electronic capture must originate from the second term on the right of Eq. 42*l*. In another paper by Gerjuoy this amplitude is derived.[20]

This definition stems from the outgoing probability current of the rearranged states, the state formed by electronic capture here. (The difficulties of these problems are discussed in detail by Gerjuoy.) The transition matrix element, which is proportional to the scattering amplitude, is derived from the quantity

$$T = \lim_{R \to \infty} \int d\tau_{cd} \phi^*(\tau_{cd}) G(\tau; \tau'; E)(V_i - U_i)\chi_i^{(+)} \tag{42m}$$

in which $\phi^*(\tau_{cd})$ represents the product of the final atomic and ionic wave functions, and τ_{cd} denotes the set of relative coordinates. The resolvent Green's function G is a function of the set of coordinates τ, the primed counterpart τ', and the energy of the system E; the variable $\mathbf{R}$ is the radius vector in the many-dimensional hyperspherical coordinates used in that analysis.[24]

The integral equation for G in terms of the final-state Green's function G_f—in abstract form—is written

$$G = G_f + G_f V_f G, \tag{42n}$$

and G_f is expressed in terms of the free-space Green's function G_{f0} and the internal states

$$G_f = \sum G_{f0}(\mathbf{r}_{cd}; \mathbf{r}'_{cd}; E)\phi(\tau_{cd})\phi^*(\tau'_{cd}). \tag{42o}$$

In this expansion the variables $\mathbf{r}_{cd}$, $\mathbf{r}'_{cd}$ represent the distances between the centers of masses of the final aggregates. Equations 42*n*, 42*o*, and 42*d* and the asymptotic expansion of G_{f0} are used to evaluate T of Eq. 42*m*, but only the final result is given:

$$\begin{aligned} T &= \eta_{cd}(\rho)\left[\Phi_f^*(\tau') + \int d\tau'' \Phi_f^*(\tau'') V_f(\tau'') G(\tau''; \tau'; E)\right] \\ &= \eta_{cd}(\rho)\Psi_f^{(-)}(\tau')^*. \end{aligned} \tag{42p}$$

$\eta_{cd}(\rho)$ consists of the diverging spherical wave in ρ, which is related to R, together with other parameters of G_{f0} (see Gerjuoy[20] for details). Equation 42*p*, and the relation between the transition amplitude and the scattering amplitude,[25] provide the information needed to define the differential cross section which we sought:

$$\frac{d\sigma}{d\Omega_f}(i \to f) = \frac{v_f}{v_i}\left(\frac{\mu_f}{2\pi\hbar^2}\right)^2 |(\Psi_f^{(-)}, (V_i - U_i)\chi_i^{(+)})|^2$$

$$= \frac{2\pi}{\hbar v_i}|T_{i\to f}|^2 \rho(E_f) = \frac{v_f}{v_i}|f_i^f(\theta_f, \phi_f)|^2, \tag{42q}$$

$$\rho(E_f) = \frac{\mu_f{}^2 v_f}{8\pi^3\hbar^3}$$

Equation 42*q* is displayed in these several forms to clarify the relation between the scattering-amplitude and the product of the transition-amplitude with the density of final states $\rho(E_f)$. The next problem confronted is the determination of U_i for practical calculations.

In the distorted-wave method an interaction,[26]

$$U_i(dw) = \int \{d(\mathbf{x}_i)\}\phi_i(\mathbf{x}_i)V_i(\mathbf{x}_i, \mathbf{R}_i)\phi_i(\mathbf{x}_i) \tag{42r}$$

is used, in which the relative position of the centers of masses, $\mathbf{R}_i$ of Eq. 2*i*, is held constant; the notation $\{d\mathbf{x}_i\}$ signifies the set of initial atomic coordinates, and includes the $\{\mathbf{x}_k\}$ and $\{\mathbf{y}_j\}$ of Eq. 2*i*. It is clear that $U_i(dw)$ is a function of $\mathbf{R}_i$ only, and it thus fulfills the requirement of affecting elastic scattering only. In applications to electronic capture from simple and complex atoms by protons, we can show that the nucleus-nucleus interaction is effectively subtracted from V_i by $U_i(dw)$. More on this topic, however, is deferred until the first Born approximation is defined and discussed.

To conclude this section, attention is directed to the paper by Grant and Shapiro.[27] In this paper the cross section of Eq. 42*q* is expressed in terms of $\chi_f^{(-)}$, $\Psi_i^{(+)}$, and $V_f - U_f$, the alternative formulation. This expression for the differential cross section is expanded so that terms up to the third order in the interaction are included, and this result was applied to charge transfer between protons and H(1*s*). The reader is referred to the original paper for the calculational details.

3.3 FIRST BORN APPROXIMATION

Although the first Born approximation is explained in many textbooks on quantum mechanics, a brief review of its origin as it pertains to electronic capture is given since the derivation not only is less straightforward than the corresponding derivation for direct processes, but the validity of the approximation also is questionable.[26,28] This approximation for high velocities of relative motion of the colliding systems is a first-order procedure, and this statement is clarified through the use of the following integral equation for scattering by a center of force:

$$\psi = e^{i\mathbf{k}\cdot\mathbf{x}} - \frac{\mu}{2\pi\hbar^2}\int \frac{e^{ik|\mathbf{x}-\mathbf{x}'|}}{|\mathbf{x}-\mathbf{x}'|} V(\mathbf{x}')\psi(\mathbf{x}')\, d\mathbf{x}',$$

$$\hbar\mathbf{k} = \mu\mathbf{v}.$$

If $\mathbf{v}$ is large enough so that the particle is under the influence of the interaction $V(\mathbf{x})$ for only a short time, little deflection of the particle is expected. Since the scattering is small in this circumstance, ψ should not differ much from the incident state, $\Phi_i = e^{i\mathbf{k}\cdot\mathbf{x}}$, and it is physically plausible to substitute this unperturbed term for the unknown wave amplitude, ψ, in the integral. The asymptotic form of the corresponding integral yields the first-order scattering amplitude,

$$f(\theta, \phi) = -\frac{\mu}{2\pi\hbar^2}\int e^{-i\mathbf{k}'\cdot\mathbf{x}'} V(\mathbf{x}')e^{i\mathbf{k}\cdot\mathbf{x}'}\, d\mathbf{x}', \qquad |\mathbf{k}'| = |\mathbf{k}|,$$

the spherical wave, e^{ikx}/x, having been removed from the integral. Moreover, if this substitutional process is repeated, the Neuman series for the integral equation is generated;[29] however, this procedure of iteration is unacceptable for electronic capture, and some of the objectionable features are discussed in references 1, 14, 20, and 28. In addition, the coordinates of the outgoing spherical wave of our example are incorrect for a rearranged system; therefore this course is altered to obtain the first Born amplitude for the wave version of electronic capture differently. This amplitude is obtained from the approximations, $\Psi_f^{(-)} = \Phi_f$, $\Psi_i^{(+)} = \Phi_i$, to the matrix elements

$$(\Psi_f^{(-)}, V_i\Phi_i) = (\Phi_f, V_f\Psi_i^{(+)}), \tag{43a}$$

reported by Gerjuoy[20] for the prior (i) and post (f) interactions, and is

$$g_{if}(\theta, \phi) = \frac{\mu_f}{2\pi\hbar^2}\int \Phi_f^*(\mathbf{x}_f)V(\mathbf{x}_i, \mathbf{x}_f)\Phi(\mathbf{x}_i)\, d\mathbf{R}_i\, d\mathbf{x}_i. \tag{43b}$$

The coordinates $d\mathbf{x}_i$ constitute all the prior atomic relative coordinates of Eq. 2*i*. Equation 43*b* is extended by definition to include other interactions, including the distorted-wave case defined in Eq. 42*q*.

An interesting property of the prior and post interactions can be obtained from the Kohn variational method.[26] Here the use of the unperturbed wave functions of the initial and final states of the system for trial wave functions in the Kohn variational amplitude yields Eq. 43*b* again with $V = V_f$, the post Born amplitude. Although this may be a poor approximation, it does suggest that it is an inconsistent approximation to use another V with these same wave functions. The last remark does not apply to the interaction $V = V_i$ since the prior and post amplitudes are equal if exact atomic wave functions are used.[30] (As a point of historical interest the previously noted paper contains the first calculation of the Born approximation using the prior and post interactions.)

In constructing the first Born amplitudes, interactions are restricted to electrostatic Coulomb potentials between each pair of particles. Ordinarily V_i and V_f decrease more strongly than a Coulomb interaction for large separations of the aggregates, and exceptions to this condition are noted at the appropriate places in the text. The amplitudes will be symmetrized consistent with the Pauli exclusion principle, but never are more than two different groups of indistinguishable particles, such as protons and electrons, treated in this chapter. Ordinarily the differential cross sections are independent of the frame of reference, but examples are given for which an average over the atomic frame of reference must be effected to remove this dependence upon the system of coordinates. For convenience Eqs. 10a and 10b are reproduced here, and the momentum-change vectors are defined:

$$E = \frac{\mu_i v_i^2}{2} - \varepsilon_i = \frac{\mu_f v_f^2}{2} - \varepsilon_f, \qquad v_i^2 = \frac{2E_i}{M_f},$$

$$\hbar K_i = \mu_i \mathbf{v}_i, \qquad \hbar K_f = \mu_f \mathbf{v}_f,$$

$$\mathbf{K}_i\cdot\mathbf{R}_i - \mathbf{K}_f\cdot\mathbf{R}_f = \mathbf{A}_f\cdot\mathbf{y}_1 + \mathbf{A}_i\cdot\mathbf{x}_i - \frac{m}{M_f}\sum_{j=2}^{n_p}\mathbf{A}_f\cdot\mathbf{y}_j - \frac{m}{M_{i1}}\sum_{k=2}^{n_a}\mathbf{A}_i\cdot\mathbf{x}_k, \tag{43c}$$

$$M_f = M_p + (n_p - 1)m, \qquad M_{i1} = M_a + n_a m,$$

$$\mathbf{A}_f = \frac{M_f}{M_{f1}}\mathbf{K}_f - \mathbf{K}_i, \qquad \mathbf{A}_i = \frac{M_i}{M_{i1}}\mathbf{K}_i - \mathbf{K}_f.$$

Reference is made to Eqs. 2i and 2f for some of the notation, and the reader is reminded that the atomic target is at rest initially.

A word of caution is pertinent here. The ratios multiplying $\mathbf{K}_i$ and $\mathbf{K}_f$ in the defining relations for $\mathbf{A}_i$ and $\mathbf{A}_f$ are nearly unity; nevertheless, these factors must be retained since the main contributions to the cross sections depend sensitively upon the values of $\mathbf{A}_i$ and $\mathbf{A}_f$ for the condition of approximate parallelism of $\mathbf{K}_i$ and $\mathbf{K}_f$. Failure to retain these ratios may lead to large errors; as one example shows, the variation of the cross section with impact energy is altered from E^{-6} to E^{-1}.[31] At low impact energies, however, the retention of this momentum transfer to the electron becomes less important. The Russell-Saunders (LS-coupling) scheme is always used for the construction of atomic wave functions, and the LSM_1M_s-representation is employed.[32] As suggested by the scalar products defining the momentum-charge vectors, a transformation of coordinates to the $\{\mathbf{x}_i\}\mathbf{y}_1$-set (the $\mathbf{x}_k$ and $\mathbf{y}_j$ of Eq. 2i and

$\mathbf{y}_1$ of Eq. 2*f*) is convenient, and integrations are greatly simplified as a result of this change of variables. As the first application of the first Born approximation the active electronic interaction is used for V in Eq. 43*b*.

3.4 ACTIVE ELECTRONIC INTERACTION (OBK)

The abbreviation OBK included in the heading of this section originated from a paper by Bates and Dalgarno[33] and signifies Oppenheimer-Brinkman-Kramers, the names of the originators.[34,35] Another nomenclature, *ne* (nucleus-electron) interaction, originates from problems in which the incident particle is a bare nucleus;[36] but in this book the interaction is designated V_a and is generalized to include the sum of all electrostatic interactions between the singly charged, incident many-electron ion, P^+, and the *active* electron of A, the process being represented by Eq. 1. If the incident ion is multiply charged, mention of this alteration will be noted since associated alterations of the unperturbed wave functions may be necessary. As a point of historical interest, Brinkman and Kramers correctly accounted for the translational motion of the electron in their impact-parameter analysis of electronic capture; moreover, they originated the transformation of coordinates that was described at the end of the preceding section. This pioneering calculation was restricted to electronic capture between atomic *s*-states, however, and it was Saha and Basu who first effected the analysis for capture from an *s*-state into a *p*-state.[37] An important feature of electronic capture either into or from atomic states that are not *s*-states relates to the orientation of the atomic frame of reference; in the preceding paper the atomic Z-axis is parallel to the rectilinear path of the incident particle. We return to this point after the OBK amplitudes and cross sections are defined.

Dimensionless variables are used, and the corresponding units differ from atomic units only in the unit of energy since $\varepsilon = me^4/2h^2$ is used here instead of 2ε. The interaction is

$$V_a = \sum_{j=2}^{n_p} |\mathbf{y}_j - \mathbf{y}_1|^{-1} - n_p|\mathbf{y}_1|^{-1}, \tag{44a}$$

and the meaning of this interaction is now explained. In these problems that contain a number of indistinguishable electrons the amplitude is first defined treating the electrons as distinguishable particles, and antisymmetrization is accomplished as the final task. It is mentioned in passing that the relative coordinates of Eqs. 2*i* and 2*f*—and any other system of relative coordinates involving electronic capture from one nucleus to another—change for each permutation of the electronic coordinates in the laboratory system. We now imagine that we know which electrons are in each atom and atomic ion initially, and we proceed to calculate the amplitude for the capture of electron

1 from A into P^+ to form P; this specification of the electrons in the atoms and atomic ions is labeled a *configuration*. This device enables us to define V_a of Eq. 44*a*, in which the first $(n_p - 1)$ terms are the ionic electron-electron (1) interactions, and the last term represents the ionic nucleus-electron (1) interaction. Next we assume that each of the four aggregates P^+, A, P, and A^+ are represented by wave functions that are antisymmetrical in the electronic coordinates; that is, we know which electrons are in each aggregate, but the electrons are indistinguishable. Note that the relative coordinates $\mathbf{R}_i$ and $\mathbf{R}_f$ of Eqs. 2_i and 2_f do not change by a permutation of the electronic coordinates within the aggregates. With this information the unsymmetrized amplitude can be constructed, and we turn to this task next.

The wave functions of the initial and final states of the system are given by

$$\begin{aligned} \Phi_f &= {}^{(2s+1)}[\mathrm{PA}^+] \exp(-i\mathbf{A}_f \cdot \mathbf{y}_1), & \Phi_i &= {}^{(2s+1)}[\mathrm{P}^+\mathrm{A}] \exp(i\mathbf{A}_i \cdot \mathbf{x}_1), \\ \mathrm{P} &= \psi_p(\mathbf{y}_1 \cdots \mathbf{y}_{n_p}; S_p M_p), & A &= \psi_a(\mathbf{x}_1 \cdots \mathbf{x}_{n_a}; S_a, M_a), \\ \mathrm{P}^+ &= \psi_{p+}(\mathbf{y}_2 \cdots y_{n_p}; S_{p^+}, M_{p^+}), & \mathrm{A}^+ &= \psi_{a+}(\mathbf{x}_2 \cdots \mathbf{x}_{n_a}; S_{a^+}, M_{a^+}). \end{aligned} \tag{44b}$$

The quantities $\mathbf{A}_{f,i}$ are defined in Eq. 43*c*, and it is supposed that the change of variables described subsequent to Eq. 43*c* has been effected. The other momentum-change variables, which are labeled *passive* since they are conjugate to *passive* electronic coordinates in the unsymmetrized amplitude, are omitted since each contain the factor, the ratio of the electronic to the nuclear masses, and thus contribute negligibly to the value of the integral.[38] (Again, since the omission of these factors leads to erroneous results in exceptional circumstances, this exception is explained in the section on asymptotic variations of cross sections at high impact velocities.) The superscript $(2s + 1)$ refers to the multiplicity of the system, and the utility of this designation originates from Wigner's spin-conservation rule.[39] According to this rule, if S_1 and S_2 are the two initial spins of the colliding aggregates with corresponding multiplicities $(2S_1 + 1)$ and $(2S_2 + 1)$, only transitions to final states with total spin in the range

$$|S_1 - S_2| \leq S \leq S_1 + S_2$$

are possible in the absence of spin-dependent forces. This rule also means that the initial and final total spins are equal. This conservation of multiplicity is expressed mathematically using the vector-coupling rules for two angular momenta.[40,41] These relations corresponding to the states in Eq. 44*b* are given here for clarification:

$$\begin{aligned} {}^{(2S+1)}[\mathrm{P}^+\mathrm{A}] &= \sum_{M_{S_1} M_{S_2}} \psi(\mathrm{A}; S_1 M_{S_1}) \psi(\mathrm{P}^+; S_2 M_{S_2}) |(S_1 S_2 M_{S_1} M_{S_2} | S_1 S_2 S_i M_{S_1}), \\ {}^{(2S+1)}[\mathrm{PA}^+] &= \sum_{M'_{S_1} M'_{S_2}} \psi(\mathrm{A}^+; S'_1 M'_{S_1}) \psi(\mathrm{P}; S'_2 M'_{S_2}) (S'_1 S'_2 M'_{S_1} M'_{S_2} | S'_1 S'_2 S_f M_{S_f}), \\ |S_1 - S_2| &\leq S_i \leq S_1 + S_2, \qquad |S'_1 - S'_2| \leq S_f \leq S'_1 + S'_2. \end{aligned} \tag{44c}$$

Contributions to the amplitude are nonvanishing only for $S_i = S_f$, $M_{si} = M_{sf}$; this follows from the orthogonality of the wave functions with respect to the spin. Since the investigations of this chapter are limited to capture from those atoms whose subshells contain only *s*- and *p*-orbitals, some mathematical properties of the relevant amplitude can be established. This coupling rule for spins rests on the assumption that *LS*-coupling is valid; that is, the electrostatic interactions dominate the spin-dependent interactions, and consequently these *LS*-states are good approximations. This is usually the case for the ground states of neutral atoms and singly charged ions for atomic numbers that are not too large. The breakdown of the coupling scheme greatly complicates calculations.[42,43]

For this purpose the wave functions are assumed to be appropriately symmetrized, linear combinations of products of one-electron orbitals which consist of a spatial-part and a spin-part; these wave functions are often conveniently represented by a linear combination of determinantal wave functions. Moreover, the one-electron spatial orbitals of each different subshell of the atom and its residual atomic ion are assumed orthogonal, and the spatial orbitals of the same subshells, atom and atomic ion, are normalized to unity. This is called the "orthonormal approximation." This approximation is not needed, however, to modify the unsymmetrized amplitude to accommodate the indistinguishability of the electrons. We recall that V_a of Eq. 44*a* and the wave functions of Eq. 44*c* refer to a given *configuration* of the electrons, and these functions are used in Eq. 43*b* to obtain the unsymmetrized OBK amplitude for the capture of electron 1, whose initial and final coordinates are $\mathbf{x}_1$ and $\mathbf{y}_1$, respectively.

Equation 43*b* refers to a given angle of scattering in the center-of-mass system, and this discussion always refers to the *same* angle. Now electron 1 can be captured into the state representing P in n_p different ways; furthermore, the relative coordinate $\mathbf{R}_f$ is unchanged for the capture of electron 1 in each of these n_p ways; consequently, each of these constituent amplitudes, together with the outgoing spherical wave in $\mathbf{R}_f$, is part of the asymptotic value of the wave function in a certain sense which is not stressed here. It is evident from the structure of Eqs. 44*a* and 44*b* that the *value* of each of these constituents amplitudes is the same. Next note that

$$N_i = \frac{(n_p + n_a - 1)!}{(n_p - 1)!\, n_a!}$$

products $[P^+A]$ and

$$N_f = \frac{(n_p + n_a - 1)!}{n_p!\,(n_a - 1)!}$$

products $[PA^+]$ constitute the antisymmetrical initial and final electronic states, respectively. (The wave function can be written in the antisymmetrical form if the electronic relative coordinates refer to the center of mass of the two nuclei as the origin.)

Although the amplitude for each of the N_f final states is associated with a different outgoing coordinate R_f, in a certain sense, the *values* of the amplitudes are equal. For each of these n_p ways of P^+ capturing the electron designated by $\mathbf{x}_1$, different coordinates $\mathbf{y}_i$ of P are involved since there must be no preferential $\mathbf{y}_i$-coordinate for the captured electron; nevertheless, the value and sign of each of these constituent unsymmetrized amplitudes is the same. In order to form the N_f products $[PA^+]$ the electrons are permuted between A^+ and P leaving the coordinates in P^+ and A unaltered, and this obviously exhausts the $n_a N_i$ terms formed from $[P^+A]$. Since this amounts to $n_p N_f$ different ways of capturing an electron into P, and since the additional normalizing factor $(N_i N_f)^{-1/2}$ is required for this counting, the effective amplitude is

$$n_p N_f (N_i N_f)^{-1/2} = (n_p n_a)^{1/2} \tag{44d}$$

times the corresponding unsymmetrized amplitude, the final result being

$$g_{if}(\theta, \phi) = \frac{(n_p n_a)^{1/2} \mu_f a_0}{2\pi m} \int {}^{2S+1}[PA^+]^* \exp(i\mathbf{A}_f \cdot \mathbf{y}_1)$$
$$\times \left(\sum_{j=2}^{n_p} |\mathbf{y}_j - \mathbf{y}_1|^{-1} - n_p |\mathbf{y}_1|^{-1} \right) {}^{(2S+1)}[P^+A] \exp(i\mathbf{A}_i \cdot \mathbf{x}_i)\, d\tau,$$
$$d\tau = d\mathbf{x}_1 d\mathbf{x}_2 \cdots d\mathbf{x}_{n_a}\, d\mathbf{y}_1 d\mathbf{y}_2 \cdots d\mathbf{y}_{n_p}. \tag{44e}$$

A slightly different version of this derivation is given elsewhere.[44] Only V_a has to be changed to get the corresponding amplitudes for the other interactions V_i, V_f, and $V(dw)$. We now study some applications of Eq. 44*e*.

Even with this simplest of the interactions, V_a, Eq. 44*e* represents a laborious calculational task, and a particular source of this difficulty is the functional form of the one-electron spatial orbitals. (Hereafter, we shall assume that *orbital* refers to the *spatial part*.) For example, the 2*s*- and 2*p*-orbitals calculated by Roothaan and Kelly[45] for *N* and *O* require at least three and two exponentials, respectively, to adequately represent these orbitals. Another source of complexity originates from the functional representation of subshells of equivalent electrons; this feature is particularly troublesome if P^+ in Eq. 44*b* contains terms of equivalent electrons with the same multiplicity because the passive electronic coordinates $\mathbf{y}_j$ of V_a mixes these terms, which are orthogonal in the spatial variables. It is adequate to say that use of the other interactions, V_i, V_f, and $V(dw)$, complicates the calculation, and hereafter for calculational convenience the atomic ion P^+ will represent a proton unless otherwise stated.

Although only atoms A containing subshells of *s*- and *p*-orbitals are treated here, capture from atoms containing electrons with orbital angular momentum $l \geq 2$ can be analyzed using the same procedure provided that *LS*-coupling adequately represents the atom. A very useful aid for studying capture from subshells of equivalent *p*-electrons is Racah's method of constructing the corresponding wave functions.[46,47] In this method the same radial wave function is used for each *p*-orbital of a group of equivalent *p*-electrons, and the wave function of a group of *n* such electrons is given by

$$\begin{aligned}\Psi(p^n SL; 1 \cdots n) &= \sum_{(S'L')} \sum_{m_l'' M_l''} \sum_{m_s'' M_s''} \Psi[p^{n-1}(S'L'M_s''M_l''; n \cdots 2)] \\ &\quad \times p_1(1, m_l'') s_1(\tfrac{1}{2}, m_s'')(L'M_l'' l m_l'' | L'1LM_l) \\ &\quad \times (S'M_s'' \tfrac{1}{2} m_s'' | S'\tfrac{1}{2} SM_s)(p^{n-1}(S'L')pSL|\}p^n SL) \qquad (44f) \\ &= \sum_{(S'L')} \sum_{(S''L'')} \Psi[p^{n-2}(S''L'')p(S'L')pSL] \\ &\quad \times (p^{n-2}(S''L'')pS'L'|\}p^{n-1}S'L')(p^{n-1}S'L'pSL|\}p^n SL)\end{aligned}$$

In the first coupling formula the sums are taken over the terms of the $p^{(n-1)}$ group, the orbital and spin components of these terms $(M_l'' M_s'')$, and the orbital and spin components of electron $1(m_l'' m_s'')$. In addition to the associated vector-coupling coefficient (Clebsch-Gordan coefficient) in Eq. 44*f* there is (the last factor) the *fractional parentage coefficient.** This is abbreviated "fpc," and the $(S'L')$-sum includes those terms for which the associated fpcs do not vanish.[48] The other representation of $\psi(p^n SL)$ is included (vector-coupling coefficients omitted) since this formula for coupling two *p*-orbitals is useful in the analysis of matrix elements of interactions that contain coordinates of both the active and passive electrons. Of practical importance is the orthonormality of the different terms of the same group in both orbital and spin spaces. Equation 44*f* is applied after the formula for the cross section is obtained.

Since the cross section is also a transition probability, we can use the formula for this quantity to define the cross section for capture from a subshell of degeneracy g_i into the 1*s* state of P, leaving the residual atomic ion in a state of degeneracy g_f,[49]

$$\begin{aligned}\sigma_{if;1s}(\theta, \phi) &= \frac{n_p n_a v_f}{g_i v_i} \sum_{M_l'} \sum_{M_l} \sum_{M_s} |g_{M_l}^{M_l'}(\theta, \phi)|^2, \\ g_i = (2L_i + 1)(2S + 1), &\qquad g_f = (2L_f + 1)(2S_f + 1), \\ -L_f \leq M_l' \leq L_f, \qquad &-L_i \leq M_l \leq L_i, \qquad -S \leq M_s \leq S, \qquad (44g) \\ Q(if; 1s) &= \int_0^{2\pi} d\phi \int_0^{\pi} d\theta \sin\theta \sigma_{if;1s}(\theta, \phi).\end{aligned}$$

* The standard equation for the fractional parentage coefficient (fpc), as used in Eq. 44*f*, is $(p^{(n-1)}(S'L')pSL|p^n SL)$.

The amplitude used here is given by $(n_a n_p)^{-1/2}$ times the one of Eq. 44*e*, and it is clear from Eq. 44*c* that the sum over M_s introduces a factor $(2S + 1)$ which cancels the same factor in g_i; the other sums include only the allowable M_l'- and M_l-values. (In this type of calculation the internal energy of these degenerate states is the energy of the center of gravity of the multiplet.[50]) The total cross section is the sum of the cross sections for capture from each subshell of A into all excited states of P, leaving the atomic residual ion in each of its allowable excited states for each combination of subshell of A and excited state of P. Again it is evident from Eq. 44*c* that the allowable excited states of A^+ are restricted by the subshell of A and the state of P. In our present application P is always a doublet, so the S_f-values of A^+ are equal to $(S_i \pm \frac{1}{2})$. Moreover, there is another sum needed to accommodate capture into excited states of P if the corresponding orbital angular momentum is $l \geq 1$. This calculation can be rather tedious to effect, but fortunately a small number of these possible cross sections provides the major contribution to the total cross section. It is instructive to see the results for special cases of Eq. 44*g*.

The first example given represents *s*-orbital capture into H(1*s*). If the one-electron orbitals of the type calculated by Roothaan and Kelly[45] are used, the cross section is

$$Q = 2\pi a_0^2 \frac{2S_f + 1}{2S + 1} \frac{v_f}{v_i} \left(\frac{16\mu_f}{m}\right)^2 \int_0^\pi \frac{d\theta \sin \theta}{(1 + A_f^2)^2}$$

$$\times \left| \frac{cz^{5/2}}{(z^2 + A_i^2)^2} - 4\sqrt{\frac{2}{45}} \sum_{j=1}^{4} \frac{c_j z_j^{9/2}(A_i^2 - z_j^2)}{(z_j^2 + A_i^2)^4} \right|^2,$$

$$S(\mathbf{x}_1) = \frac{1}{\sqrt{\pi}} \left[cz^{3/2} \exp(-zx_1) + \sqrt{\frac{2}{45}}\, x_1^2 \sum_{j=1}^{4} c_j z_j^{7/2} \exp(-z_j x_1) \right]. \quad (44h)$$

Also included is the *s*-orbital wave function.[51] As mentioned before, at least three terms of $S(\mathbf{x}_1)$ are needed for 2*s*-orbital capture, but the first term of S suffices for 1*s*-orbital capture. (In connection with these formulas and other formulas that contain $A_{i,f}$, see Appendix 1 in which various combinations of the momentum-change variables are expressed in terms of the relevant masses, internal energies, and impact energy.) Suppose that A and A^+ represent $Mg(^1S)$ and $Mg^+(1s^2 2s^2 2p^6 3s;\, ^2S)$, respectively; it is thus clear that $S = 0$, $M_l' = M_l = 0$. It follows from Eq. 44*c* that

$$^1[PA^+] = \frac{H(1s)}{\sqrt{2}} [\alpha(1)Mg^+(-\tfrac{1}{2}) - \beta(1)Mg^+(\tfrac{1}{2})],$$

and it is evident that the effective part of $^1[PA^+]$ is

$$\frac{2}{\sqrt{2}} H(1s)Mg^+(-\tfrac{1}{2}).$$

The contribution from $Mg^+(\frac{1}{2})$ is the same as that originating from the other part of $^1[PA^+]$ since there is a compensating change of sign in the expanded part of $^1[P^+A]$ that is a factor of $\beta(1)$. Since the orthonormal approximation is used, and since V_a of Eq. 44*e* contains only one term in these examples, the integration over $d\mathbf{x}_2 \cdots d\mathbf{x}_{12}$ leaves the uncanceled factor, $n_a^{-1/2}$, stemming from the normalizing factor $(12!)^{-1/2}$ of the single determinantal wave function for $Mg(^1S)$; consequently, the factor $n_a(n_p = 1)$ of Eq. 44*g* is canceled. The other factor, $(2/\sqrt{2})^2$, is just equal to $(2S_f + 1)(2S + 1)^{-1}$ in Eq. 44*h* since $S = 0$, and $S_f = \frac{1}{2}$ in this example. The rest of Eq. 44*h* results from the spatial integration.

Another case that Eq. 44*h* represents is 1*s*- (or 2*s*-) orbital capture from $O(1s^22s^22p^4;\,^3P)$ leaving the ions $O^+(1s2s^22p^4)$ in the ^2P- or ^4P-states. In this case the ^2P-, ^4P-wave functions are constructed by vector-coupling to the initial $2p^4(^3P)$ group. If we carry out the calculation, the n_a factor is again canceled and the ratios, $(2S + 2)(2S + 1)^{-1}$, and $2S(2S + 1)^{-1}$, appear as factors in Eq. 44*h* corresponding to the residual atomic ions $O^+(^4P)$ and $O^+(^2P)$, respectively. In this case, moreover, the sum over M_l and M_l' just cancel $(2L + 1)$ from g_i since each ionic wave function has the same M_l' value as the atom from which it evolves. The factor $2S_f + 1$ in this example thus originates from the vector-coupling coefficients used to construct the ionic wave functions.

The next examples involve *p*-orbital capture into H(1*s*), and the cross sections and wave functions are

$$Q = \frac{2\pi v_f}{v_i}\left(\frac{64\mu_f a_0}{m}\right)^2 \frac{n}{3}\,(p^{n-1}(S_fL_f)p^nSL|\}p^nSL)^2 \times \int_0^{\pi} \frac{d\theta \sin\theta A_i^2}{(1 + A_f^2)^2}\left|\sum_{j=1}^{3} \frac{c_j z_j^{7/2}}{(z_j^2 + A_i^2)^3}\right|^2, \quad (44i)$$

$$p_{\pm 1,0}(\mathbf{x}_1) = \frac{x_1}{\sqrt{\pi}}\left[\frac{\sin\theta_1}{\sqrt{2}}\exp(\pm i\phi_1), \cos\theta_1\right]\sum_{j=1}^{3} c_j z_j^{5/2}\exp(-z_j x_1).$$

Only the wave function of the *p*-orbital is needed here by virtue of the orthonormal approximation and the fact that V_a does not contain any of the coordinates of the atom A. [In this case two terms of $p_m(\mathbf{x}_1)$ usually suffice, and the form of the spherical harmonics used simplifies the calculations.] The first form of Eq. 44*f* is used here with $p_1(1, m_l'')$ occupying $\mathbf{x}_1$ of *A*, whereas $\mathbf{y}_1$ of V_a occurs in the wave function for H(1*s*). Since the wave function for *A* is here decomposed into sums of products of $(n_a - n)$- and *n*-electron wave functions, and since the wave function for $[PA^+]$ is correspondingly decomposed into three groups of electrons, the factor $n_a n_p$ of Eq. 44g is removed and replaced by the factor *n*. [The additional group of electrons in $[PA^+]$

refers to the one in $H(^2S)$.] This is easily verified using the previous analysis of symmetrization. In this example,

$$N_i = \frac{n_a!}{n!\,(n_a - n)!}, \qquad N_f = \frac{n_a!}{1!\,(n-1)!\,(n_a - n)!},$$

and $n_p = 1$; therefore we find that

$$\frac{n_p N_f}{(N_i N_f)^{1/2}} = n$$

as asserted, and this explains the factor of n of Eq. 44i.

The integration over $d\mathbf{x}_1$ and $d\mathbf{y}_1$ are performed, and the integrated result is inserted into Eq. 44g, which still contains a fourfold sum to be evaluated:

$$S = \sum_{M_{l_f}} \sum_{M_l} \sum_{m_l} \sum_{m_{l'}} (L_f 1 M_{l_f} m_l | L_f 1 L M_l)^* (L_f 1 M_{l_f} m_{l'} | L_f 1 L M_l) A_{im_l} A_{im_{l'}} \\ (p^{n-1}(S_f L_f) pSL|\}p^n SL)^2,$$
$$-L_f \le M_{l_f} \le L_f, \qquad -L \le M_l \le L, \qquad (44j)$$
$$-1 \le m_l \le 1, \qquad -1 \le m_{l'} \le 1.$$

The square of the fpc in Eq. 44j is diagonal in M_{lf}, M_l, m_l, $m_{l'}$, M_{S_f}, and M_S, and thus is independent of the summing operations.[47] The components of $\mathbf{A}_i$ appearing here result from a p-orbital electron being captured from the atomic target A, and these components depend upon the orientation of the associated atomic frame of reference with respect to some fixed coordinate frame. In this case S is independent of this atomic frame of reference, but in other, more complicated cases the sums corresponding to Eq. 44j do not remove this dependence upon the coordinate system, and averaging is required to remove this dependence. This point is discussed further in Section 3.5.

The sums over m_l, $m_{l'}$, and M_{l_f} are first effected with the use of a property of the vector-coupling coefficients,

$$M_{l_f} + m_l = M_{l_f} + m_{l'} = M_l$$

and the value of S can be written as

$$S = \sum_{M_l} \{(L_f 1 (M_l - 1) 1 | L_f 1 L M_l)^2 |A_{+1}|^2 + (L_f 1 M_l 0 | L_f 1 L M_l)^2 |A_0|^2 \\ + (L_f 1 (M_l + 1) - 1 | L_f 1 L M_l)^2 |A_{-1}|^2\} (p^{n-1}(S_f L_f) p(SL)|\}p^n SL)^2. \qquad (44k)$$

By reasons of the restriction to p-orbital capture, only the three values $L_f = L, L \pm 1$ are admissible, and these sums can be evaluated with the aid of the formulas

$$\sum_{M_l=-L}^{L} 1 = 2L + 1 \qquad \text{and} \qquad \sum_{M_l=-L}^{L} M_l^2 = \frac{L}{3}(L+1)(2L+1)$$

and a table of vector-coupling coefficients.[52] The sum

$$\sum_{M_l=-L}^{L} (L_f 1(M_l - m_l)m_l|L_f 1 L M_l)^2 = \frac{2L+1}{3}$$

is the same for each of the nine coefficients in Eq. 44*k*, and we finally get the asserted Eq. 44*i* since each of the $A_{im}{}^2$ have the same common factor and

$$\sum_{m=-1}^{1} A_{im}{}^2 = A_i^2.$$

The fact that Q of Eq. 44*i* is proportional to the number of equivalent electrons and the square of the fpc is the interesting feature of this result. Equations 44*h* and 44*i*, it is recalled, accommodate capture into H(1*s*) only, and we should like to generalize these formulas for capture into all states of H with principal quantum number *n*.

The formulas needed for this purpose have been derived by May[53] in connection with his studies of charge transfer and ionization and also by Fock[54] in his pioneering paper in which he proved the equivalence of the Schrödinger equation in momentum space for the hydrogen atom to the integral-equation for spherical harmonics of four-dimensional potential theory. May starts with the integral

$$f_{nlm}(\mathbf{A}) = \int d\mathbf{x}\phi_{nlm}(\mathbf{x})e^{i\mathbf{A}\cdot\mathbf{x}}$$

and proves that

$$\sum_{l=0}^{n-1} \sum_{m=-l}^{l} \left| f_{nlm}(\mathbf{A}) \right|^2 = \frac{64\pi a_0{}^{-5}}{n^3[A^2 + (1/na_0)^2]^4}. \tag{44l}$$

In the paper by Fock one of the results obtained is

$$\sum_{l=0}^{n-1} \sum_{m=-l}^{l} \left| \phi_{nlm}(\mathbf{A}) \right|^2 d\mathbf{A} = \frac{8a_0{}^{-5}\, d\mathbf{A}}{\pi^2 n^3 \hbar^3 [A^2 + (na_0)^{-2}]^4},$$

and if the volume element $d\mathbf{A}$, together with the square of the normalizing factor $(2\pi\hbar)^{-3/2}$, is removed from the last result, we recapture Eq. 44*l*. [The normalizing factor removed is the factor commonly used to define symmetrical spatial, and momentum transforms, and thus the $\phi_{nlm}(\mathbf{A})$ without this factor agrees with f_{nlm} of Eq. 44*l*.] The next result established by May is

$$[A^2 + (na_0)^{-2}]f_{nlm}(\mathbf{A}) = 2a_0{}^{-1}g_{nlm}(\mathbf{A}),$$

$$g_{nlm}(\mathbf{A}) = \int d\mathbf{x}\, \frac{\phi_{nlm}(\mathbf{x})e^{i\mathbf{A}\cdot\mathbf{x}}}{x}. \tag{44m}$$

This result is proven by taking the Fourier transform of the equation satisfied by $\phi_{nlm}(\mathbf{x})$. The desired result emerges from the substitution of Eq. 44*m* in 44*l*:

$$\sum_{l=0}^{n-1} \sum_{m=-l}^{l} \left| g_{nlm}(\mathbf{A}) \right|^2 = \frac{16\pi a_0^{-3}}{n^3[A^2 + (na_0)^{-2}]^2} \tag{44n}$$

since, according to Eq. 44*e*, $g_{nlm}(\mathbf{A})$ represents the integral over $d\mathbf{y}_1$ if $\mathbf{A} = \mathbf{A}_f$. We here refer to the interaction $|\mathbf{y}_1|^{-1}$ of Eq. 44*e*. Equations 44*h* and 44*i* are modified to represent capture into all states of H with principal quantum number *n* by replacing

$$(1 + A_f^2)^{-2}$$

with

$$(n^3\pi)^{-1}(A_f^2 + n^{-2})^{-2}$$

if dimensionless variables are used. The factor $(16\pi^2)$ is removed from Eq. 44*n* since this factor occurs in the factor outside the integrals. Note that alterations also must be made in the equation expressing conservation of energy.

Although no mathematical basis has been given in support of the approximation of the OKB cross section, and even though it seems well established that OBK cross sections predict values larger than the experimentally determined values,[30,55] by empirical adjustment of some of the atomic parameters it is sometimes possible to reproduce the measured values remarkably well. Such approaches are in fact occasionally appealing since other, more refined approximating procedures not only involve more elaborate calculations but are additionally restricted by the paucity of information on analytic representations of atomic wave functions, a limitation that is gradually vanishing.

Another reason for interest in these cross sections arises from the observation that the ratios

$$R(\mathrm{A}) = \frac{Q_i(1s)}{Q_{\mathrm{OBK}}(1s)}, \tag{44o}$$

are strikingly similar for the atoms (A) = H(1*s*) and He($1s^2$).[56] In this formula $Q_i(1s)$ and $Q_{\mathrm{OBK}}(1s)$ denote the cross sections calculated in the first Born approximation using V_i and V_a, respectively, for electronic capture into H(1*s*) by H^+ from the ground state of the atom, leaving the residual atomic ion in its ground state. This similarity suggests using the ratio *R*(H) and the OBK cross sections calculated for the same impact energy *E* to estimate the cross sections for electronic capture by protons from complex atoms. These estimated total cross sections are given by the formula

$$Q(\mathrm{A}) = \sum_{m,n} Q_{mn}(\mathrm{OBK}; \mathrm{A}; E)R(\mathrm{H}; E). \tag{44p}$$

Here m and n denote, respectively, the mth subshell of the atom and the ns-state of H; no allowance is made for capture into other states of H or for simultaneous excitation of the residual atomic ion, which should be relatively small at high impact velocities. (Equation 44p can be altered to include the modification given by Eq. 44n.)

This estimating method has been moderately successful in predicting total cross sections for electronic capture by protons. By "successful" we mean that the calculated and measured values do not disagree by a factor in excess of three. It is doubtful, however, that the constituent parts of Eq. 44p would predict the corresponding cross section with the same accuracy. [In this connection the recent paper by Gaily compares prior and post cross sections (Section 3.5) with measured values for capture into H(2s) by incident protons.[57] In one case the calculated values underestimate, and in the other case they overestimate the experimentally determined cross sections.] It is not known, moreover, why Eq. 44p is as successful as it is; an explanation of this phenomenon may be derived in the future from a correct mathematical description of electronic capture. (OBK cross sections also are being calculated to get estimates of the cross sections into highly excited states of the hydrogen atom, methods which have been developed principally by Hiskes.[58])

3.5 PRIOR, POST, AND DISTORTED WAVE INTERACTIONS

Most of the mathematical details needed for this section have been developed in Sections 3.2, 3.3, and 3.4; therefore the chief task of this section is to supply the necessary revisions to the preceding equations so that the interactions of this section can be used in place of V_a. (The reader is again reminded that it is not known in what sense, if any, the first Born approximation represents the solution to the problem of electronic capture.) Only the case of an incident, positively charged nucleus capturing from an atom (usually neutral) to form a hydrogenic atom (or ion) is treated here since most measured processes are in this category. We simplify additionally to capture by protons for the first task, and proceed to record the associated $V_{i,f}$ for a neutral n-electron atomic target; dimensionless variables and the coordinates defined by Eqs. 2i, 2f, and 44e are used:

$$V_i = n|\mathbf{x}_1 - \mathbf{x}_{n+1}|^{-1} - |\mathbf{x}_{n+1}|^{-1} - \sum_{j=2}^{n} |\mathbf{x}_1 - \mathbf{x}_{n+1} - \mathbf{x}_j|^{-1},$$

$$\mathbf{x}_1 = \mathbf{x}_i, \qquad \mathbf{x}_{n+1} = \mathbf{y}_1 = \mathbf{x}_f, \tag{45a}$$

$$V_f = V_i + |\mathbf{x}_{n+1}|^{-1} - n|\mathbf{x}_1|^{-1} + \sum_{j=2}^{n} |\mathbf{x}_1 - \mathbf{x}_j|^{-1}.$$

The prior and post amplitudes are given by Eq. 44*e* in which V_a is replaced by V_i or V_f since the procedure used to handle the indistinguishability of the electrons is also applicable to the interactions of this section. Use of this revised formula means also that only the two momentum-change vectors $\mathbf{A}_i$ and $\mathbf{A}_f$ are retained; the purpose of this approximation is to simplify the calculation, as noted before.

With the use of the second form of Eq. 44*f*, we can demonstrate that the passive interactions,

$$|\mathbf{x}_i - \mathbf{x}_f - \mathbf{x}_j|^{-1},$$

mix certain contributions arising from terms p^{n-1} of the same multiplicity of the atomic ion.[59] This occurs for both *s*- and *p*-orbital capture for processes such as

$$H^+ + O(^3P{:}2p^4) \rightarrow H(1s) + \begin{cases} O^+(^2P{:}2p^3) \\ O^+(^2P,\ ^4P{:}ms2p^4), \end{cases}$$

but this mixing does not occur in the process

$$H^+ + O(^3P{:}2p^4) \rightarrow H(1s) + O^+(^4S{:}2p^3).$$

This mixing is absent in the last process since only one term of p^3 is a quartet. If mixing occurs, the cross section depends upon the orientation of the atomic target unless some parts of the amplitude are omitted, or unless the orientation of the atomic target is averaged over all directions. This problem has been studied before, and only a sketch of that analysis is presented here.[60]

In problems of this type, the amplitude is proportional to a combination of integrals (the Feynman variety discussed subsequently) and trigonometric functions of (θ_i, ϕ_i) and (θ_f, ϕ_f), which are the spherical angles between the atomic frame of reference and the wave vectors $\mathbf{K}_i$ and $\mathbf{K}_f$, respectively, as in Eq. 43*c*. If (α, β, γ) are the Euler angles of the atomic frame of reference with respect to the *XYZ*-axes—a frame of reference fixed in the center-of-mass system—and if the direction of incidence is parallel to the direction of the positive *Z*-axis, the identifications of θ_i, ϕ_i to

$$\beta = \theta_i, \qquad \gamma = \frac{\pi}{2} - \phi_i$$

suffice, and it is also convenient to replace the angle α by

$$\phi_z = \alpha + \frac{\pi}{2}.$$

The matrix that represents these rotations through the three Euler angles is used to relate θ_f and ϕ_f to (α, β, γ).[61] The average value of a function is defined by

$$\overline{F(\theta_i, \phi_i, \theta_f, \phi_f, \theta, \phi)} = \frac{1}{8\pi^2} \int_0^{\pi} d\theta_i \sin\theta_i \int_0^{2\pi} d\phi_i \int_0^{2\pi} d\phi_z F, \qquad (45b)$$

and this formula is used to remove the dependence of the cross section upon the relative orientation of the atomic target. [If the cross section specifies the state with quantum numbers nlm ($l \geq 1$) for the outgoing H atom, it is convenient to use another set of Euler angles to specify the orientation of the H atom.] This averaging feature, of course, complicates and lengthens numerical procedures used to calculate cross sections.

The numerical evaluation of cross sections for which averaging is unnecessary, nevertheless, has been made more feasible by using the Feynman-integral representations.[30,62,63] Two representations commonly employed are

$$(ab)^{-1} = \int_0^1 dx[b + (a - b)x]^{-2},$$

$$(abc)^{-1} = 2\int_0^1 dx \int_0^1 dy\{c(1 - y) + [b + (a - b)x]y\}^{-3}.$$

These two representations, together with the standard representations,

$$I(\mathbf{x}, \mathbf{x}'; \alpha) = \frac{e^{-\alpha|\mathbf{x}-\mathbf{x}'|}}{|\mathbf{x} - \mathbf{x}'|} = (2\pi^2)^{-1} \int_0^1 \frac{d\mathbf{k} e^{i\mathbf{k}\cdot(\mathbf{x}-\mathbf{x}')}}{k^2 + \alpha^2},$$

$$|\mathbf{x} - \mathbf{x}'|^{-1} = \lim_{\alpha\to 0} I(\mathbf{x}, \mathbf{x}'; \alpha), \qquad |\mathbf{x} - \mathbf{x}'| = \lim_{\alpha\to 0} \frac{\partial^2}{\partial\alpha^2} I(\mathbf{x}, \mathbf{x}'; \alpha), \qquad (45c)$$

suffice to reduce the Born cross sections to a numerically calculational level. In these calculations, the approximations discussed in Section 3.4 are commonly employed. Except in special cases the $(\mathrm{fpc})^2$ cannot be factored, nor can the sums occurring in Eqs. 44*j* and 44*k* be isolated, as is accomplished to get Eq. 44*i*.

The prior version of the distorted wave interaction,

$$V_i(dw) = V_i - U_i,$$

as shown in the paper by Bassel and Gerjuoy,[14] may be approximated for most purposes.[64] For this purpose V_i is the first expressed in terms of $\mathbf{R}_i$ and $\mathbf{x}_j$ eliminating $\mathbf{x}_{n+1}$, again using Eqs. 2*i* and 2*f*:

$$V_i = n\left|\frac{m}{M_{i1}}\mathbf{x}_1 + \frac{m}{M_{i1}}\sum_{k=2}^{n}\mathbf{x}_k + \mathbf{R}_i\right|^{-1} - \left|\frac{M_i}{M_{i1}}\mathbf{x}_1 - \frac{m}{M_{i1}}\sum_{k=2}^{n}\mathbf{x}_k - \mathbf{R}_i\right|^{-1}$$

$$- \sum_{j=2}^{n}\left|\mathbf{R}_i - \mathbf{x}_j + \frac{m}{M_{i1}}\mathbf{x}_1 + \frac{m}{M_{i1}}\sum_{k=2}^{n}\mathbf{x}_k\right|^{-1} \qquad (45d)$$

$$\approx n|\mathbf{R}_i|^{-1} - |\mathbf{x}_1 - \mathbf{R}_i|^{-1} - \sum_{j=2}^{n}|\mathbf{R}_i - \mathbf{x}_j|^{-1}.$$

An approximate interaction used in the wave version of the distorted wave method is given by Eq. 42*r* using for V_i the approximate expression in Eq. 45*d*. This interaction is labeled U_i, and it is clear that it is a function of $\mathbf{R}_i$ only; moreover, since $\mathbf{R}_i$ is approximately

$$\mathbf{R}_i = \mathbf{x}_1 - \mathbf{x}_{n+1} - \frac{m}{M_{i1}} \sum_{k=1}^{n} \mathbf{x}_k \approx \mathbf{x}_1 - \mathbf{x}_{n+1},$$

the calculational procedures pertaining to V_i and V_f clearly apply to cross sections derived from $V_i(dw)$.

If in Eq. 42*q* we approximate the wave functions by

$$\Psi_f^{(-)} = \Phi_f, \qquad \chi_i^{(+)} = \Phi_i, \tag{45e}$$

the first Born approximation for $V_i(dw)$ results. The corresponding amplitude for the process

$$H^+ + D(1s) \rightarrow H(1s) + D^+ \tag{45f}$$

is recorded for comparative and illustrative purposes (the nuclear binding forces in D and D^+ are assumed to be infinite here):

$$\Phi_i = \frac{b^{3/2}}{(\pi a_0{}^3)^{1/2}} \exp\left(-\frac{bx_i}{a_0} + i\mathbf{K}_i \cdot \mathbf{R}_i\right) = \phi_i \exp(i\mathbf{K}_i \cdot \mathbf{R}_i),$$

$$\Phi_f = \frac{a^{3/2}}{(\pi a_0{}^3)^{1/2}} \exp\left(-\frac{ax_f}{a_0} + i\mathbf{K}_f \cdot \mathbf{R}_f\right) = \phi_f \exp(i\mathbf{K}_f \cdot \mathbf{R}_f),$$

$$\begin{aligned} V_i - U_i = e^2\Bigg(\left[\frac{1}{|\mathbf{x}_f - \mathbf{x}_i|} - \frac{1}{|\mathbf{x}_f|}\right] - \Bigg\{\frac{1}{R_i} - \exp\left[-\left(\frac{2b}{1-b}\right)\right]\left(\frac{R_i}{a_0}\right) \\ - \frac{b}{a_0(1-b)} \exp\left[-\frac{2b}{(1-b)a_0} R_i\right] \\ - \frac{1}{R_i}\frac{\exp(-2R_i/a_0)}{R_i} + \frac{\exp(-2R_i/a_0)}{a_0}\Bigg\}\Bigg), \end{aligned} \tag{45g}$$

$$a = \frac{M}{M_1} = \frac{M_f}{M_{f1}}, \qquad b = \frac{M_D}{M_{D1}} = \frac{M_i}{M_{i1}}.$$

Since $b/(1 - b)$ is large, it is a good approximation to discard the corresponding terms, and from previous discussions $\mathbf{R}_i$ can be approximated by

$$\mathbf{R}_i \approx \mathbf{x}_i - \mathbf{x}_f.$$

The resulting approximate amplitude for capture can be derived from Eq. 44*e* and is

$$g(\theta, \phi) = \frac{\mu_f a_0}{2\pi m} \int \phi_f^*(\mathbf{x}_f) \exp(i\mathbf{A}_f \cdot \mathbf{x}_f) \left[\frac{1}{|\mathbf{x}_f - \mathbf{x}_i|} - \frac{1}{|\mathbf{x}_f|} - \frac{\exp(-2|\mathbf{x}_i - \mathbf{x}_f|)}{|\mathbf{x}_i - \mathbf{x}_f|} - \exp(-2|\mathbf{x}_i - \mathbf{x}_f|) \right] \phi_i(\mathbf{x}_i) \exp(i\mathbf{A}_i \cdot \mathbf{x}_i)\, d\mathbf{x}_i\, d\mathbf{x}_f,$$

$$\mathbf{A}_i = b\mathbf{K}_i - \mathbf{K}_f, \qquad \mathbf{A}_f = a\mathbf{K}_f - \mathbf{K}_i. \tag{45h}$$

Dimensionless variables are used in Eq. 45*h* and *a* and *b* are equated to unity except in $\mathbf{A}_i$ and $\mathbf{A}_f$.

A study of Eq. 45*h* at large impact energies, or at large angles of scattering, reveals that effectively only the matrix element of $V_a = x_f^{-1}$ contributes, and $g(\theta, \phi)$ tends to the associated OBK amplitude. This pattern of variation is appealing since the nucleus-nucleus interaction is also absent from the impact-parameter version for rectilinear motion of the nuclei, a feature discussed in detail in Section 3.1. Moreover, a study of Eqs. 40*k* and 41*h* reveals the similarity of the three amplitudes recorded in Eqs. 40*k*, 41*h*, and 45*h*. A final comment is made about the last result.

This process exemplifies accidentally resonant electronic capture since the internal energy of H(1*s*) and D(1*s*) are nearly equal; however, the *masses* of H and D are different. If the angular integration of the differential cross section is replaced by an integration over one of the momentum-change variables, a study of the constituent parts of the cross section reveals that the cross section is independent of the mass of the atomic target at moderately high impact velocities. Thus, if the interactions and other atomic parameters are the same, the cross sections calculated in the first Born approximation, $[V_a, V_i, V(dw)]$, are independent of the atomic nuclear mass for incident singly charged nuclei at moderately high impact velocities. [The interaction V_f is excluded since $Q(f) = Q(i)$ if the atomic wave functions are exact.] It should be mentioned in concluding these sections on the first Born approximation that no simple variation of the cross section with impact energy occurs except for atomic targets containing only one subshell or for 1*s*-orbital capture from complex atoms at very high impact velocities.

3.6 IMPULSE APPROXIMATION

The impulse approximation is discussed at this point in the chapter in order to contrast this approximation with other first-order approximations already treated and to clarify the connection with the second Born approximation. It is also the purpose of this section to elucidate certain difficulties that emerge from a derivation using time-independent techniques. This approximation was originally developed for application to nuclear scattering, a process that usually involves short-range forces.[65,66] Pradhan,[67] however, first employed these methods of formal scattering theory to derive the impulse approximation for electronic capture from atomic hydrogen by protons.[17,68] An attractive feature of this approximation is the absence of the effect of the nucleus of the atomic target (parent nucleus) in the determination of the wave function of the system except that the Coulomb force between the parent nucleus and the electron determines the initial distribution in electronic velocity. This interpretation is appealing, moreover, by virtue of its analogy to the classical impulse approximation where the interaction between the electron and the parent nucleus is neglected during the time that the impulsive force acts, and the classical electronic motion about the parent nucleus enters only in the determination of the initial distribution in electronic-velocity.[69] (This point is discussed further in the sections on classical methods.)

A simplified derivation of the impulse matrix element, I_{if}, is presented next, and this particular deduction elicits the connection with the sum of the corresponding first and second Born matrix elements at the outset.[70] * Proton 2 is incident on the hydrogen atom consisting of electron 3 bound in the 1s-state to proton 1. After the collision the electron is bound in the 1s-state to proton 2. This process and specialized notation is recorded in Eq. 46a:

$$\begin{aligned}
&H^+(2) + H(1s{:}13) \to H(1s{:}23) + H^+(1),\\
&H\Psi = E\Psi, \qquad H = H_i + V_i = H_f + V_f,\\
&H_i = T + U_{13}, \qquad H_f = T + U_{23},\\
&V_i = U_{12} + U_{23}, \qquad V_f = U_{12} + U_{13},\\
&H_i\Phi_i = E\Phi_i, \qquad G_f^{(+)} = (E + i\varepsilon - H_f)^{-1}\\
&H_f\Phi_f = E\Phi_f,\\
&T\chi_m = E_m\chi_m,\\
&\omega_{23}^{(+)} = 1 + (E_m + i\varepsilon - H_f)^{-1}U_{23}\\
&G^{(+)} = (E + i\varepsilon - H)^{-1},\\
&\Psi = (1 + GV_i)\Phi_i.
\end{aligned} \tag{46a}$$

(The superscript (+) is omitted hereafter.) In this equation T is the operator of kinetic energy, and the subscript m denotes a six-dimensional continuum, or momentum space, and χ_m is a plane wave that is labeled by these variables; for example, $m = (\mathbf{K}_1, \mathbf{K}_2)$, $\chi_m = (2\pi)^{-3} \exp[i(\mathbf{K}_1\cdot\mathbf{x}_1 + \mathbf{K}_2\cdot\mathbf{x}_2)]$. Assume it is possible to represent G by the Born series,

$$G = G_f + G_fV_fG + \cdots = G_f + G_fV_fG_f + \cdots, \tag{46b}$$

* Professor M. R. C. McDowell derived this same result (private communication).

and retain only the first term of this series so that Ψ' of Eq. 46*a* is approximated by

$$\Psi' \approx (1 + G_f V_i)\Phi_f. \tag{46c}$$

If Eq. 46*c* is used in Eq. 43*a*, we get

$$R = (\Phi_f, V_f \Psi') \approx (\Phi_f, V_f(1 + G_f V_i)\Phi_i), \tag{46d}$$

which is one form for the sum of the first and second Born matrix elements. If only the interaction U_{23} of V_i in Eq. 46*a* is retained, we obtain the final approximation that we want to use,

$$R_{if} = (\Phi_f, V_f(1 + G_f U_{23})\Phi_i). \tag{46e}$$

The last step approximates the part of Eq. 46*d* belonging only to the second Born approximation. (A practical reason for deleting U_{12} from V_i is a calculational one. The function given by ω_{23} of Eq. 46*a* operating on Φ_i is known; the replacement of U_{23} by U_{12} in ω_{23}, however, yields an operator ω_{23} whose associated function, $\omega_{23}\Phi_i$, is unknown. The following *proof* obviously lacks mathematic rigor.)

The identity of Eq. 42*i*,

$$(P - Q)^{-1} - P^{-1} = (P - Q)^{-1}QP^{-1},$$

is first invoked to get the relation

$$\begin{aligned}&[(E + i\varepsilon - H_f)^{-1} - (E_m + i\varepsilon - H_f)^{-1}]U_{23}\chi_m(\chi_m, \Phi_i)\\ &\quad = -[(E + i\varepsilon - H_f)^{-1}(E_m + i\varepsilon - H_f)^{-1}]U_{23}\chi_m(\chi_m, U_{13}\chi_n)(\chi_n, \Phi_i),\end{aligned} \tag{46f}$$

where repetition of the dummy indices m and n signifies an integration over the appropriate domain. The right member of Eq. 46*f* is obtained from its original form with the use of Eq. 46*a*, the Hermitian property of T, and the expansibility of Φ_i in the set of function χ_n, a six-dimensional Fourier transform; the separate steps are

$$\begin{aligned}&(E_m - E)(\chi_m, \Phi_i)\\ &\quad = (T\chi_m, \Phi_i) - (\chi_m, H_i\Phi_i) = -(\chi_m, U_{13}\Phi_i) = -(\chi_m, U_{13}\chi_n)(\chi_n, \Phi_i).\end{aligned} \tag{46g}$$

(The Hermitian property of T is readily proven.) Next Eq. 46*f* is expressed anew in terms of ω_{23} and G_f,

$$\begin{aligned}&[E + i\varepsilon - H_f]^{-1}U_{23}\Phi_i\\ &\quad = (\omega_{23} - 1)\chi_m(\chi_m, \Phi_i) - (E + i\varepsilon - H_f)^{-1}(\omega_{23} - 1)\chi_m(\chi_m, U_{13}\chi_n)(\chi_n, \Phi_i).\end{aligned} \tag{46h}$$

In the original derivation of the impulse approximation, the right-hand member of

$$1 + G(U_{12} + U_{23}) = \omega_{23} + GU_{12}\omega_{23} + G[U_{13}(\omega_{23} - 1) - (\omega_{23} - 1)U_{13}]$$

(summations over dummy variables understood) was approximated by ω_{23}, and in particular, the term

$$-G(\omega_{23} - 1)U_{13}$$

was discarded.[67] This term, we note, corresponds to one of the terms in Eq. 46*h*. Thus we also discard the same term, and substitute the resulting approximation to $G_f U_{23}\Phi_i$ of Eq. 46*h* in Eq. 46*e* to get

$$\begin{aligned} I_{if} &= (\Phi_f, V_f \omega_{23}\Phi_i) \\ \omega_{23}\Phi_i &= \omega_{23}\chi_m(\chi_n, \Phi_i). \end{aligned} \tag{46i}$$

The first line of Eq. 46*i* is Eq. 30 of the paper by Pradhan.[67] The operation of ω_{23} on Φ_i is repeated in Eq. 46*i* to emphasize how ω_{23} is defined to operate on a function. Hereafter, this spectral decomposition of the function is understood in an expression such as $\omega_{23}\Phi$.

Next the evaluation of the approximate wave function, $\omega_{23}\Phi_i$, is analyzed similarly to previous studies.[71–73] The operator ω_{23}, according to Eq. 46*a*, is $(1 + G_f U_{23})$ for the variable energy E_m. This Green's function and its defining differential equation are

$$\begin{aligned} &\left[\nabla_{R_f}{}^2 + \frac{2\mu}{\hbar^2}\left(\frac{\hbar^2}{2m}\nabla_{x_f}{}^2 + \frac{e^2}{x_f}\right) + K_t^2 + i\delta'\right] G_f(E_m) = \delta(\mathbf{x}_f - \mathbf{x}_f')\delta(\mathbf{R}_f - \mathbf{R}_f') \\ &G_f = (2\pi)^{-3}\Sigma_n \int \frac{\phi_n(\mathbf{x}_f)\phi_n^*(\mathbf{x}_f')}{K_t^2 + (2\mu/\hbar^2)\varepsilon_n - U^2 + i\delta'} \exp\left[iU\cdot(\mathbf{R}_f - \mathbf{R}_f'\right) d\mathbf{U}, \\ &\mu = \frac{MM_i}{M_t}, \qquad E_m = \frac{\hbar^2 K_t^2}{2\mu} = \frac{\hbar^2 K_1{}^2}{2\mu} + \frac{\hbar^2}{2m}K^2, \qquad U_{23} = \frac{-e^2}{x_f}, \\ &\Phi_i = (2\pi)^{-3/2}\phi_i(\mathbf{x}_i)\exp(i\mathbf{K}_i\cdot\mathbf{R}_i), \qquad \Phi_f = (2\pi)^{-3/2}\phi_f(\mathbf{x}_f)\exp(i\mathbf{K}_f\cdot\mathbf{R}_f), \\ &\chi_m = (2\pi)^{-3/2}\exp\left[i(-\mathbf{K}_1\cdot\mathbf{R}_f + \mathbf{K}\cdot\mathbf{x}_f)\right]. \end{aligned} \tag{46j}$$

Some of the apparent redundancy in Eq. 46*j* serves to emphasize that most of the plane waves are normalized to the delta function. In the present analysis the operator,

$$\lim_{\varepsilon\to 0} i\varepsilon G_f, \tag{46k}$$

which is equivalent to $(1 + G_f U_{23})$, is used; thus, the intermediate result obtained from integrating over $\mathbf{U}$ and $\mathbf{R}_f'$ yields

$$\begin{aligned} \psi_m &= (2\pi)^{-3/2}\exp(-i\mathbf{K}_1\cdot\mathbf{R}_f)F, \\ F &= \lim_{\varepsilon\to 0}\Sigma_n \int \frac{\phi_n(\mathbf{x}_f)\phi_n^*(\mathbf{x}_f')\,d\mathbf{x}_f'}{K^2 + (2m/\hbar^2)\varepsilon_n + i\varepsilon}\exp(i\mathbf{K}\cdot\mathbf{x}_f'). \end{aligned} \tag{46l}$$

The sum in Eq. 46*l* includes not only the discrete states of hydrogen but also an integration of the associated continuum; thus F is

$$\lim i\varepsilon \int G_c(\mathbf{x}_f, \mathbf{x}'_f; \mathbf{K})\, d\mathbf{x}'_f \exp(i\mathbf{K}\cdot\mathbf{x}'_f)$$
$$= \int d\mathbf{x}'_f[\delta(\mathbf{x}_f - \mathbf{x}'_f) + G_c(\mathbf{x}_f, \mathbf{x}'_f; \mathbf{K})U_{23}(\mathbf{x}'_f)] \exp(i\mathbf{K}\cdot\mathbf{x}'_f), \quad (46m)$$

in which G_c is the Coulomb-Green's function for outgoing boundary conditions. According to formal scattering theory, the solution of the equation

$$\psi = \phi + (E + i\varepsilon - H_0)^{-1}U\psi, \qquad (E - H_0)\phi = 0, \qquad (46n)$$
$$(E - H_0 - U)\psi = 0$$

is

$$\psi = [1 + (E + i\varepsilon - H_0 - U)^{-1}U]\phi = \lim_{\varepsilon\to 0} i\varepsilon(E + i\varepsilon - H_0 - U)^{-1}\phi \qquad (46o)$$

provided, in particular, that

$$\lim_{R\to\infty} RU \to 0. \qquad (46p)$$

In this last equation R represents the distance between the center of masses of the outgoing aggregates. Since ψ of Eq. 46*p* represents the solution to both the integral equation and the associated differential equation for the same *boundary conditions*, the solution to the differential equation and Eq. 46*o* must be identical. Moreover, the asymptotic form of ψ,

$$\lim_{R\to\infty} \psi,$$

calculated from Eq. 46*o* contains ϕ as the incident state, and ϕ is a plane wave for this two-body example provided that Eq. 46*p* is satisfied; however, if U is a single Coulomb potential and H_0 is the corresponding kinetic-energy operator, we know that the asymptotic form of the scattering solution to the differential equation in Eq. 46*n* has Coulomb, or distorted, plane and spherical waves.[74] In Eq. 46*m*, nevertheless, the incident state is an *undistorted plane wave* which disagrees with the *incident state* of the associated differential equation. Equation 46*m* has been evaluated using both forms of the solution,[75] and the two calculations both yield the same nonunique, incorrectly normalized function.[72] In a subsequent study of Eqs. 46*m* and 46*o*, Eq. 46*m* was evaluated using an improved limiting procedure; and Eq. 46*o* was also calculated for the three-dimensional attractive square-well potential, a form of U that satisfies Eq. 46*p*.[73] The latter solution was the unique, correctly normalized scattering solution for this square-well interaction, whereas the result for Eq. 46*m* was unchanged.

This result for the single Coulomb potential is not surprising, because we know that the plane wave in Eq. 46*m* is not the correct incident state for the

Coulomb scattering problem. It is important to realize that standard, time-independent, formal scattering solutions are used here, for the Coulomb-operator of Eq. 46*m can be modified* using *time-dependent* scattering theory so that the plane wave is projected into the correct scattering solution.[76] Other related research on this subject is devoted both to the derivation of different representations of the Coulomb-Green's function and to the investigation of different aspects of scattering solutions for the single Coulomb potential.[77–87]

In practice the correct solution to the Coulomb scattering problem for the attractive interaction,

$$F = \exp\left(\frac{\pi\alpha}{2}\right)\Gamma(1 - i\alpha)\exp(i\mathbf{K}\cdot\mathbf{x}_f)\,{}_1F_1(i\alpha, 1; i[Kx_f - \mathbf{K}\cdot\mathbf{x}_f]),$$
$$\alpha = \frac{M}{(M_1 a_0 K)} \tag{46q}$$

is used in place of the nonunique result of Eq. 46*m*. This substitution amounts to a renormalization of the calculated result; moreover, this renormalizing factor approaches unity as the impact energy increases. This renormalized solution[72] is now used to complete the calculation of $\omega_{23}\Phi_i$:

$$(\chi_m, \Phi_i) = (2\pi)^{-3}\int d\mathbf{x}_i\, d\mathbf{x}_f \exp\{i[\mathbf{K}_1\cdot(\mathbf{x}_i - a\mathbf{x}_f) - \mathbf{K}\cdot\mathbf{x}_f]\}$$
$$\times \frac{\exp(-x_i/a_0)}{(\pi a_0{}^3)^{1/2}}\exp[i\mathbf{K}_i\cdot(a\mathbf{x}_i - \mathbf{x}_f)]$$
$$= \frac{8(\pi a_0{}^3)^{1/2}\delta(a\mathbf{K}_1 + \mathbf{K} + \mathbf{K}_i)}{(1 + a_0{}^2|a\mathbf{K}_i + \mathbf{K}_1|^2)^2},$$

$$\omega_{23}\Phi_i = (2\pi)^{-3/2}\int d\mathbf{K}_1\, d\mathbf{K}\exp\{i[-\mathbf{K}_1\cdot(\mathbf{x}_i - a\mathbf{x}_f) + \mathbf{K}\cdot\mathbf{x}_f]\}$$
$$\times \exp\left(\frac{\pi\alpha}{2}\right)\Gamma(1 - i\alpha)\,{}_1F_1[i\alpha, 1; i(Kx_f - \mathbf{K}\cdot\mathbf{x}_f)] \tag{46r}$$
$$\times \frac{8(\pi a_0{}^3)^{1/2}\delta(a\mathbf{K}_1 + \mathbf{K} + \mathbf{K}_i)}{(1 + a_0{}^2|a\mathbf{K}_i + \mathbf{K}_1|^2)^2}$$
$$= \left(\frac{8a_0{}^3}{\pi^2 a^6}\right)^{1/2}\int d\mathbf{K}\exp\left\{i\left[\left(\frac{\mathbf{K} + \mathbf{K}_i}{a}\right)\cdot\mathbf{R}_f + \mathbf{K}\cdot\mathbf{x}_f\right]\right\}\Gamma(1 - i\alpha)$$
$$\times \exp\left(\frac{\pi\alpha}{2}\right)\frac{{}_1F_1[i\alpha, 1; i(Kx_f - \mathbf{K}\cdot\mathbf{x}_f)]}{[1 + a_0{}^2|a\mathbf{K}_i - a^{-1}(\mathbf{K} + \mathbf{K}_i)|^2]^2},$$
$$\mathbf{R}_i = a\mathbf{x}_i - \mathbf{x}_f, \qquad \mathbf{R}_f = \mathbf{x}_i - a\mathbf{x}_f.$$

As might be expected, $\omega_{23}\Phi_i$ is a complicated function since it contains some contribution from the second Born approximation. Although the remaining calculation for I_{if} is conceptually simple, numerical methods are required for this evaluation.

Pradhan originally showed that the contribution from the nucleus-nucleus interaction, U_{12}, vanished if the ratio m/M vanished.

In Cheshire's theoretical and numerical study of the cross section calculated using I_{if} of Eq. 46*i*, he found that contributions originating from

$$(\Phi_f, V_{12}\omega_{23}\Phi_i)$$

became decreasingly important with increasing impact velocity.[88,89] This quality is understandably appealing by reason of the connection with the impact-parameter version for uniform rectilinear motion of the nuclei, a connection already noted several times. Thus, in spite of the formal difficulties and objectionable features that harass the impulse approximation, we see that it possesses some desirable qualities. Bransden and Cheshire in their analysis of electronic capture from helium by protons used the IP version of the impulse approximation to calculate the probability for capture $|c_f(\rho)|^2$ as a function of impact parameter[90,91] (see Eq. 41*m* and subsequent discussion in Section 1.1). At an impact energy of 700 keV, for example, they found that $|c_f(\rho)|^2$ never exceeded unity, but $|c_f(\rho)|^2$ calculated using V_i or V_f in the first Born approximation often exceeded unity. The angular distribution calculated in the wave version, moreover, varied similarly to the results obtained using the method of Bassel and Gerjuoy as explained in Section 3.2, but both of these angular distributions disagreed markedly with the corresponding distributions obtained using V_i or V_f with the first Born approximation.

The foregoing comparisons emphasize how different approximating procedures may yield total cross sections agreeing closely with the experimentally determined values for a restricted range of impact velocities, the situation here, although not only do the angular distributions disagree significantly, but the theoretical basis of the approximations is also questionable. Note also that OBK cross sections have been used with the impulse approximation for the purpose of estimating the total impulse-approximation cross sections.[92] We next turn to another approximating procedure that also shares some of the qualities of the second Born approximations.

3.7 CONTINUUM DISTORTED WAVE APPROXIMATION OF CHESHIRE

As noted by Cheshire,[93] approximating methods that attempt to include transitions to intermediate states of the continuum are loosely termed "second order." In a previous paper Cheshire derived his *continuum distorted wave* approximation, and the associated amplitude for electronic capture contained transitions into and from intermediate states of the continuum.[94] In this respect the method is similar to the impulse approximation, and both can be interpreted as second order. We now study the results obtained from this 1964 paper by Cheshire.

The impact-parameter version is used, and initially the problem of interest consists of an incident ion P^+ of nuclear charge Z_p capturing an electron originally bound to a nucleus of charge Z_a to form an atom A. Uniform rectilinear motion of the nuclei is assumed. [The charges on P^+ and A, initially, are Z_p and $(Z_a - 1)$, respectively, but this refinement in notation is ignored.] Although Eqs. 4, 5, and 15*b* contain the equations and coordinates for this analysis, these are repeated here for more convenient reference:

$$\begin{aligned} \mathbf{R} = \boldsymbol{\rho} + \mathbf{v}t, \qquad \mathbf{x}_i &= \mathbf{s} + \tfrac{1}{2}\mathbf{R}, & \nabla_s &= \nabla_{x_i} \\ t_i &= t, & \frac{\partial}{\partial t} &= \frac{\partial}{\partial t_i} + \frac{\mathbf{v}\cdot\nabla_{x_i}}{2}; \\ \mathbf{x}_f &= \mathbf{s} - \tfrac{1}{2}\mathbf{R}, & \nabla_s &= \nabla_{x_f}, \\ t_f &= t, & \frac{\partial}{\partial t} &= \frac{\partial}{\partial t_f} - \frac{\mathbf{v}\cdot\nabla_{x_f}}{2}, \end{aligned} \tag{47}$$

The electronic motion is governed by the equation (in atomic units)

$$H\Psi(\mathbf{s}, t) = i\frac{\partial \Psi}{\partial t}$$

$$H = -\tfrac{1}{2}\nabla_s^2 - \frac{Z_p}{x_f} - \frac{Z_A}{x_i} \tag{47a}$$

and the initial condition is

$$\lim_{t \to -\infty} \Psi(\mathbf{s}, t) = \Phi_i = \phi_i(\mathbf{x}_i)e^{-i\gamma_i} \tag{47b}$$

which initial atomic function $\phi_i(\mathbf{x}_i)$ satisfies the equation

$$\left(\tfrac{1}{2}\nabla_{x_i}^2 + \frac{Z_A}{x_i} + \varepsilon_i\right)\phi_i(\mathbf{x}_i) = 0. \tag{47c}$$

The phase factor γ_i is customarily determined by the requirement that ϕ_i of Eq. 47*b* satisfies Eq. 47*a*, ignoring any effect of the Coulomb perturbing potential $-Z_p/x_f$ for large values of $\mathbf{R}(\mathbf{x}_f = \mathbf{x}_i - \mathbf{R})$. In this problem the associated boundary condition is

$$\left(\tfrac{1}{2}\nabla \mathbf{s}^2 + \frac{Z_A}{x_i} + i\frac{\partial}{\partial t}\right)\Phi_i = 0. \tag{47d}$$

Cheshire noted that the Coulomb interactions have a well-known effect at infinity, and he proceeded to revise the pair of equations, 47*a* and 47*d*, in order to remove this troublesome feature. First, the arbitrariness of an interaction (see Section 3.1) that is a function of

$$\mathbf{R} = \mathbf{r}_p - \mathbf{r}_a$$

only is invoked to modify Eq. 47*a* to

$$\begin{aligned} H_\lambda \Psi_\lambda &= i \frac{\partial}{\partial t} \Psi_\lambda, \\ H_\lambda &= H + \frac{Z_p}{R}, \end{aligned} \tag{47e}$$

but the associated initial condition is again

$$\lim_{t \to -\infty} \Psi_\lambda \to \Phi_i = \phi_i(\mathbf{x}_i)^{-i\gamma_i}. \tag{47f}$$

The perturbing potential, however, is now

$$Z_p\left(\frac{1}{x_f} - \frac{1}{R}\right)$$

instead of the original single Coulomb interaction, and the boundary equation (47*d*) can be used with confidence to determine γ_i since

$$\lim_{R \to \infty} R\left(\frac{1}{x_f} - \frac{1}{R}\right) \to 0.$$

Consequently, as γ_i is determined only up to an arbitrary additive constant, this constant can be discarded since the perturbing interaction does *not* influence γ_i. However, there is another way to attain the same goal since the interaction λ/R can be removed by a unitary transformation.* It is readily verified that the operator

$$-i \frac{\partial}{\partial t}$$

applied to the function

$$\begin{aligned} \Psi_\lambda &= \Psi \exp\left[i\lambda v^{-1} \log\left(vR - v^2 t\right)\right] \\ \Psi &= \Psi_{\lambda=0}, \qquad \lambda = Z_p \end{aligned} \tag{47g}$$

yields

$$-\lambda R^{-1} \Psi_\lambda$$

so that the transformation represented by Eq. 47*g* is also a device for removing the nucleus-nucleus interaction.

* Attention to this possibility for problems in electronic capture apparently originated from G. C. Wick, as noted in a footnote to the paper by Jackson and Schiff.[30]

In order not to have an unwanted phase factor initially, an additional condition

$$\lim_{t\to-\infty} \Psi \to \Phi_i \exp\left[-i\lambda v^{-1} \log\left(vR - v^2 t\right)\right] \tag{47h}$$

is imposed since the initial condition Φ_i is correct. Alternatively, the equivalent conditions

$$\lim_{t\to-\infty} \Psi \to \Phi_i \exp\left[-i\lambda v^{-1} \log\left(vx_f + \mathbf{v}\cdot\mathbf{x}_f\right)\right] \tag{47i}$$

can be used since

$$\lim_{t\to-\infty} \frac{vR - v^2 t}{vx_f + \mathbf{v}\cdot\mathbf{x}_f} = \lim \frac{vR - v^2 t}{v|\mathbf{x}_i - \mathbf{R}| + \mathbf{v}\cdot(\mathbf{x}_i - \mathbf{R})} = 1,$$

and this limit is correct since $\mathbf{x}_i$ is usually finite initially according to Eq. 47*b*. It is now evident that we can use Eq. 47*a* together with Eq. 47*h* since $i(\partial/\partial t)$ operating on this initial state inserts $-Z_p R^{-1}$ into Eq. 47*a*, and the resulting perturbing potentials $Z_p(x_f^{-1} - R^{-1})$ can be discarded to get Eq. 47*d*, which is again free from unwanted long-range effects. Nevertheless, this procedure suffers from the defect that the incorrect initial condition in Eq. 47*h* must be used in place of the correct condition of Eq. 47*b*. Thus both Eqs. 47*h* and 47*g* must be satisfied to fulfill Eq. 47*f*, the correct condition. In contrast (for emphasis) Eq. 47*e* is used together with the initial condition of Eq. 47*f*. This method is now specialized to protons and atomic hydrogen.

Equation 47*e*, together with Eqs. 47*c* and 47*d* ($Z_p = Z_A = 1$), is first supplemented with Eqs. 48*a* through 48*d*:

$$\left(\tfrac{1}{2}\nabla_s^2 + \frac{1}{x_f} + i\frac{\partial}{\partial t}\right)\Phi_f = 0, \tag{48a}$$

$$\left(\tfrac{1}{2}\nabla_{x_f}^2 + \frac{1}{x_f} + \varepsilon_f\right)\phi_f(\mathbf{x}_f) = 0, \tag{48b}$$

$$\Phi_i = \phi_i(\mathbf{x}_i)e^{-i\gamma_i}, \qquad \gamma_i = \mathbf{v}\cdot\mathbf{s} + \frac{v^2}{8}t + \varepsilon_i t, \tag{48c}$$

$$\Phi_f = \phi_f(\mathbf{x}_f)e^{i\gamma_f}, \qquad \gamma_f = \tfrac{1}{2}\mathbf{v}\cdot\mathbf{s} - \frac{v^2 t}{8} - \varepsilon_f t. \tag{48d}$$

It has been demonstrated many times already that γ_i and γ_f are the correct choices for the phases. Let Ψ_i and Ψ_f be solutions of Eqs. 47*e* subject to the initial conditions

$$\lim_{t\to-\infty} \Psi_i \to \Phi_i, \qquad \lim_{t\to\infty} \Psi_f \to \Phi_f; \tag{48e}$$

furthermore, assume that these solutions can be decomposed as

$$\Psi_i = \Phi_i L_i, \tag{48f}$$

$$\Psi_f = \Phi_f L_f. \tag{48g}$$

The corresponding equations and initial conditions satisfied by L_i and L_f are (Eq. 47)

$$\left(\tfrac{1}{2}\nabla_s^2 + \frac{1}{x_f} - \frac{1}{R} + i\frac{\partial}{\partial t} - \frac{i\mathbf{v}\cdot\nabla_s}{2}\right)L_i = \frac{\mathbf{x}_i}{x_i}\cdot\nabla_s L_i, \qquad \lim_{t\to-\infty} L_i \to 1 \tag{48h}$$

$$\left(\tfrac{1}{2}\nabla_s^2 + \frac{1}{x_i} - \frac{1}{R} + i\frac{\partial}{\partial t} + \frac{i\mathbf{v}\cdot\nabla_s}{2}\right)L_f = \frac{\mathbf{x}_f}{x_f}\cdot\nabla_s L_f, \qquad \lim_{t\to\infty} L_f \to 1. \tag{48i}$$

Approximate solutions to Eqs. 48*h* and 48*i* are obtained by neglecting the right-hand member of each equation.

The two approximate solutions are rewritten as

$$\begin{aligned} L_i' &= F_i \exp\left[iv^{-1}\log(vR - v^2t)\right] = F_i e^{i\beta_i}, \\ L_f' &= F_f \exp\left[-iv^{-1}\log(vR + v^2t)\right] = F_f e^{-i\beta_f}, \end{aligned} \tag{48j}$$

and these results are substituted into the approximate forms of Eqs. 48*h* and 48*i* to get

$$\left(\tfrac{1}{2}\nabla_{x_f}^2 + \frac{1}{x_f} - i\mathbf{v}\cdot\nabla_{x_f}\right)F_i = 0, \tag{48k}$$

$$\left(\tfrac{1}{2}\nabla_{x_i}^2 + \frac{1}{x_i} + i\mathbf{v}\cdot\nabla_{x_i}\right)F_f = 0. \tag{48m}$$

In these equations use was made of the transformations in Eq. 47. These two equations are common to scattering theory, and the solutions for L_i' and L_f' are[95]

$$L_i' = e^{\pi/2v}\Gamma\left(1 - \frac{i}{v}\right){}_1F_1\left[\frac{i}{v}, 1; i(vx_f + \mathbf{v}\cdot\mathbf{x}_f)\right]e^{i\beta_i}, \tag{48n}$$

$$L_f' = e^{\pi/2v}\Gamma\left(1 + \frac{i}{v}\right){}_1F_1\left[-\frac{i}{v}, 1; -i(vx_i + \mathbf{v}\cdot\mathbf{x}_i)\right]e^{-i\beta_f}. \tag{48o}$$

The positive signs are used in the arguments of the two confluent hypergeometric functions because the initial conditions in Eqs. 48*h* and 48*i* must be satisfied. With the positive signs in the arguments each argument tends to infinity corresponding to the initial condition, and the asymptotic form of F_i and F_f is $e^{-i\beta_i}$ and $e^{i\beta_f}$, respectively.

At this point distorted wave functions are introduced as solutions to the equations

$$\left(\tfrac{1}{2}\nabla_x{}^2 + \frac{1}{x_i} + i\frac{\partial}{\partial t} + U_i\right)\chi_i = 0, \tag{48p}$$

$$\left(\tfrac{1}{2}\nabla_x{}^2 + \frac{1}{x_f} + i\frac{\partial}{\partial t} + U_f\right)\chi_f = 0 \tag{48q}$$

with the initial conditions

$$\lim_{t\to-\infty} \chi_i \to \Phi_i, \qquad \lim_{t\to\infty} \chi_f \to \Phi_f. \tag{48r}$$

The functions χ_i and χ_f are defined by the equation

$$\chi_i = L_i'\Phi_i, \qquad \chi_f = L_f'\Phi_f, \tag{48s}$$

and with the use of Eqs. 48*h* and 48*i*, right-hand members equaling zero, we can easily find that

$$U_i\chi_i = \left(\frac{1}{x_f} - \frac{1}{R}\right)\chi_i + \Phi_i\frac{\mathbf{x}_i}{x_i}\cdot\nabla_s L_i', \tag{48t}$$

$$U_f\chi_f = \left(\frac{1}{x_i} - \frac{1}{R}\right)\chi_f + \Phi_f\frac{\mathbf{x}_f}{x_f}\cdot\nabla_s L_f'. \tag{48u}$$

The last step of the problem is to determine the cross sections for electronic capture. The amplitude for electronic capture, as we now know, is given by

$$b_{if} = \lim_{t\to\infty}\int d\mathbf{s}\Phi_f^*\Psi_i,$$

and according to Eq. 48*r* the equivalent expression is

$$b_{if} = \lim_{t\to\infty}\int d\mathbf{s}\chi_f^*\Psi_i. \tag{49a}$$

This equation for b_{if} is transformed into another form,

$$b_{if} = \int_{-\infty}^{\infty} dt \int d\mathbf{s}\frac{\partial}{\partial t}(\chi_f^*\Psi_i) + \lim_{t\to-\infty}\int d\mathbf{s}\chi_f^*\Psi_i. \tag{49b}$$

If the integrated term of Eq. 49*b* vanishes, Eq. 48*q* for χ_f and Eq. 47*e* for Ψ_i can be used to transform Eq. 49*b* into

$$b_{if} = i\int_{-\infty}^{\infty} dt \int d\mathbf{s}\left[\chi_f^*\left(\frac{1}{x_i} - \frac{1}{R} - U_f\right)\Psi_i + \tfrac{1}{2}\nabla_s\cdot(\chi_f^*\nabla_s\Psi_i - \Psi_i\nabla_s\chi_f^*)\right]. \tag{49c}$$

With the use of the second form of Green's theorem, the integral of the divergence is transformed into a surface integral on a large sphere in **s**-space, and it is physically plausible to expect this integral to vanish as $s \to \infty$; consequently, this integral is conjectured to vanish. (An expansion of the functions of Eq. 49c in terms of the eigenfunctions of atomic hydrogen is instructive. Contributions to the surface integral from the discrete part of the expansion vanish, and since the contribution from the continuum contains an integral over the associated spectral variables, we can use partial integration to show that this part also vanishes.) The final expression for the amplitude b_{if} thus is given by

$$b_{if} = i\int_{-\infty}^{\infty} dt \int d\mathbf{s}\chi_f^*\left(\frac{1}{x_i} - \frac{1}{R} - U_f\right)\Psi_i. \tag{49d}$$

If the time-reversed reaction is used (see Section 2.5), the associated amplitude is

$$a_{if} = \lim_{t\to-\infty} \int d\mathbf{s}\Psi_f^*\Phi_i = \lim_{t\to-\infty} \int d\mathbf{s}\Psi_f^*\chi_i. \tag{49e}$$

With the use of Eqs. 48p and 47e we proceed as in the last example to establish that

$$a_{if} = i\int_{-\infty}^{\infty} dt \int d\mathbf{s}\Psi_f^*\left(\frac{1}{x_f} - \frac{1}{R} - U_i\right)\chi_i, \tag{49f}$$

provided that the associated surface integrals and

$$\lim_{t\to\infty} \int d\mathbf{s}\Psi_f^*\chi_i$$

both vanish. The cross section for electronic capture (atomic units) is

$$Q_{if} = 2\pi \int_0^{\infty} d\rho\rho|a_{if}|^2 = 2\pi \int_0^{\infty} d\rho\rho|b_{if}|^2. \tag{49g}$$

In the actual calculation, Cheshire used a_{if}, and with the relation in Eq. 48t transformed Eq. 49f into

$$a_{if} = -i\int_{-\infty}^{\infty} dt \int d\mathbf{s}\Psi_f^*\Phi_i \frac{\mathbf{x}_i}{x_i} \cdot \nabla_s L_i'. \tag{49h}$$

The additional approximation $\Psi_f^* = \chi_f^*$ was used in illustrative calculations, and the correct initial condition is maintained in this approximation according to Eqs. 48s and 48r.

The approximation used to get the equation satisfied by L_i', however, does not contain coupling to the initial atomic state. This is evident since the term discarded in Eq. 48h originates from the term

$$e^{-i\gamma_i}\nabla_s L_i \cdot \nabla_s \phi_i(\mathbf{x}_i),$$

which is part of the coupling terms; the other part involving the translational motion of the electron (see Eq. 48*f*) is canceled. Note also that the similar approximate equation for L_f does not contain its associated coupling,

$$e^{i\gamma_f}\nabla_s L_f \cdot \nabla_s \phi_f(\mathbf{x}_f).$$

As Cheshire remarks, these omissions affect the accuracy of the approximation at low impact velocities. Another noteworthy point is the agreement of these solutions at high impact velocities with one form of the second Born approximation—a detailed discussion which is reserved for the sections on the second Born approximation and the asymptotic variations of amplitudes at high impact velocities. It is pertinent here to direct attention to another second-order approximation proposed by Cheshire.[93] Equations 37*a* and 37*b* of his paper are clearly derived and require no additional elucidation; however, since illustrative calculations had not been effected using this method (according to this author's knowledge) the method is not presented. We turn instead to a study of the second Born approximation.

REFERENCES

1. D. R. Bates, *Proc. Roy. Soc. (London)A*, **247**, 294 (1958).
2. K. Takayanagi, *Sci. Rept. Saitama Univ., Ser. A*, **2**, 33 (1955).
3. N. C. Sil, *Proc. Phys. Soc. (London)*, **75**, 194 (1960).
4. M. H. Mittleman, *Phys. Rev.*, **122**, 499 (1961).
5. D. R. Bates and R. McCarroll, *Phil. Mag. Suppl.*, **11**, 39 (1962).
6. D. R. Bates, "Theoretical Treatment of Collisions between Atomic Systems," in *Atomic and Molecular Processes* (D. R. Bates, ed.), Academic Press, New York (1962), pp. 585–588.
7. B. H. Bransden, "Atomic Rearrangement Collisions," in *Advances in Atomic and Molecular Physics*, Vol. 1, (D. R. Bates and I. Estermann, eds.), Academic Press, New York (1965), pp. 98–102.
8. D. R. Bates, *Quantum Theory*, Academic Press, New York (1961), p. 253.
9. N. F. Mott and H. S. W. Massey, *The Theory of Atomic Collisions* (3rd ed.), Oxford University Press, London (1965), p. 432.
10. *Ibid.*, Chapter 4.
11. J. W. Frame, *Proc. Cambridge Phil. Soc.*, **27**, 511 (1931).
12. A. M. Arthurs, *Proc. Cambridge Phil. Soc.*, **57**, 904 (1961).
13. R. McCarroll, *Proc. Roy. Soc. (London)A*, **264**, 547 (1961).
14. R. H. Bassel and E. Gerjuoy, *Phys. Rev.*, **117**, 749 (1960).
15. D. F. Gallaher and L. Wilets, *Phys. Rev.*, **169**, 139 (1968).
16. S. E. Lovell and M. B. McElroy, *Proc. Roy. Soc. (London)A*, **283**, 100 (1965).
17. M. L. Goldberger and K. M. Watson, *Collision Theory*, Wiley, New York (1964), Chapter 5.
18. Ta-You Wu and T. Ohmura, *Quantum Theory of Scattering*, Prentice-Hall, Englewood Cliffs, N.J. (1962), pp. 26–36.

19. A. Messiah, *Quantum Mechanics*, Vol. 2, (J. Potter, trans.) (1962); reprint ed., North-Holland, Amsterdam, and Interscience, New York (1963), Chapter 19.
20. E. Gerjuoy, *Ann. Phys.* (*N. Y.*), **5**, 58 (1958).
21. Mott and Massey, *op. cit.*, pp. 79–85.
22. P. M. Morse and H. Feshbach, "Solutions of the Schröedinger Equations," in *Methods of Theoretical Physics*, McGraw-Hill, New York (1953), Part II, Section 12.3.
23. B. Z. Vulikh, *Introduction to Functional Analysis for Scientists and Technologists*, Addison-Wesley, Reading, Mass. (1963).
24. A. Sommerfeld, *Partial Differential Equations in Physics* (E. G. Straus, trans.), Academic Press, New York (1949), pp. 227–235.
25. Messiah, *op. cit.*, Vol. 2, pp. 806–807; 836–839.
26. Mott and Massey, *op. cit.*, pp. 424–428.
27. J. Grant and J. Shapiro, *Proc. Phys. Soc.* (*London*), **86**, 1007 (1965).
28. Bates, *Atomic and Molecular Processes*, pp. 570–578.
29. R. Courant and D. Hilbert, *Methods of Mathematical Physics*, Vol. 1, trans. ed., Interscience, New York (1953), pp. 140–142.
30. J. D. Jackson and H. Schiff, *Phys. Rev.*, **89**, 359 (1953).
31. Bates, *Atomic and Molecular Processes*, p. 574.
32. E. U. Condon and G. H. Shortley, *Theory of Atomic Spectra* (1935); reprint corr. ed., Cambridge University Press, New York (1953), Chapter 7.
33. D. R. Bates and A. Dalgarno, *Proc. Phys. Soc.* (*London*)*A*, **66**, 972 (1953).
34. J. R. Oppenheimer, *Phys. Rev.*, **31**, 349 (1928).
35. H. C. Brinkman and H. A. Kramers, *Proc. Acad. Sci. Amsterdam*, **33**, 973 (1930).
36. Bates, *Atomic and Molecular Processes*, p. 573.
37. M. N. Saha and D. Basu, *Indian J. Phys.*, **19**, 121 (1945).
38. B. H. Bransden, A. Dalgarno, and N. M. King, *Proc. Phys. Soc.* (*London*)*A*, **67**, 1075 (1954).
39. H. S. W. Massey and E. H. S. Burhop, *Electronic and Ionic Impact Phenomena*, Oxford University Press, London (1952), pp. 427–432.
40. Condon and Shortley, *op. cit.*, Chapter 2.
41. R. A. Mapleton, *Phys. Rev.*, **130**, 1829 (1963).
42. C. C. Lin and R. G. Fowler, *Ann. Phys.* (*N. Y.*), **15**, 461 (1969).
43. C. C. Lin and R. M. St. John, *Phys. Rev.*, **128**, 1749 (1962).
44. R. A. Mapleton, *Proc. Phys. Soc.* (*London*), **85**, 1109 (1965).
45. C. C. J. Roothaan and P. S. Kelly, *Phys. Rev.*, **131**, 1177 (1963).
46. G. Racah, *Phys. Rev.*, **62**, 438 (1942).
47. G. Racah, *Phys. Rev.*, **63**, 367 (1943).
48. A. de-Shalit and I. Talmi, *Nuclear Shell Theory*, Academic Press, New York (1963).
49. G. Herzberg, *Molecular Spectra and Molecular Structure:* I. *Spectra of Diatomic Molecules*, Van Nostrand, Princeton, N.J. (1951), p. 21.
50. Condon and Shortley, *op. cit.*, p. 195.
51. R. A. Mapleton, "Classical and Quantal Calculations on Electron Capture," Air Force Cambridge Research Laboratories Report AFCRL 67–0351, Physical Science Research Papers, No. 328, 1967, App. D, Part I.
52. Condon and Shortley, *op. cit.*, Table 23, p. 76.
53. R. M. May, *Phys. Rev.*, **136**, A669 (1964).
54. V. Fock, *Z. Phys.*, **98**, 145 (1936).
55. Bates, *Atomic and Molecular Processes*, p. 578.
56. R. A. Mapleton, *Phys. Rev.*, **126**, 1477 (1962).

57. T. D. Gaily, *Phys. Rev.*, **178**, 207 (1969).
58. J. R. Hiskes, *Phys. Rev.*, **180**, 146 (1969).
59. R. A. Mapleton, AFCRL 67-0351, App. E, Part. I.
60. R. A. Mapleton, *J. Phys. B*, **1**, 850 (1968).
61. H. Goldstein, *Classical Mechanics*, Addison-Wesley, Reading, Mass. (1950), pp. 107–109.
62. R. P. Feymann, *Phys. Rev.*, **76**, 769 (1949).
63. E. Corinaldesi and L. Trainor, *Nuovo Cimento*, **9**, 940 (1952).
64. Mott and Massey, *op. cit.*, p. 427.
65. G. F. Chew and G. C. Wick, *Phys. Rev.*, **85**, 636 (1952).
66. G. F. Chew and M. L. Goldberger, *Phys. Rev.*, **87**, 778 (1952).
67. T. Pradhan, *Phys. Rev.*, **105**, 1250 (1957).
68. M. Gell-Mann and M. L. Goldberger, *Phys. Rev.*, **91**, 398 (1953).
69. Wu and Ohmura, *op. cit.*, pp. 280–285.
70. R. A. Mapleton, *Proc. Phys. Soc.* (*London*), **83**, 35 (1964).
71. R. A. Mapleton, *J. Math. Phys.*, **2**, 478 (1961).
72. R. A. Mapleton, *J. Math. Phys.*, **2**, 482 (1961).
73. R. A. Mapleton, *J. Math. Phys.*, **3**, 297 (1962).
74. Mott and Massey, *op. cit.*, Chapter III.
75. S. Okubo and D. Feldman, *Phys., Rev.*, **117**, 292 (1960).
76. J. D. Dollard, *J. Math. Phys.*, **5**, 729 (1964).
77. E. A. Wichmann and C. H. Woo, *J. Math. Phys.*, **2**, 178 (1961).
78. L. Hostler and R. H. Pratt, *Phys. Rev. Letters*, **10**, 469 (1963).
79. K. Mano, *J. Math. Phys.*, **5**, 505 (1964).
80. L. Hostler, *J. Math. Phys.*, **5**, 591 (1964).
81. L. Hostler, *J. Math. Phys.*, **5**, 1235 (1964).
82. J. Schwinger, *J. Math. Phys.*, **5**, 1606 (1964).
83. G. B. West, *J. Math. Phys.*, **8**, 942 (1967).
84. R. Finklestein and D. Levy, *J. Math. Phys.*, **8**, 2147 (1967).
85. G. L. Nutt, *J. Math. Phys.*, **9**, 796 (1968).
86. W. F. Ford, *Phys. Rev.*, **133**, B1616 (1964).
87. W. F. Ford, *J. Math. Phys.*, **7**, 626 (1966).
88. I. M. Cheshire, *Proc. Phys. Soc.* (*London*), **82**, 113 (1963).
89. Bransden, *op. cit.*, pp. 115–119.
90. B. H. Bransden and I. M. Cheshire, *Proc. Phys. Soc.* (*London*), **81**, 820 (1963).
91. H. Schiff, *Canadian J. Phys.*, **32**, 393 (1954).
92. J. P. Coleman and S. Trelease, *J. Phys. B.*, **1**, 172 (1968).
93. I. M. Cheshire, *Phys. Rev.*, **138**, A992 (1965).
94. I. M. Cheshire, *Proc. Phys. Soc.* (*London*), **84**, 89 (1964).
95. Mott and Massey, *op. cit.*, pp. 55–57.

CHAPTER 4

SECOND BORN APPROXIMATION

4.1 INTRODUCTION

In recent years attention has been directed anew to the second Born approximation partly by reason of interest in the asymptotic variation of the associated cross section at high impact velocities. Other interest, however, originates from comparisons of cross sections calculated using the first Born approximation with different interactions, the impulse approximation, continuum distorted wave approximation, and approximations to the second Born approximation. Further interest stems from similarities of certain classical cross sections to the cross sections derived from the second Born approximation. Moreover, there are two forms of the second Born approximation for electronic capture from one charged nucleus by another charged nucleus, and a criterion is needed to determine which amplitude should be used, for there are apparent mathematical differences between these two amplitudes, particularly at high impact velocities. Furthermore, there may be more fundamental mathematical problems involved.

According to an analysis by Aaron, Amado, and Lee the Born series for rearrangement collisions may diverge.[1] The conclusion is based on the study of two particles bound to a fixed center of force and depends upon the sign of the Fourier components of the attractive interaction not changing sign in a specified domain of the associated variables. In possible disagreement with this conjecture, however, is the result of another study.[2] In this study the Born series for a one-dimensional model of a three-body collision is investigated. It is concluded from the study of this counterexample that a Born series for rearrangement collisions does not necessarily diverge for all impact energies if the Fourier components of the attractive interaction do not change sign. It is thus clear why the interest in the second Born approximation continues. In the present chapter, however, the goal is descriptive, explanatory, and calculational, and no attempt is made to support or reject the use of the Born series to represent the process of electronic capture. The approximation of interest is now defined and described.

4.2 THE TWO EXPANSIONS AND CONTRIBUTIONS FROM THE DISCRETE INTERMEDIATE STATES

The processes studied consist of proton 2 incident upon an atom comprising an electron 3 bound in the $1s$-state to a singly charged nucleus 1. Two processes are treated that together represent symmetrical (exact) resonance and accidental resonance (see Section 1.4.). In the center-of-mass system the energy is denoted here by E; the kinetic-energy operators are represented by T; the separate interactions are designated U_{if}, and the other notation is now standard to this book.[3,4]

The Hamiltonian of the system is decomposed into three forms,

$$H = H_i + V_i = H_f + V_f = H_0 + V,$$
$$H_i = T_i + U_{13}, \qquad H_f = T_f + U_{23}, \qquad H_0 = T_0,$$
$$V = U_{12} + U_{13} + U_{23}, \tag{50a}$$

and the corresponding Green's functions needed are

$$G = (E + i\varepsilon - H)^{-1}, \qquad G_f = (E + i\varepsilon - H_f), \qquad G_0 = (E + i\varepsilon - H_0). \tag{50b}$$

The formal solution in terms of the initial state Φ_i and the resolvent Green's function G is[5,6]

$$\Psi = (1 + GV_i)\Phi_i, \qquad (E - H_i)\Phi_i = 0, \tag{50c}$$

and it is assumed that G can be expanded in the two series

$$G = G_0 + G_0 V G_0 + \cdots,$$
$$G = G_f + G_f V_f G_f + \cdots. \tag{50d}$$

Only the first term of each series is retained and substituted for G in Eq. 50c to obtain the sum of the amplitudes corresponding to the first and second Born approximations; these amplitudes are

$$f_0 = \frac{\mu}{2\pi\hbar^2}(\Phi_f, V_f(1 + G_0 V_i)\Phi_i),$$

$$f = \frac{\mu}{2\pi\hbar^2}(\Phi_f, V_f(1 + G_f V_i)\Phi_i), \tag{50e}$$

$$\mu = \mu_f = \frac{M_i M_{f1}}{M_t}, \qquad M_f = M, \qquad M_i = M, 2M.$$

In Eq. 50*e* the mass of the atomic nucleus M_i ($= M_a$) can have one of two values corresponding to exact $M_i = M$ or accidental $M_i = 2M$ resonance. (The atomic target, deuterium, is selected to represent the process of accidental resonance, and the proton-neutron binding energy is assumed infinite in this application.) The second Born amplitude using G_f will be described in detail, and at the appropriate place the second Born amplitude employing G_0 will be introduced; this is convenient since there are no discrete states in the spectrum of G_0 as there are in the spectrum of G_f. We now construct G_f and the associated second Born amplitude.

The relation between E, the impact energy E_i, and conservation of energy is given in Eqs. 9*b*, 10*a*, and 10*b*, and an additional notation is presented in Eq. 50*f*:

$$\epsilon_i = b\epsilon_1, \qquad \epsilon_f = a\epsilon_1, \qquad \epsilon_1 = \frac{me^4}{2\hbar^2}, \qquad b = \frac{2M}{M + M_1}, \qquad a = \frac{M}{M_1},$$

$$\mu_1 = am, \qquad \mu_2 = mb, \qquad \Delta\epsilon = (b - a)\epsilon_1, \qquad \frac{2\mu}{\hbar^2}\,\epsilon_n = \frac{\mu\mu_1}{n^2 m^2 a_0},$$

$$K_i^2(\text{a.u.}) = K_i^2 = \frac{M_{i1}\mu_i E_i}{mM_t\epsilon_1}, \tag{50f}$$

$$K_f^2(\text{a.u.}) = K_f^2 = \frac{\mu M_{i1} E_i}{mM_t\epsilon_1} - \frac{\mu}{m}\,\Delta\epsilon,$$

$$\mu_i = \frac{M_{i1}M_f}{M_t}, \qquad M_t = M_{i1} + M_f = M_i + M_{f1}.$$

At this point dimensionless units are introduced; these agree with atomic units except that the unit of energy is ϵ_1. Next, the final-state Green's function G_f is decomposed into the two parts associated with the discrete and continuous spectrum of atomic hydrogen (see Eq. 46*j*, Section 3.6):

$$G_f = G_{fd} + G_{fc},$$

$$G_{fd} = \left(\frac{1}{2\pi}\right)^3 \sum_{n=1}^{\infty} \phi_n(\mathbf{x}_f)\phi_n^*(\mathbf{x}_f') \int \frac{d\mathbf{U}\exp[i\mathbf{U}\cdot(\mathbf{R}_f - \mathbf{R}_f')]}{D_d}$$

$$D_d = K_f^2 + i\delta - \frac{\mu a}{m}\left(1 - \frac{1}{n^2}\right) - U^2, \tag{50g}$$

$$G_{fc} = \left(\frac{1}{2\pi}\right)^6 \sum_{l=0}^{\infty} \sum_{m=-l}^{l} \int \frac{d\mathbf{U}\exp[i\mathbf{U}\cdot(\mathbf{R}_f - \mathbf{R}_f')]}{D_c}\, d\mathbf{k}\phi_{klm}(\mathbf{x}_f)\phi_{klm}^*(\mathbf{x}_f'),$$

$$D_c = K_f^2 + i\delta - \frac{\mu a}{m} - \frac{\mu}{\mu_1}k^2 - U^2.$$

The functions ϕ_n of atomic hydrogen are normalized to the Kronecker delta function, $(\phi_m, \phi_n) = \delta_{mn}$, whereas the corresponding functions of the continuum are normalized to the three-dimensional delta function,

$$\int d\mathbf{x}_f \phi_{k'}(\mathbf{x}_f)^* \phi_k(\mathbf{x}_f) = \sum_{l=0}^{\infty} \sum_{m=-l}^{l} \sum_{p=0}^{\infty} \sum_{q=-p}^{p} \int d\mathbf{x}_f \phi_{k'pq}(\mathbf{x}_f)^* \phi_{klm}(\mathbf{x}_f) = \delta(\mathbf{k} - \mathbf{k}').$$

The result for G_{fc} is readily simplified by replacing the corresponding sums in Eq. 50*g* by the equivalent sums,[7]

$$\begin{aligned}\int d\Omega_k \sum_{l=0}^{\infty} \sum_{m=-l}^{l} & i^l(2l+1)L_l(kx_f)N_{lm}P_l^{|m|}(\cos\theta_{x_f})P_l^{|m|}(\cos\theta_k) \\ & \times \exp[im(\phi_{x_f} - \phi_k)] \sum_{p=0}^{\infty} \sum_{q=-p}^{p} i^{-p}(2p+1)L_p(kx_f')^* N_{pq} \\ & \times P_p^{|q|}(\cos\phi_k)P_p^{|q|}(\cos\theta_{x_f'}) \exp[iq(\phi_k - \phi_{x_f'})] \\ = \frac{\alpha}{4\pi^2(1 - e^{-2\pi\alpha})} & \exp[i\mathbf{k}\cdot(\mathbf{x}_f - \mathbf{x}_f')]\,{}_1F_1(i\alpha, 1; i[kx_f - \mathbf{k}\cdot\mathbf{x}_f]) \\ & \times {}_1F_1(-i\alpha, 1; -i[kx_f' - \mathbf{k}\cdot\mathbf{x}_f']);\end{aligned}$$

$$\int d\Omega_k N_{lm} P_l^{|m|}(\cos\theta_k) P_p^{|q|}(\cos\theta_k) \exp[i(q-m)\phi_k] = \frac{4\pi}{2l+1}\delta_{lp}\,\delta_{mq}, \qquad (50h)$$

$$N_{lm} = \frac{(l - |m|)!}{(l + |m|)!},$$

$$L_l(kx) = \frac{\Gamma(l + 1 - i\alpha)}{(2l+1)!}(2kx)^l e^{ikx}\,{}_1F_1(l + 1 - i\alpha, 2l + 2; -i2kx),$$

$$\alpha = k^{-1},$$

$$\begin{aligned}\phi_{klm}(\mathbf{x}_f) = \frac{1}{2\pi\Gamma(1 - i\alpha)}\left(\frac{\alpha}{1 - e^{-2\pi\alpha}}\right)^{1/2} & i^{-l}(2l+1)L_l(kx_f)N_{lm} \\ & \times P_l^{|m|}(\cos\theta_{x_f})P_l^{|m|}(\cos\theta_k)\exp[im(\phi_{x_f} - \phi_k)].\end{aligned}$$

Now G_{fc} is represented compactly by

$$\begin{aligned}G_{fc} = \left(\frac{1}{2\pi}\right)^5 \int \frac{\alpha}{1 - e^{-2\pi\alpha}} & \frac{d\mathbf{k}\, d\mathbf{U} \exp[i\mathbf{U}\cdot(\mathbf{R}_f - \mathbf{R}_f')]}{D_c} \exp[i\mathbf{k}\cdot(\mathbf{x}_f - \mathbf{x}_f')] \\ & \times {}_1F_1(i\alpha, 1; i[kx_f - \mathbf{k}\cdot\mathbf{x}_f])\,{}_1F_1(-i\alpha, 1; -i[kx_f' - \mathbf{k}\cdot\mathbf{x}_f']). \qquad (50i)\end{aligned}$$

The amplitudes originating from the constituent parts of the Green's functions are represented by

$$f = \frac{\mu a_0}{2\pi m}[f^{(1)} + f_i^{(2)}], \qquad i = d, c, \qquad (50j)$$

and contributions to f_d are investigated first.

The independent variables $\mathbf{x}_i$, $\mathbf{x}_f$ are used in the actual calculation, and we assume that this change of variables has been accomplished:

$$\frac{\mu a_0}{2\pi m} f^{(1)} = \frac{\mu a_0}{2\pi m} \int \Phi_f^* V_f \Phi_i \, d\mathbf{x}_i \, d\mathbf{x}_f,$$

$$\Phi_i = \frac{\exp(-x_i)}{\sqrt{\pi}} \exp\{i[\mathbf{K}_i \cdot (b\mathbf{x}_i - \mathbf{x}_f)]\};$$

$$\Phi_f^* = \frac{\exp(-x_f)}{\sqrt{\pi}} \exp\{-i[\mathbf{K}_f \cdot (\mathbf{x}_i - a\mathbf{x}_f)]\},$$

$$\begin{aligned} V_i &= (U_{12} + U_{23}) = |\mathbf{x}_i - \mathbf{x}_f|^{-1} - x_f^{-1}; \qquad \mathbf{R}_f = \mathbf{x}_i - a\mathbf{x}_f, \\ V_f &= U_{12} + U_{13} = |\mathbf{x}_i - \mathbf{x}_f|^{-1} - x_i^{-1}; \qquad \mathbf{R}_f' = \mathbf{x}_i' - a\mathbf{x}_f'. \end{aligned} \tag{51a}$$

$$f^{(2)} = \frac{2\mu}{(2\pi)^3 m} \sum_{n=1}^{\infty} \int d\mathbf{x}_i \, d\mathbf{x}_f \Phi_f^*(\mathbf{x}_i, \mathbf{x}_f)(|\mathbf{x}_i - \mathbf{x}_f|^{-1} - x_i^{-1})\phi_n(\mathbf{x}_f)$$

$$\times \exp(i\mathbf{U}\cdot\mathbf{R}_f) \frac{d\mathbf{U} \exp(-i\mathbf{U}\cdot\mathbf{R}_f') \, d\mathbf{x}_i' \, d\mathbf{x}_f' \phi_n^*(\mathbf{x}_f')(|\mathbf{x}_i' - \mathbf{x}_f'|^{-1} - x_f'^{-1})\Phi_i(\mathbf{x}_i', \mathbf{x}_f')}{D_d}.$$

A word of explanation about the origin of the factor of $2\mu/m$ is in order. Before the change to dimensionless variables was effected there was the additional factor $2\mu e^2\hbar^{-2}$ associated with the second Born amplitude. The change to dimensionless variables introduces another factor a_0, and the resultant final factor is $2\mu/m$, as asserted. The contribution from $n = 1$ is evaluated first, and two terms of D_d cancel according to Eq. 50g:

$$\begin{aligned} F(12 - 13; 12) &= \frac{512\mu}{m\pi^3} \int \frac{d\mathbf{U}}{(K_f^2 + i\delta - U^2)|\mathbf{U} - \mathbf{K}_f|^2} \\ &\times \left[\frac{1}{(4 + (1-a)^2|\mathbf{U} - \mathbf{K}_f|^2)^2} - \frac{1}{(4 + a^2|\mathbf{U} - \mathbf{K}_f|^2)^2}\right] \\ &\times \frac{d\mathbf{V}}{(\mathbf{V}^2 + \varepsilon^2)(1 + |\mathbf{V} + a\mathbf{U} - \mathbf{K}_i|^2)^2(1 + |\mathbf{V} + \mathbf{U} - b\mathbf{K}_i|^2)^2}, \end{aligned} \tag{51b}$$

$$\begin{aligned} G(13 - 12; 23) &= \frac{512\mu}{m\pi} \int \frac{d\mathbf{U}}{(K_f^2 + i\delta - U^2)|\mathbf{U} - \mathbf{K}_f|^2} \\ &\times \left[\frac{1}{(4 + a^2|\mathbf{U} - \mathbf{K}_f|^2)^2} - \frac{1}{(4 + (1-a)^2|\mathbf{U} - \mathbf{K}_f|^2)^2}\right] \\ &\times \frac{1}{(1 + |a\mathbf{U} - \mathbf{K}_i|^2)(1 + |\mathbf{U} - b\mathbf{K}_i|^2)^2}. \end{aligned} \tag{51c}$$

The symbols (12 − 13; 12) denotes the contribution from $V_f U_{12}$ in accordance with the notation of Eq. 51*a*; similarly, (13 − 12; 23) signifies the product $V_f U_{23}$. These integrals are difficult to evaluate without the aid of simplifying approximations, but before turning to this effort we must emphasize that the integrals cannot be simplified by equating a and b to unity; the values of these integrals are very sensitive to these small alterations since the associated amplitudes usually are sharply peaked in the forward direction. (For more details on this subject see the discussions following Fig. 3 of Section 1.2 and Eq. 43*c* of Section 3.3.)

Since the first two terms of the Born series can scarcely be expected to represent the scattering solution accurately at low impact energies, it will be assumed that the energy is large enough so that a form of the Riemann-Lebesgue theorem can be used to obtain a good approximation to the integrals.[8] The three-dimensional form of this theorem can be conveniently represented by the frequently used formula

$$\lim_{\mathbf{A}\to\infty} \int \frac{d\mathbf{U} f(\mathbf{U})}{(\lambda^2 + |\mathbf{U} - \mathbf{A}|^2)^2} = \frac{\pi^2}{\lambda} \int \delta(\mathbf{U} - \mathbf{A}) f(\mathbf{U})\, d\mathbf{U} = \frac{\pi^2}{\lambda} f(\mathbf{A}), \tag{51d}$$

if $f(\mathbf{U})$ satisfies the restrictions of the theorem. [See reference 8 for the details of these limitations; as a practical rule, Eq. 51*d* is safe to apply if $f(\mathbf{U})$ does not vary strongly near $\mathbf{U} = \mathbf{A}$, is not strongly peaked at $\mathbf{U} = \mathbf{A}$, and if $f(\mathbf{U})$ is absolutely integrable.] Without recourse to this (or other) simplifying approximations, the integrated expressions are so complicated that the pattern of variation with angle and impact energy is extremely difficult to analyze. The present approximate course is thus selected, and Eq. 51*d* is applied to Eq. 51*c*.

According to Eq. 50*f*, K_i^2 and K_f^2 both are proportional to E_i since $\Delta\epsilon$ either vanishes or is relatively small, and it is consequently clear that pronounced peaks occur at

$$U = \mathbf{K}_f, \qquad b\mathbf{K}_i.$$

The value of G at the second peak, designated G_2 is

$$G_2(b\mathbf{K}_i) = \frac{512\pi\mu}{mA_i^2(K_f^2 - b^2K_i^2)[1 + (1 - ab)^2K_i^2]} \times ([4 + a^2A_i^2]^{-2} - [4 + (1 - a^2)A_i^2]^{-2}), \tag{51e}$$

$$\mathbf{A}_i = b\mathbf{K}_i - \mathbf{K}_f, \qquad \mathbf{A}_f = a\mathbf{K}_f - \mathbf{K}_i.$$

The evaluation at the other peak requires caution since the factor

$$|\mathbf{U} - \mathbf{K}_f|^{-2}$$

diverges at $\mathbf{U} = \mathbf{K}_f$. Moreover, if the last two factors of the integrand are removed, it is readily proven that the remaining two integrals diverge. The situation of exact resonance is additionally complicated by the fact that $a = b$ and $\mathbf{A}_i(\pi) = -\mathbf{A}_f(\pi)$ since $\mathbf{K}_f(\pi) = -\mathbf{K}_i$; therefore we might anticipate contributions from the angular region near $\theta = \pi$. We shall return to this consideration after the integrals of Eq. 51c are evaluated.[9] (Dettman and Leibfried[9] have obtained similar results using their linear model for rearrangement collisions.) It is conjectured that a good approximation to the remaining integrals is obtained by replacing the last two factors of the integrand by their values at $\mathbf{U} = \mathbf{K}_f$. This is done, and for purposes of consistency (as explained later) the integrals are made convergent by replacing K_f^2 with K^2 in D_d, and the limiting value of the difference of the two integrals for $K = K_f$ is effected after inserting the limits of integration in the following single integrals:

$$I(\Gamma_i) = \frac{\pi}{2K_f}\int_0^1 dx\,x\left[\frac{-i}{(\beta + S_i)^2} + \frac{i}{(4K_f^2 + S_i)^2} - \frac{K_f}{S_i^{3/2}(4K_f^2 + S_i)} - \frac{2K_f}{S_i^{1/2}(4K_f^2 + S_i)^2}\right] \qquad (51g)$$

$$\beta = (K - K_f)^2, \qquad S_i = \Gamma_i^2 x, \qquad \Gamma_1^2 = 4a^{-2}, \qquad \Gamma_2^2 = 4(1 - a)^{-2}.$$

If $\beta = 0$, the first term of the integrand yields an integral that diverges logarithmically; nevertheless, the difference of the two relevant integrals is finite and is equal to

$$\lim_{\beta\to 0}\frac{512\mu}{m}(1 + A_f^2)^{-1}(1 + A_i^2)^{-2}\left[\left(\frac{\Gamma_1}{2}\right)^4 I(\Gamma_1) - \left(\frac{\Gamma_2}{2}\right)^4 I(\Gamma_2)\right]$$

$$= \frac{512\mu\pi}{16K_f m}(1 + A_f^2)^{-1}(1 + A_i^2)^{-2}\left(\frac{i}{2}\left\{\log\left(\frac{a}{1 - a}\right)^2 + \log\frac{1 + (K_f a)^{-2}}{1 + [K_f(1 - a)]^{-2}} + \frac{K_f^2[a^2 - (1 - a)^2]}{[1 + (K_f a)^2][1 + K_f^2(1 - a)^2]}\right\}\right.$$

$$+ \tan^{-1}[K_f(1 - a)]^{-1} - \tan^{-1}[K_f a]^{-1}$$

$$\left. + \frac{K_f(2a - 1)[1 - a(1 - a)K_f^2]}{2[1 + a^2K_f^2][1 + K_f^2(1 - a)^2]}\right)$$

$$= \frac{512\mu\pi}{16K_f m}[1 + A_f^2]^{-1}[1 + A_i^2]^{-2} I = G_1(\mathbf{K}_f). \qquad (51h)$$

If the divergent parts of the integrals $\beta \to 0$ are subtracted before the limits of integration are inserted, a different result is obtained since

$$\lim_{\beta \to 0} \left[\left(\frac{\Gamma_1}{2}\right)^4 (\beta + S_1)^{-2} - \left(\frac{\Gamma_2}{2}\right)^4 (\beta + S_2)^{-2} \right] = 0, \tag{51i}$$

and the term $\log (a/1 - a)^2$ would not occur in Eq. 51h. Consequently, the procedure used insures that the result for nonresonance, $\beta \neq 0$, goes continuously into the case for exact and accidental resonance in the limit $\beta \to 0$. The approximate value of Eq. 51c is given by

$$G(13 - 12; 23) \approx G_1(\mathbf{K}_f) + G_2(b\mathbf{K}_i), \tag{51j}$$

with G_1 and G_2 defined by Eqs. 51h and 51e, respectively. Two instructive points are made before the calculation is continued.

First, the two terms in square brackets of Eqs. 51b and 51c can be combined and in this form a factor $|\mathbf{U} - \mathbf{K}_f|^2$ occurs which cancels the corresponding external factor. Although the resulting integrals are complicated, they are bounded, consistent with Eq. 51h. The second point is that the separate divergences exhibited in Eqs. 51b and 51c also occur in the second Born approximation for elastic scattering and thus are not some peculiarity emerging from the problem of electronic capture. We now return to Eq. 51b and its evaluation.

The calculational approximations are identical to those just employed. First, the Riemann-Lebesgue theorem, referred to as the *peaking approximation,* is invoked to evaluate the integral over $d\mathbf{V}$; this is designated I_V:

$$I_V = \pi^2(|a\mathbf{U} - \mathbf{K}_i|^{-2} + |\mathbf{U} - b\mathbf{K}_i|^{-2})[1 + |(1 - a)\mathbf{U} + (1 - b)\mathbf{K}_i|^2]^{-2}. \tag{51k}$$

The resulting integral over $d\mathbf{U}$ of Eq. 51b has peaks at

$$\mathbf{U} = \mathbf{K}_f, \qquad -\frac{1 - b}{1 - a}\mathbf{K}_i. \tag{51l}$$

These peaks are separated sufficiently in $\mathbf{U}$-space provided that $b \neq a$, but if $b = a$, $\mathbf{K}_f$ can equal $-\mathbf{K}_i$ at $\theta = \pi$, and the peaking approximation is invalid, strictly speaking; nevertheless, if the dependence on impact energy is sought irrespective of the correct numerical factors, this method still gives the correct dependence on energy for *ns*-intermediate states. Thus, with this limitation on accuracy understood, we continue using this approximation, and the remaining integrations over $d\mathbf{U}$ are accomplished as follows: (1) $\mathbf{U}$ is equated to $\mathbf{K}_f$ in I_V of Eq. 51k and the remaining integration over $d\mathbf{U}$ is effected; (2) the value of $\mathbf{U}$ corresponding to the peak in Eq. 51k is used in

$|\mathbf{U} - \mathbf{K}_f|^{-2}$ and in the bracketed terms of Eq. 51*b*, and the rest of the integration over $d\mathbf{U}$ is performed. The result is

$$F(12 - 13; 12) = -\frac{512\mu\pi}{m}\Bigg(\{(4 + R_1^2)^{-2} - (1 - a)^4[4(1 - a)^2 + a^2R_1^2]^{-2}\}$$

$$\times \frac{(1 - a)^2(1 - ab)^{-2}K_i^{-2}R_i^{-2}}{(1 - a)^{-1} + i2K_f + (1 - a)\{[(1 - b)K_i/(1 - a)]^2 - K_f^2\}}$$

$$+ \frac{1}{16K_f}[A_f^{-2}(1 + R^2)^{-2} + A_i^{-2}(1 + R^2)^{-2}]I\Bigg), \tag{51m}$$

$$I = I(\text{Eq. } 51h), \qquad R_1^2 = |\mathbf{A}_i + \mathbf{A}_f|^2, \qquad R^2 = (1 - a)^2|\mathbf{K}_i + \mathbf{K}_f|^2.$$

None of the remaining matrix elements originating from the discrete intermediate states will be evaluated since these additional calculations more appropriately belong in an appendix (see Appendix 3); instead a description of the results of these (unpublished) calculations is given.

First the cases of nonresonance at arbitrary scattering angles and exact resonance at small scattering angles are described. As was just explained (Eqs. 51*h* and 51*m*), the constituent 1*s*-amplitude varies as $E_i^{-3.5}$ for large impact energies, and we can demonstrate by a lengthy calculation that the other *ns*-amplitudes ($n \geq 2$) all vary as E_i^{-4}. Again it can be demonstrated that the *np*-amplitudes also vary as E_i^{-4}, but all other *nl*-amplitudes decrease more strongly than E_i^{-4}. For exact resonance the situation is different at large scattering angles.[4]

In this case some of the constituent amplitudes increase near $\theta = \pi$. This variation originates from $\mathbf{R}_1 = \mathbf{R}(b = a)$ vanishing at $\theta = \pi$. In this instance the two terms inside the first brace of Eq. 51*m* can be combined, and the resultant numerator contains a factor R^2 that cancels R^{-2} from the external factor. As a consequence, this part of $F(12 - 13; 12)$ varies as $E_i^{-1.5}$ at high impact energies, and another lengthy calculation reveals that each *ns*-intermediate state contributes to this $E_i^{-1.5}$-variation. Each of the *nl*-intermediate states, however, contributes a part that varies as $E_i^{-1.5} \log E_i$ and thus dominates the *ns*-contribution for sufficiently large E_i. No effort has been made to determine whether these infinite sums converge, and in this connection another item of interest emerges. Only contributions from the 1*s*-intermediate state are needed here.

As has been abundantly demonstrated, the separate integrals of Eqs. 51*b* and 51*c* diverge. Since we use Eq. 50*d*, it is equally valid to use the series

$$G_f = G_0 + G_0U_{23}G_0 + \cdots, \tag{51n}$$

and we might be tempted to use this expansion to analyze the matrix element,

$$(\Phi_f, V_fG_fV_i\Phi_i)$$

of Eq. 50e. Now each of the matrix elements of this second Born element diverges for the 1s-intermediate state, but each matrix element of the second Born matrix element using G_0 is bounded for small scattering angles, and they are probably bounded for large angles also, as we shall learn in the next section. (Exact resonance is always understood in discussions of the variation at large scattering angles since otherwise the angular distribution is dominant in the forward direction.) This is the first point we sought. The fact that the matrix elements in G_f are unbounded suggests that either Eq. 51n is invalid or else that matrix elements in G_0 of order greater than the second also are unbounded. However, the fact that the second Born terms $(\Phi_f, V_f G_f U_{12} \Phi_i)$ and $(\Phi_f, V_f G_f U_{23} \Phi_i)$ are finite suggests that cancellations also occur for the corresponding higher order matrix elements in G_0. We can scarcely rely upon results calculated in the second Born approximation, however, until this puzzling situation is clarified. The other contribution to the amplitude containing G_{fc} is studied next.

4.3 FREE-STATE GREEN'S FUNCTION AND CONTRIBUTIONS FROM THE CONTINUOUS INTERMEDIATE STATES OF G_f

The continuous spectrum of G_f, denoted by G_{fc}, is given by Eq. 50i. Suppose that the impact energy is large enough so that contributions from G_{fc} to the second Born amplitude are negligible if k^2 departs much from K_f^2. This is a plausible assumption since D_c of Eq. 50g is very large otherwise except for a special range of $\mathbf{U}$. We thus imagine k to be sufficiently large so that the functions ${}_1F_1$ of Eq. 50i differ negligibly from unity for most $\mathbf{x}_f$. In this case the factor, $\alpha(1 - e^{-2\pi\alpha})^{-1}$, is closely equal to $(2\pi)^{-1}$, and this approximate G_{fc} would be the free-state Green's function G_0 if the integration over $d\mathbf{k}$ were not restricted to large values of $\mathbf{k}$; however, since D_c is large over the omitted domain of $d\mathbf{k}$, all of $\mathbf{k}$-space may be restored with small error in the relevant matrix elements. With this approximation we have thus established a relation between the two amplitudes in Eq. 50e at high impact energies. Nevertheless, this deceptively simple derivation of G_0 should be used only with matrix elements whose values are well-defined, and in particular, the associated integrals should be uniformly convergent with respect to interchanging ${}_1F_1$ with unity. If this simple rule is violated, ambiguous results may ensue. Examples of such nonuniformity will be encountered in this section, but matrix elements containing G_0 in Eq. 50e do not suffer from this defect, and they are calculated first.

These matrix elements were originally calculated by Drisko, and much of the present approach is either patterned after or strongly influenced by that original work.[10] The denominator D_c is first transformed into new variables

$$\mathbf{U} = \mathbf{P} + \mathbf{K}_f, \qquad \mathbf{k} = a\mathbf{P} + \mathbf{Q} + \mathbf{A}_f,$$

$$-D_c = \frac{\Delta}{2\mu_1} = \frac{1}{2\mu_1}[|\mathbf{P} + \mathbf{Q}|^2 - 2(1-a)\mathbf{P}\cdot\mathbf{Q} - 2\mathbf{A}_i\cdot\mathbf{P} + 2\mathbf{A}_f\cdot Q \tag{52a}$$

$$+ A_f^2 + a^2 - i\delta - \gamma(P^2 + 2\mathbf{P}\cdot\mathbf{Q})],$$

$$\gamma = (1-a)(1 - MM_d^{-1}), \qquad M_d = 2M.$$

The program is to calculate the matrix elements for the accidentally resonant process, and then determine whether the mathematical approximations used are also valid for the case $b = a$. After this case of exact resonance has been analyzed, the final calculation will be effected for the partial amplitudes stemming from G_{fc} in Eq. 50*e*. Equations 50*g*, 51*a*, and 52*a* contain the material needed for the evaluation of $f_c^{(2)}$ of Eq. 50*j*, which is readily accomplished to give

$$M_0(13, 23) + M_0(12, 23) = -\frac{32\mu_1}{m\pi^3}\int \frac{d\mathbf{U}\,d\mathbf{k}}{\Delta}[I_0(13) - I_0(12)]F_0(23),$$

$$M_0(12, 12) + M_0(13, 12) = -\frac{32\mu_1}{m\pi^3}\int \frac{d\mathbf{U}\,d\mathbf{k}}{\Delta}[I_0(12) - I_0(13)]F_0(12),$$

$$I_0(13) = |\mathbf{U} - \mathbf{K}_f|^{-2}[1 + q^2]^{-2}, \qquad \mathbf{q} = \mathbf{k} - a\mathbf{U} + a\mathbf{K}_f, \tag{52b}$$

$$I_0(12) = |\mathbf{U} - \mathbf{K}_f|^{-2}(1 + p^2)^{-2},$$

$$m = \mathbf{k} + (1-a)(\mathbf{U} - \mathbf{K}_f) = \mathbf{q} + \mathbf{U} - \mathbf{K}_f;$$

$$F_0(23) = (q')^{-2}(1 + |\mathbf{U} - b\mathbf{K}_i|^2)^{-2},$$

$$\mathbf{q}' = -\mathbf{k} + a\mathbf{U} - \mathbf{K}_i; \qquad F_0(12) = (\mathbf{q}')^{-2}(1 + |b\mathbf{K}_i - \mathbf{U} + \mathbf{q}'|^2)^{-2}.$$

The matrix element, $M_0(13, 23)$, is evaluated first. The change of variables used in Eq. 52*a*, together with another change of variables, are used to transform the integral of interest into

$$M_0(13, 23) = -\frac{32}{\pi^3}\int \frac{d\mathbf{K}_1\,d\mathbf{K}_2}{|\mathbf{K}_1 + \mathbf{A}_i|^2|\mathbf{K}_2 - \mathbf{A}_f|^2(1 + K_1^2)^2(1 + K_2^2)^2\Delta},$$

$$K_1 = \mathbf{P} - \mathbf{A}_i, \qquad \mathbf{K}_2 = \mathbf{Q} + \mathbf{A}_f, \tag{52c}$$

$$\Delta = |\mathbf{K}_1 + \mathbf{K}_2|^2 + 2\mathbf{A}_i\cdot\mathbf{K}_2 - 2\mathbf{A}_f\cdot\mathbf{K}_1 - \mathbf{A}_i^2 - 2\mathbf{A}_i\cdot\mathbf{A}_f + a^2 - i\delta.$$

The approximation to Δ in Eq. 52*c* is obtained by omitting terms with the factor $(1 - a) \approx m/M$. At high impact energies both $\mathbf{A}_i$ and $\mathbf{A}_f$ are large; therefore the preceding integrand decreases strongly for small departures of K_1 and K_2 from zero, or, in the usual description, there are sharp peaks about $\mathbf{K}_1 = \mathbf{K}_2 = 0$. Consequently, the Fourier transforms of the interactions may be replaced by their values at these peaks, to a very good approximation. Moreover, the term $|\mathbf{K}_1 + \mathbf{K}_2|^2$ may be omitted from Δ since this term is negligible in comparison to the terms in $\mathbf{A}_i$ and $\mathbf{A}_f$ for $\mathbf{K}_1$ and $\mathbf{K}_2$ near the peaks. An examination of Δ reveals that this latter approximation improves with increasing impact energy. With this explanatory digression completed, we return to the integration. Cylindrical coordinates are introduced; the integrations over the angular and radial variables are effected, and the paths of integration over dK_{1z} and dK_{2z} are closed in the upper and lower half-planes, respectively, since there are no zeros of Δ in their respective half-spaces. The result is

$$M_0(13, 23) = \frac{-32\pi}{A_i^2 A_f^2[-A_i^2 - 2\mathbf{A}_i\cdot\mathbf{A}_f - i2(A_i + A_f)]} \tag{52d}$$

$$\mathbf{A}_i = b\mathbf{K}_i - \mathbf{K}_f, \qquad \mathbf{A}_f = a\mathbf{K}_f - \mathbf{K}_i.$$

In Eq. 52d the term $(a^2 - i\delta)$ is omitted since it is negligible. We next use Eqs. 50f and 51e to get the following calculationally convenient relations:

$$A_i^2 \approx A_f^2 = E_i(1 + \lambda); \qquad \lambda = \frac{2\sqrt{ab}}{(1 - \sqrt{ab})^2}(1 - \cos\theta),$$

$$-A_i^2 - 2\mathbf{A}_i\cdot\mathbf{A}_f = E_i(\lambda - 3); \qquad E_i = \left(\frac{v_i}{2v_0}\right)^2,$$

$$v_0 = \frac{e^2}{\hbar}, \qquad d\theta\sin\theta = \frac{(1 - \sqrt{ab})^2}{2\sqrt{ab}}\,d\lambda\left(\approx \frac{1}{2}\left(\frac{m}{M}\right)^2 d\lambda, \qquad b = a\right). \tag{52e}$$

(The small-angle approximation, $E_i\cos\theta \approx E_i$, is used to derive the approximations to $A_{i,f}^2$. The impact energy E_i is here expressed in units of 100 keV.) These relations are used to facilitate the integration of the differential cross section over angle, but the remaining parts of Eq. 50e are needed to get the cross section.

First we use Eq. 51a and the peaking approximation to get the contributions from the first Born approximation; these constituent amplitudes are[11]

$$\int d\mathbf{x}_i\, d\mathbf{x}_f \Phi_f^*(|\mathbf{x}_i - \mathbf{x}_f|^{-1} - \mathbf{x}_i^{-1})\Phi_i$$

$$= \frac{32}{\pi}\int \frac{d\mathbf{K}}{(K^2 + \varepsilon^2)(1 + |\mathbf{K} - \mathbf{A}_f|^2)^2(1 + |\mathbf{K} + \mathbf{A}_i|^2)^2} - \frac{32\pi}{(1 + A_f^2)^2(1 + A_i^2)}$$

$$= 32\pi\left[\frac{1}{(1 + R_1^2)^2}\left(\frac{1}{A_f^2} + \frac{1}{A_i^2}\right) - \frac{1}{(1 + A_f^2)^2(1 + A_i^2)}\right], \tag{52f}$$

$$\mathbf{R}_1 = \mathbf{A}_i + \mathbf{A}_f.$$

The other two terms stemming from G_{fc} that contribute at high impact energies are $M_0(12, 23)$ and $M_0(13, 12)$, since Drisko has shown that $M_0(12, 12)$ is negligible.[10] In addition to the change of variables in Eq. 52*a*, an additional transformation (whose corresponding Jacobian is unity in all these cases) is used to simplify the integration of $M_0(12, 23)$:

$$M_0(12, 23) = \frac{32}{\pi^3} \int \frac{d\mathbf{K}_1\, d\mathbf{K}_2}{|\mathbf{K}_1 + \mathbf{A}_i|^2 |\mathbf{K}_1 - \mathbf{K}_2 + \mathbf{R}_1|^2 (1 + \mathbf{K}_1{}^2)^2 (1 + \mathbf{K}_2{}^2)^2 \Delta},$$
$$K_1 = \mathbf{P} - \mathbf{A}_i, \qquad \mathbf{K}_2 = \mathbf{P} + \mathbf{Q} + \mathbf{A}_f, \tag{52g}$$
$$\Delta = K_2{}^2 - 2\mathbf{K}_1 \cdot \mathbf{R}_1 - 2A_i{}^2 - 2\mathbf{A}_i \cdot \mathbf{A}_f.$$

The terms containing the factor $(1 - a)$ and the term $(a^2 - i\delta)$ (Eq. 52*a*) are omitted with negligible error. There are no zeros in Δ, and the peaking approximation is applicable, which gives

$$M_0(12, 23) = \frac{32\pi}{A_i{}^2 R_1{}^4}. \tag{52h}$$

For the evaluation of $M_0(13, 12)$ another change of variables is effected, and the peaking approximation is again applicable. The result is

$$M_0(13, 12) = \frac{32}{\pi^3} \int \frac{d\mathbf{K}_1\, d\mathbf{K}_2}{|\mathbf{K}_2 - \mathbf{K}_1 + \mathbf{R}_1|^2 |\mathbf{K}_1 - \mathbf{A}_f|^2 (1 + K_1{}^2)^2 (1 + K_2{}^2)^2 \Delta},$$
$$\mathbf{K}_1 = \mathbf{Q} + \mathbf{A}_f, \qquad \mathbf{K}_2 = \mathbf{P} + \mathbf{Q} - \mathbf{A}_i, \qquad \Delta = K_2{}^2 + 2\mathbf{K}_1 \cdot \mathbf{R}_1 - R_1{}^2,$$
$$M_0(13, 12) = \frac{32\pi}{A_i{}^2 R_1{}^4} = M_0(12, 23). \tag{52i}$$

Since $A_i{}^2$, $A_f{}^2$, and $R_1{}^2$ all greatly exceed unity, it is clear from Eqs. 52*i* and 52*f* that the part of Eq. 52*f* originating from U_{12}, the nucleus-nucleus interaction, is canceled by $M_0(12, 23)$ and $M_0(13, 12)$, a fact originally discovered by Drisko. Thus the amplitude f_0 of Eq. 50*e* and the associated cross section are

$$f_0 = -\frac{16\mu a_0}{m} \left\{ \frac{1}{A_i{}^2 A_f{}^2 [-A_i{}^2 - 2\mathbf{A} \cdot \mathbf{A}_f - i2(A_i + A_f)]} - \frac{1}{A_f{}^4 A_i{}^2} \right\}$$
$$= -\frac{16\mu a_0}{mE_i^3} \left\{ \frac{1}{(1 + \lambda)^2 \left[\lambda - 3 - i4 \left(\dfrac{1 + \lambda}{E_i} \right)^{1/2} \right]} + \frac{1}{(\lambda + 1)^3} \right\} \tag{52j}$$
$$Q_0 = \frac{64\pi a_0{}^2}{E_i{}^6} \int_0^\infty \frac{d\lambda}{(1 + \lambda)^4} \left| \frac{1}{\lambda + 1} + \frac{1}{\lambda - 3 - i4 \left(\dfrac{1 + \lambda}{E_i} \right)^{1/2}} \right|^2.$$

The variable of integration λ was obtained from the transformation in Eq. 52*e* and the upper limit is used with negligible error since the integrand decreases strongly with increasing λ. Note that Q_0 is independent of mass in this system of units, notwithstanding the fact that the masses of the atomic target and incident particle are different. This independence of the cross section upon the mass of the atomic target is generally true for a given incident particle provided that the change in internal energy is negligible relative to the impact energy in the defining relations for the momentum-change variables. Since $E_i = (v/2v_0)^2$, the important parameter at high impact energies is the velocity because the only other parameter in Eq. 52*j* is a_0. (In this example all nuclear charges are equal to one, but in the next section this restriction will be removed.) We return to the evaluation of the integral in Eq. 52*j* and write the result of this lengthy calculation:*

$$Q_0 = \frac{64\pi a_0{}^6}{E_i{}^6}(I_1 + I_2 + I_3) = Q(\mathrm{OBK})\left(0.3011 + \frac{5\pi}{2^{11}}\sqrt{E_i}\right),$$

$$Q(\mathrm{OBK}) = \frac{64\pi a_0{}^2}{5E_i{}^6},$$

$$I_1 = \int_1^\infty \frac{dU}{U^6} = \frac{1}{5}; \tag{52k}$$

$$I_2 = 3\int_1^\infty \frac{dU}{U^4}\left[U^2 - 8\left(1 - \frac{2}{E}\right)U + 16\right]^{-1}$$

$$= 0.1574 + 3\pi\frac{\sqrt{E_i}}{2^{11}};$$

$$I_3 = -8\int_1^\infty \frac{dU}{U^5}\left[U^2 - 8\left(1 - \frac{2}{E}\right)U + 16\right]^{-1} = -0.2972 - 2\pi\frac{\sqrt{E_i}}{2^{11}},$$

Only the terms in $E_i{}^{-6}$ and $E_i^{-11/2}$ are retained in this calculation, but the important feature of Eq. 52*k* is the $E_i^{-11/2}$-variation, although this term does not become dominant until very high impact energies. Recently the Fadeev equations together with the Coulomb two-body *T*-matrix have been used to evaluate the asymptotic cross section for the $U_{13}U_{23}$ interaction of this section. Drisko's result of Eq. 52*k* was obtained. Furthermore, the contributions from the associated infinity of two-body bound states were explicitly summed in the asymptotic limit.[12] A rough calculation reveals that for an

* This result agrees with the result of Drisko only to the first two significant figures in the first term, the original number being 0.295. This difference probably arises from the different methods used to effect the integrations.

impact energy of 830 MeV the factor multiplying Q(OBK) is equal to unity. The theory, of course, is not valid at these relativistic energies of the proton; nevertheless, the result is important since it shows that $f_0^{(2)}$ is not negligible in the mathematical sense.

Another particularly important feature of Q_0 is its independence of the nucleus-nucleus interaction as are the asymptotic cross sections obtained from the distorted wave method, the impulse approximation, and the continuum distorted wave method. Moreover, the continuum distorted wave method of Cheshire[13] reproduces Eq. 52k exactly at high impact velocities although the asymptotic form of the impulse approximation gives a cross section whose V_i^{-11}-term exceeds the one of Eq. 52k by a factor of two.[14] (This difference had not been explained at the time of this writing.) This $E_i^{-11/2}$ variation of the cross section, furthermore, has been a subject of much interest since Thomas deduced this same dependence from classical mechanics many years ago, but this comparison is postponed until the classical methods have been described.[15] A final point to note is that $M_0(12, 12)$ varies as

$$R_1^{-5}A_i^{-2} \log A_i,$$

according to the calculations of Drisko, and clearly does not contribute at high impact energies.

Next we inquire about the corresponding results for exact resonance. In this case $b = a$, and $\mathbf{R}_1$ of Eq. 52f becomes

$$R_1^2 = R^2 = (1 - a)^2|\mathbf{K}_i + \mathbf{K}_f|^2 = 2E_i(1 + \cos\theta), \qquad (52l)$$

with E_i again expressed in units of 100 keV. In this example of exact resonance, $\mathbf{R}$ vanishes at $\theta = \pi$, for which angle $\mathbf{A}_i = -\mathbf{A}_f$; consequently, the terms previously discarded from Δ were larger than the terms retained in Δ of Eqs. 52g and 52i. Moreover, the corresponding approximate amplitudes diverge at $\theta = \pi$! We digress now to emphasize an important new feature of this apparent backward scattering.

If the dominant contribution to the amplitude originates from the angular region near $\theta = \pi$, center-of-mass system, the associated cross section is undefined unless it dominates the cross section for elastic scattering! This is easily understood. Since the hydrogen atom formed is deflected nearly 180°, the scattered proton, the recoil particle, emerges near the forward direction; thus the wave amplitudes for capture and elastic scattering interfere, and by reason of the indistinguishability of the two protons, the cross sections for the separate processes are undefined unless one of them is dominant. In practice elastic scattering is dominant at high impact velocities, and as a consequence the cross section for electronic capture is undefined. Nevertheless, the cross section may still be defined if a better approximation is used since this type of angular variation may not appear. Here we mean only that the corresponding cross section is not defined if calculated using this approximation. We return now to the calculations for θ near π.

The partial amplitude $M_0(13, 23)$, it is first noted, does not peak in this angular region and thus is not needed. The exact values of Δ for $M_0(12, 23)$ and $M_0(13, 12)$ are

$$\Delta(12, 23) = K_2^2 - 2\mathbf{K}_1 \cdot \mathbf{R} - R^2 + a^2 - i\delta - 2(1 - a)C,$$
$$C = (\mathbf{K}_1 + \mathbf{A}_i)\cdot(\mathbf{K}_2 - \mathbf{K}_1 - \mathbf{R})$$

and

$$\Delta(13, 12) = K_2^2 + 2\mathbf{K}_f \cdot \mathbf{R} + a^2 - i\delta - 2(1 - a)C,$$
$$C = (\mathbf{K}_2 - \mathbf{K}_1 + \mathbf{R})\cdot(\mathbf{K}_1 - \mathbf{A}_f).$$

At the peaks of the associated integrands of Eqs. 52*g* and 52*i*, these quantities each equal

$$\Delta(\mathbf{K}_1 = \mathbf{K}_2 = 0) = a(a - R^2) - i\delta. \tag{52m}$$

The sign of Δ changes to a positive value in the angular region for which $R^2 < a$; therefore the cancellation noted prior to Eq. 52*j* does not occur; however, this cancellation does occur over most of the angular region, and the results for exact and inexact resonance are essentially the same for impact energies of the proton below 40 MeV. An approximate answer for the sum of these two contributions is obtained by using Eq. 52*m* for Δ and by integrating the remaining approximate integrals of Eqs. 52*g* and 52*i*. These calculated results are

$$M_0(13, 12) + M_0(12, 23) \approx \frac{64\pi}{a(a - R^2)(1 + A_i^2)(4 + R^2)}\cdot \tag{52n}$$

This procedure should give the correct dependence upon impact energy E_i^{-1} and the correct sign at $\theta = \pi$, although the numerical coefficients probably are incorrect. As will be established in the next section, the corresponding first Born amplitude also varies as E_i^{-1} at $\theta = \pi$. Furthermore, similar results for backward capture can be obtained from an approximate classical mechanical analysis (a result explained in Section 5.6). A similar, but more complicated, calculation of $M_0(12, 12)$ reveals an $E_i^{-5/2}$ variation at $\theta = \pi$, and this contribution is negligible indeed. Equation 50*i* is consulted next for the calculation of the elements originating from G_{fc} of Eq. 50*e*.

The notation of Eq. 52*b* is modified slightly to accommodate the changes of including the confluent hypergeometric functions in the integrands. The modified integrands are given in Eq. 53*a*:

$$I(13) = I_0(13)\left\{1 + i\alpha\,\frac{2A - ikB}{B - A}\right\}\left\{\frac{B}{B - A}\right\}^{i\alpha},$$

$$A = \mathbf{k}\cdot\mathbf{q} + ik, \qquad B = 1 + q^2,$$

$$I(12) = I_0(12)\left\{1 + i\alpha\,\frac{2C - ikD}{D - C}\right\}\left\{\frac{D}{D - C}\right\}^{i\alpha},$$

$$C = \mathbf{k}\cdot\mathbf{p} + ik, \qquad D = 1 + p^2, \tag{53a}$$

$$F(23) = F_0(23)\left\{\frac{(q')^2 + 2\mathbf{K}\cdot\mathbf{q}'}{(q')^2}\right\}^{i\alpha},$$

$$F(12) = \lim_{\lambda\to 0} F_0(12)\left\{1 + \frac{i}{2k}\right\}\left\{\frac{2ik}{\lambda}\right\}^{i\alpha}.$$

These modified quantities replace the associated quantities with the subscript zero in Eq. 52*b*, and the corresponding matrix elements are designated simply $M(ij, kl)$. The spatial integrations leading to these results were accomplished with the aid of the integral representation,

$$_1F_1(i\alpha, 1; z) = (2\pi i)^{-1} \oint dt\, t^{(i\alpha - 1)}(t - 1)^{-i\alpha} e^{zt}, \tag{53b}$$

used by Pradhan in his calculations on the impulse approximation.[16] The contour in the t-plane encircles the points 0 and 1, which are the endpoints of the branch cut on the real t-axis. The results for the $I(ij)$ in Eq. 53*a*, exclusive of the phase factors, each can be interpreted as a decomposition into two parts, the first term being the answer for $\alpha = 0$ and the other term being the addition needed to give the correct answer for $\alpha \neq 0$. The last term, $F(12)$, was evaluated with the use of the convergence factor

$$e^{-\lambda x_f},$$

and the last factor of $F(12)$ is a divergent phase factor; moreover, at $\alpha = 0$, ($k \to \infty$) the value of this factor depends upon the order that the limiting procedures are effected, and thus the corresponding integrals are probably nonuniformly convergent. As a consequence, the values of the two matrix elements $M(13, 12)$ and $M(12, 12)$ are possibly ambiguous, and it is not known whether an exact calculation would yield results in agreement with those obtained using G_0. Although no effort is made to evaluate these integrals, the important fact has been elicited that the two approximate amplitudes of Eq. 50*e* are different, although these two amplitudes may be equal at high impact velocities. In an actual numerical calculation at intermediate impact velocities, nevertheless, we might find noticeably different results since Eq. 50*e* does represent two different expansions for the amplitude.

To summarize, for high impact energies approximate answers to the integrals representing Eq. 50*e* have been obtained for the discrete spectrum and the free-state Green's function G_0, and although the same approximations are used for the integrals arising from G_{fc}, it is not known whether or not such results are unique. In the region of small scattering angles the sum of the first and second Born amplitudes is given by Eq. 52*j* for both exact and inexact resonance if the expansion containing G_0 is used. Moreover, the corresponding cross section is also given by this equation for inexact resonance. Nevertheless, in the case of exact resonance a peak occurs at the scattering angle $\theta = \pi$, and this contribution apparently dominates the scattering at small angles with the consequence that interference with elastic scattering occurs, and the cross section for capture is not defined in this approximation. This effect near $\theta = \pi$ also occurs if G_{fc} is used, but these results may be ambiguous. Contributions from the discrete intermediate states are dominated by those from the associated continuum; nevertheless, by reason of the difficulties mentioned in connection with the expansion of Eq. 51*n* relevant to the separate matrix elements of the 1*s*-intermediate state, the second Born approximation should be used with reservation until these, and other presently unacceptable features are removed.

4.4 WAVE-MECHANICAL PREDICTION OF ASYMPTOTIC CROSS SECTIONS AND AMPLITUDES FOR ELECTRONIC CAPTURE AT HIGH NONRELATIVISTIC VELOCITIES

This section is devoted to the study of the variation with impact velocity, or energy, for the amplitudes and cross sections calculated according to some of the approximating schemes that are discussed in this book. Only the wave mechanical version of the asymptotic forms are calculated, and in the remainder of this section the qualification *asymptotic* always refers to the limiting value of a quantity as the impact velocity, or energy, becomes large. By *nonrelativistic* we mean that the kinetic energy of the incident particle is approximated well by the nonrelativistic kinetic energy,

$$M_pC^2\left[1 - \left(\frac{v_i}{c}\right)^2\right]^{-1/2} - M_pC^2 \approx \tfrac{1}{2}M_pv_i^2,$$

and this approximation is satisfactory if $(v_i/c)^2 < 0.1$. For incident protons this requires the impact energy not exceed 100 MeV. However, in the case of capture from the $1s^2$ subshell of atoms the asymptotic form usually is not achieved for nonrelativistic impact velocities; in these cases the corresponding limiting forms refer to *mathematical, not physical,* solutions. Nearly all attention is devoted to hydrogenic systems, and to atoms whose wave functions are approximated by appropriate linear combinations of one-electron Slater type orbitals (STOs) since these forms comprise most of the analytic representations of wave functions in current use. As a first item in the program for this section, Eq. 4 is generalized to allow for arbitrary positive values of the two nuclear charges:

$$\begin{aligned}
(E - H_i - V_i) &= E + \left(\frac{\hbar^2}{2\mu_i}\nabla_{R_i}{}^2 + \frac{\hbar^2}{2\mu_1}\nabla_{x_i}{}^2 + \frac{Z_a e^2}{x_i} - \frac{Z_p(Z_a - 1)e^2}{R_i}\right) \\
&\quad + Z_p e^2\left(\frac{1}{x_f} - \frac{Z_a}{|\mathbf{x}_i - \mathbf{x}_f|} + \frac{Z_a - 1}{R_i}\right), \\
(E - H_f - V_f) &= E + \left(\frac{\hbar^2}{2\mu_f}\nabla_{R_f}{}^2 + \frac{\hbar^2}{2\mu_2}\nabla_{x_f}{}^2 + \frac{Z_p e^2}{x_f} - \frac{Z_a(Z_p - 1)e^2}{R_f}\right) \\
&\quad + Z_a e^2\left(\frac{1}{x_i} - \frac{Z_p}{|\mathbf{x}_i - \mathbf{x}_f|} + \frac{(Z_p - 1)}{R_f}\right), \\
\mathbf{x}_f &= -\mathbf{R}_i + \frac{M_a}{M_{a1}}\mathbf{x}_i, \qquad \mathbf{x}_i - \mathbf{x}_f = \mathbf{R}_i + \frac{m}{M_{a1}}\mathbf{x}_i, \\
\mathbf{x}_i &= \mathbf{R}_f + \frac{M_p}{M_{p1}}\mathbf{x}_f, \qquad \mathbf{x}_i - \mathbf{x}_f = \mathbf{R}_f - \frac{m}{M_{p1}}\mathbf{x}_f.
\end{aligned} \tag{54a}$$

With the use of the relations between the appropriate coordinates in Eq. 54*a*, it is clear that

$$\lim_{R_i\to\infty} R_i(V_i)_{x_i} \to 0, \qquad \lim_{R_f\to\infty} R_f(V_f)_{x_f} \to 0, \tag{54b}$$

and the terms in R_i and R_f are added to H_i and H_f in order that V_i and V_f each decreases more strongly than the Coulomb potential for large distances between the centers of masses of the two aggregates. (The subscripts x_i and x_f in Eq. 54*b* signify that the corresponding atomic coordinate is kept constant.) This modification is made so that the perturbing prior and post interactions satisfy one of the requirements (Eq. 54*b*) of standard scattering theory; however, as a result of these Coulomb potentials the initial and final states are also modified, but this alteration is explained in connection with the OBK cross section, which is analyzed first.

In this case the active electronic interaction is

$$Z_p e^2 x_f{}^{-1},$$

and Section 3.4 is consulted next. The initial and final wave functions associated with Eq. 44*b* and 54*a*[17] (in the original coordinate system) are

$$\Phi_i = \phi_i(\mathbf{x}_i) \exp\left(-\frac{\pi\alpha_i}{2}\right)\Gamma(1 + i\alpha_i) \exp(i\mathbf{K}_i\cdot\mathbf{R}_i)\,{}_1F_1(-i\alpha_i; 1, iW_i),$$

$$W_i = K_iR_i - \mathbf{K}_i\cdot\mathbf{R}_i,\ \cos\Theta_i = \frac{\mathbf{K}_i\cdot\mathbf{R}_i}{K_iR_i},\qquad \alpha_i = \frac{Z_p(Z_a - 1)v_0}{v_i}, \tag{54c}$$

$$v_0 = \frac{e^2}{\hbar}$$

and

$$\Phi_f = \phi_f(\mathbf{x}_f) \exp\left(-\frac{\pi\alpha_f}{2}\right)\Gamma(1 - i\alpha_f)$$
$$\times \exp[-iK_fR_f\cos(\pi - \theta_f)]\,{}_1F_1(i\alpha_f, 1; -iW_f),$$
$$W_f = K_fR_f[1 - \cos(\pi - \theta_f)],\qquad \cos\theta_f = \frac{\mathbf{K}_f\cdot\mathbf{R}_f}{K_fR_f}, \tag{54d}$$
$$\alpha_f = \frac{Z_a(Z_p - 1)v_0}{v_f}.$$

(We do not write the relation between K_i, K_f and impact energy each time.) Now in the matrix element defining the scattering amplitude the complex conjugate of this converging (incoming) wave solution in $\mathbf{R}_f$ is used; thus in the corresponding integral the diverging wave solution (outgoing) appears as it should.[18]

These solutions are designated *converging* and *diverging* by reason of their asymptotic form in R_f, and in this connection, the initial wave function is a diverging wave solution in R_i. The parameters α_i and α_f both approach zero with increasing $\mathbf{v}_i$ and $\mathbf{v}_f$, respectively; therefore in the asymptotic region (again we refer to impact velocity) the solutions in $\mathbf{R}_i$ and $\mathbf{R}_f$ differ negligibly from a plane wave, and we shall use this plane wave approximation for all cases of the first Born approximation of this section. It is important to observe that larger values of Z_a and Z_p require correspondingly larger values of $\mathbf{v}_i$ and $\mathbf{v}_f$ to maintain a given smallness in α_i and α_f, and this is the effect that operates to move the asymptotic region toward larger impact velocities for capture from inner subshells. (Investigations of the quality of the plane wave approximation—or, equivalently, the neglect of the Coulomb repulsive term in R_i and R_f—have been effected for ionization and excitation[19] and electronic capture for exact resonance,[20] and these findings reveal that the effect of the Coulomb terms decrease with increasing impact velocity; however, it is not known whether these conclusions are also applicable to systems containing Coulomb attractive potentials.) Of incidental interest is the fact that α_i and α_f both vanish identically if Z_a and Z_p each is equal to one, the situation most commonly analyzed.

We now use the coordinates of Eq. 44*e* to describe the amplitude; dimensionless units are used:

$$g_{if}(\theta) = \frac{\mu_f a_0}{2\pi m} \frac{Z_a^{3/2}}{\pi} \int d\mathbf{x}_i \, d\mathbf{x}_f \, \frac{Z_p}{x_f} \exp{(i\mathbf{A}_f \cdot \mathbf{x}_f)} \mathrm{H}(\mathbf{x}_f \cdot ns) \exp{(i\mathbf{A}_i \cdot \mathbf{x}_i - Z_a x_i)}. \tag{54e}$$

The symbol "H" in Eq. 54*e* denotes a final hydrogenic *ns*-atomic state, and the dominant contribution to the associated amplitude in the asymptotic region is readily obtained by using a general term of the radial function, a Laguerre polynomial.[21] The integral is

$$g_{if}(\theta) = \frac{\mu_f a_0}{2\pi m} \frac{(Z_a Z_p)^{5/2}}{\pi} (-1)^m \frac{\partial^m}{\partial (Z_p/n)^m} \frac{32\pi^2 N_m}{[(Z_p/n)^2 + A_f^2][Z_a^2 + A_i^2]^2}, \tag{54f}$$

and since both A_i^2 and A_f^2 are proportional to E_i (impact energy) at high impact energies, it is clear that the dominant term occurs for $m = 0$.

An examination of the radial wave functions reveals that

$$N_0 = n^{-3/2}, \qquad n = 1, 2, \ldots, \tag{54g}$$

a relation to the principal quantum number n first noted by Oppenheimer[22] and later reported by Jackson and Schiff.[11] This is often known as the n^{-3} law for the contribution of *ns*-states to the total cross section. Since the Riemann zeta function represents the sum,

$$\zeta(3) = \sum_{n=1}^{\infty} n^{-3} = 1.202,$$

this number is occasionally employed as a corrective factor to the cross section for capture into H(1*s*) in order to estimate the effect of the omitted *ns*-states $n \geq 2$. It is clear that this rule strictly applies only to the asymptotic region. [It is readily established with the use of Eqs. A1.1*d* and A1.1*e* that $A_f^2 + (Z_p/n)^2 = A_i^2 + Z_a^2$ with negligible error for one-electron systems. However, in accord with previous remarks, implicit dependence upon n only is negligible in the asymptotic region for the resulting factor, $(A_i^2 + Z_a^2)^{-3}$ of Eq. 54*f*.] Thus the cross section for capture from *s*-states into *s*-states varies as E_i^{-6} and as the fifth power of the product $Z_a Z_p$, a fact discovered by the originators.[23] A less well known fact, however, is that the square of the integral over $d\mathbf{x}_f$ corresponding to $m = 0$ in Eq. 54*f*—normalizing factors appropriate to this case included—is exactly equal to Eq. 44*n* of Section 3.4. Thus the sum over all $l \geq 1$ in Eq. 44*n* cancels all but one of the terms originating from the *ns*-state ($l = 0$), and this remainder is exactly the contribution that is dominant at high impact energies. Separate contributions from the H(*nl*)-states $l > 0$ decrease more strongly than E^{-6}; thus these cross sections do not contribute asymptotically, and it is readily demonstrated that the sum of the cross sections for capture into all p-states and d-states, for a given n, vary as E_i^{-7} and E_i^{-8}, respectively, in agreement with a conjecture of Bates and McCarroll.[24]

We next investigate the other atomic parameters upon which the asymptotic OBK cross section depends. For this purpose attention is first directed to the asymptotic expansions for A_i^2 and A_f^2 and to an associated change of variable:

$$A_i^2 \approx A_f^2 = \frac{8M}{M_p}\left(\frac{\mu_f}{m}\right)^2 E_i(1 - \cos\theta) + \frac{M}{M_p} E_i \cos\theta, \tag{54h}$$

$$d\theta \sin\theta = \frac{2A_i\, dA_i}{(M/M_p)[8(\mu_f/m)^2 - 1]E_i}.$$

The asymptotic cross section corresponding to Eqs. 54f and 54g is thus approximately equal to

$$Q(1s \to ns) = \frac{64\pi a_0^2}{n^3}(Z_a Z_p)^5 x^{-2} \int_x^\infty \frac{2A_i\, dA_i}{[A_f^2 + (Z_p/n)^2]^2 (A_i^2 + Z_a^2]^4},$$

$$E_i(100\text{ keV}) = \frac{M_p}{M} x^2, \qquad x = \frac{v_i}{2v_0}. \tag{54i}$$

At sufficiently high impact velocities the integral in Eq. 54i depends only upon the ratios of velocities $(v_i/2v_0)$; here M and M_p denote the respective masses of a proton and the incident ion. If Z_p and Z_a are neglected in the integrand, only the E_i^{-6} term of the cross section remains. Only in this system of units is the external factor free from a ratio of masses.

Another interesting feature of the physical predictions of this cross section is deduced from the factor of Eq. 54f,

$$Z_a^{-4}\left[1 + \left(\frac{A_i}{Z_a}\right)^2\right]^{-2},$$

which is proportional to the Fourier transform of the initial state. In the asymptotic region, $A_i \gg Z_a$, and the associated cross section is larger for larger Z_a, which suggests that the more tightly bound electron is more readily captured at a *given impact energy*. Moreover, accurate OBK cross sections require accurate values of the Fourier transform corresponding to large values of A_i, and according to the theory of the Fourier transform, this means that the initial atomic wave function must be known accurately near the nucleus.[25,26]

One more example of the OBK cross section is studied to illustrate the effect of the passive momentum-change vector in the asymptotic region. These quantities were defined in Eq. 43c, Section 3.3, and were labeled *passive* following Eq. 44*b*, Section 3.4. If the one-parameter approximate wave function,

$$Z^3\pi^{-1}\exp\left[-Z(x_1 + x_3)\right],$$

is used to represent atomic helium (He $1s^2$)], the corresponding OBK amplitude for electronic capture into H(1*s*) by protons leaving the residual ion He^+ (1*s*), is

$$g_{if}(\theta) = \frac{256Z^4(2 + Z)\sqrt{2}\,\mu_f a_0}{m(1 + A_f^2)(Z^2 + A_i^2)^2[(2 + Z)^2 + A_1^2]^2}, \qquad A_1^2 = \left(\frac{m}{M_i}\right)^2 A_i^2. \tag{54j}$$

The passive vector is $\mathbf{A}_1$, and it is evident that the amplitude in Eq. 54*j* varies asymptotically as E_i^{-5}, and the associated cross section varies as E_i^{-10} However, this variation is *not* attained for nonrelativistic impact velocities; consequently, discarding A_1^2 relative to $(2 + Z)^2$ is an excellent nonrelativistic approximation, and with this alteration the E_i^{-6} variation for *s-s* capture is restored. It is consequently clear that the passive vectors usually can be omitted from the matrix elements at the outset.

If we assume that the operators H_i and H_f of Eq. 54*a* are Hermitian, standard manipulative procedures can be used to show the equality of the prior and post amplitudes:

$$(\Phi_f, V_f\Phi_i) = (\Phi_f, [H_i - H_f + V_i]\Phi_i) = ([E - H_f]\Phi_f, \Phi_i) + (\Phi_f, [H_i - E]\Phi_i) + (\Phi_f, V_i\Phi_i) = (\Phi_f, V_i\Phi_i),$$

$$(E - H_i)\Phi_i = 0, \qquad (E - H_f)\Phi_f = 0.$$

The preceding relation is established using the exact functions Φ_i and Φ_f of Eqs. 54*c* and 54*d*, but it does not necessarily follow that this equation goes uniformly into the result obtained in the plane wave approximation. A calculation of the two amplitudes in the plane wave approximation, moreover, reveals that the corresponding asymptotic amplitudes are not equal unless $Z_p = Z_a$ and the approximations, $(M_a/M_{a1} \approx (M_p/M_{p1}) \approx 1$, are used. (If the nuclei also are identical, the analysis leading to the result in Eq. 55*g* applies to the present problem.) Examples of nonuniformity with respect to the interchange of integration and limit do occur in problems involving the confluent hypergeometric, one such example having been studied by this author.[28] For purposes of consistency, therefore, the prior and post interactions are used without the terms in R_i and R_f. Only the prior interaction is needed since Jackson and Schiff[11] proved the equality of these prior and post amplitudes:

$$g_{if}(\theta) = \frac{\mu_f a_0}{2\pi m} \frac{(Z_p Z_a)^{3/2}}{\pi} Z_p \int d\mathbf{x}_i \, d\mathbf{x}_f \exp(i\mathbf{A}_f \cdot \mathbf{x}_f - Z_p x_f) \times [x_f^{-1} - Z_a |\mathbf{x}_i - \mathbf{x}_f|^{-1}] \exp(i\mathbf{A}_i \cdot \mathbf{x}_i - Z_a x_i). \tag{55a}$$

Although this amplitude is readily modified to accommodate the dominant contribution for capture into an *ns*-state, this is not done here. The integral representation for

$$|\mathbf{x}_i - \mathbf{x}_f|^{-1}$$

from Eq. 45*c*, together with the peaking approximation, is invoked to evaluate the asymptotic amplitude (see Eq. 54*i*):

$$g_{if} = \frac{16\mu_f a_0 (Z_a Z_p)^{5/2}}{m} \left(\frac{1}{(Z_p^2 + A_f^2)(Z_a^2 + A_i^2)^2} - \frac{Z_a}{A_f^2 (Z_a^2 + R^2)^2} - \frac{Z_p}{A_i^2 (Z_p^2 + R^2)^2} \right), \tag{55b}$$

$$R^2 = 4x^2 F, \qquad F = \frac{1}{(M_a + M_p)^2} [M_a^2 + M_p^2 + 2M_a M_p \cos\theta].$$

In Eq. 55*b* the approximation $R^2 = 4x^2$ (the value at $\theta = 0$)—an excellent approximation for nonresonance—is commonly used to simplify the angular integration of the differential cross section. The other equally good approximations and change of variables defined in Eq. 54*h* are used in the calculation of the asymptotic prior cross section,

$$Q(1s \to 1s) = Q(\text{OBK}) L(Z_a, Z_p, M_p),$$
$$L = 1 + \frac{5(Z_p + Z_a)^2}{256} - \frac{5(Z_p + Z_a)}{24}, \tag{55c}$$
$$Q(\text{OBK}) = \frac{64 (Z_a Z_p)^5 \pi a_0^2}{5 x^{12}}.$$

There are several noteworthy cases of Eq. 55*c*. If the incident particle is a proton and the atomic target is H($1s$: $Z_a = 1$), the factor L equals 0.661, a frequently noted result.[29] If the incident particle is again a proton, the interaction used in Eq. 55*a* can be shown to approximate the effective interaction for the capture of $1s$-electrons (the innermost subshell) from a neutral atomic target of atomic number Z_a provided that the wave function

$$Z^{3/2} \pi^{-1/2} \exp(-Z_a r)$$

is a good approximation to the one-electron orbital for the $1s$-subshell (see assumptions following Eq. 43*c* of Section 3.3). A study of $L = L(Z_a, 1, M)$ of this example reveals that L equals 0.661 for $Z_a = 1$, as just noted, and attains its smallest value for positive integers at $Z_a = 4$; L increases monotonically for increasing integral values of Z_a, and at $Z_a = 10$, L exceeds unity.[30] In contrast, the plane wave approximation used with the distorted wave interaction yields asymptotically the OBK cross section.

The prior interaction used in Eq. 55*c* is used to calculate U_i, described in connection with Eq. 45*g* of Section 3.5. The same approximations are used again, and the result of the last calculation is

$$\begin{aligned} V_i - U_i &= \{Z_p(x_f^{-1} - Z_a|\mathbf{x}_i - \mathbf{x}_f|^{-1}) + Z_p|\mathbf{x}_i - \mathbf{x}_f|^{-1} \\ &\quad \times \{Z_a - [1 - \exp(-2Z_a|\mathbf{x}_i - \mathbf{x}_f|)]\} + Z_p Z_a \exp(-2Z_a|\mathbf{x}_i - \mathbf{x}_f|)\} \\ &= Z_p\{x_f^{-1} - |\mathbf{x}_f - \mathbf{x}_i|^{-1}[1 - \exp(-2Z_a|\mathbf{x}_i - \mathbf{x}_f|)] \\ &\qquad + Z_a \exp(-2Z_a|\mathbf{x}_i - \mathbf{x}_f|)\}. \end{aligned} \tag{55d}$$

If the standard integral representations of Eq. 45*c* (Section 3.5) are used to represent the terms containing $|\mathbf{x}_i - \mathbf{x}_f|$ in Eq. 55*d*, the Riemann-Lebesgue theorem[8] can be applied to the corresponding *dw*-amplitude to show that contributions to these integrals are appreciable only in the neighborhood of $|\mathbf{x}_i - \mathbf{x}_f| \approx 0$. Thus it is evident that only the active electronic interaction of Eq. 55*d* contributes to the asymptotic *dw* cross section. The calculation is a straightforward repetition of the previous calculations, and the cross section[31] is

$$\lim_{v_i \to \infty} Q(dw\colon 1s \to 1s) = Q(\text{OBK}), \tag{55e}$$

with Q(OBK) defined in Eq. 55*c*. In all preceding calculations of this section it has been presupposed that the reaction was nonresonant; we now go on to investigate the alterations needed to accommodate exact resonance.

This situation has already been discussed in Section 4.3. In that section the process of charge transfer between H^+ and H(1*s*) is used, and it is found that contributions to the amplitude from the neighborhood of $\theta = \pi$ become important at high impact velocities. Moreover, the approximate amplitude for this process is given by Eq. 52*n* [the factor $\mu_f a_0(2\pi m)^{-1}$ is omitted in that equation] with the correct sign relative to the corresponding asymptotic prior amplitude as given in Eq. 52*f*. This result was obtained with the aid of the peaking approximation; however, since the peaks coincide at $\theta = \pi$, the same integral was also integrated exactly for θ in the neighborhood of π to find that the amplitude in this angular region is correctly given by

$$g_{if}(\theta) = \frac{32\mu_f a_0}{mA_i^2(4 + R^2)^2}, \tag{55f}$$

and this amplitude originates solely from the proton-proton interaction.[32] The associated cross section was found to be equal to

$$Q_{pp} = \frac{1}{12E_i^3}\left(\frac{m}{M}\right)^2 \pi a_0^2, \tag{55g}$$

in which E_i is again expressed in units of 100 keV. Note that the small-angle approximation used previously, $R^2 = 4E_i$, is not applicable in this case since the important angular variation originates from R^2. In a similar calculation, the exact *dw*-interaction was used, and the resulting asymptotic cross section was found to be identical to the one in Eq. 55*g*. [Exact *dw*-interaction means that the approximation, $\mathbf{R}_i \approx \mathbf{x}_i - \mathbf{x}_f$, used to simplify Eq. 45*g* in Section 3.5, *cannot* be used in this calculation. Moreover, it can be proven that cross sections for capture into excited states of H do not contribute to the asymptotic cross section for this backward capturing process[33] (see Appendix 3)]. Although it is known from the discussion relevant to Eq. 52*n* that the cross section of Eq. 55*g* is not defined if the protons are correctly represented as indistinguishable particles, this material is included for a subsequent comparison with an approximate classical mechanical analysis. Next, several asymptotic cross sections are compared.

These comparisons refer to nonresonance only. The cross sections stemming from the sum of the first and second Born approximations using the free-state Green's function, G_0, is given by Eq. 52*k* (Section 4.3). For high, yet nonrelativistic, velocities this cross section is 0.3011 Q(OBK), which is roughly a half of the asymptotic value, 0.661 Q(OBK), calculated from the first Born approximation with the prior interaction. Bransden and Cheshire[14] obtained nearly the same ($2^{-11}v_i$ instead of $2^{-12}v_i$) asymptotic cross section using the impulse approximation, but Cheshire derived the identical result from his continuum distorted wave approximation.[13] Although this cross section calculated from the impulse approximation is clearly related to the active electronic interaction, the contributing interactions are not easily determined in the continuum distorted wave method. More recently McCarroll and Salin[34] used a different approximation to derive the asymptotic result of Cheshire; however, at the time of this writing there was disagreement concerning the validity of some of the approximations.[35,36] Before a numerical comparison of some of the limits is given, a peculiar feature of the impulse approximation is described.

By reason of the function $_1F_1$, the asymptotic amplitude is not uniformly convergent with respect to the interchange of the order of integration and the limit $E_i \to \infty$. If uniform convergence is forced, the limit of the amplitude approaches the limit of the corresponding first Born approximation.[28] It is emphasized that this limit is mathematical; nevertheless, these mathematical vagaries should be noted since the asymptotic cross section obtained by Bransden and Cheshire also disagrees slightly with the results obtained directly from the sum of the first and second Born approximations, the sum of which is the source of the impulse approximation. Coleman and McDowell,[37] moreover, showed that the asymptotic cross section calculated in the impulse approximation agrees with Eq. 55*g* for resonance charge transfer between H^+ and H(1s). In conclusion to this section some numerical comparisons of the asymptotic cross sections are given.

As a first point the OBK cross section is not asymptotic unless $A_f^2 \gg Z_p^2$ and $A_i^2 \gg Z_a^2$. For the accidentally resonant process of capture from D(1s) into H(1s) (nuclear binding infinite), the asymptotic prior cross section, 0.661 Q(OBK) is still not attained at an impact energy of 100 MeV. In the case of capture from He(1s^2) by H^+ into H(1s) leaving the residual ion He^+(1s), the prior cross section has attained only 88% of its asymptotic value, 0.535 Q(OBK), at the same impact energy. For an interesting comparison, the value of the corresponding dw cross sections are equal to 0.97 Q(OBK:D) and 0.95 Q(OBK:He), respectively, at an impact energy of only 10 MeV. Another interesting result has been obtained by Mittleman and Quong,[38] who obtained the cross section (varying as E_i^{-6}),

$$Q = (23/48)Q(\text{OBK}),$$

for charge transfer between protons and hydrogen atoms in the 1s-state. Their approximation resulted from a study of the regularity of the phases and amplitudes between exact and approximate solutions to one-dimensional models of the capturing process. It is reemphasized that a simple law for the variation of the cross section with impact energy for electronic capture from complex atoms is never achieved in the nonrelativistic domain of impact energies. Even for capture by protons from the simplest atoms—atoms containing only s-electrons in their ground state configuration—the cross sections obtained from the prior and post interactions as well as the sum of the first and second Born approximations do not attain their asymptotic limit for impact energies below the region starting at 100 MeV.

REFERENCES

1. R. Aaron, R. D. Amado, and B. W. Lee, *Phys. Rev.*, **121**, 319 (1961).
2. K. Dettman and G. Leibfried, *Phys. Rev.*, **148**, 1271 (1966).
3. R. A. Mapleton, "Classical and Quantal Calculations on Electron Capture," Air Force Cambridge Research Laboratories Report AFCRL 67-0351, Physical Science Research Papers, No. 328, 1967, Sections 8–11, Part I.
4. R. A. Mapleton, *Proc. Phys. Soc.* (*London*), **91**, 868 (1967).
5. E. Gerjuoy, *Phys. Rev.*, **109**, 1806 (1958).
6. E. Gerjuoy, *Ann. Phys.* (*N. Y.*), **5**, 58 (1958).
7. N. F. Mott and H. S. W. Massey, *The Theory of Atomic Collisions* (3d ed.), Oxford University Press, London (1965), Chapter 3.
8. E. C. Titchmarsh, *The Theory of Functions*, Oxford University Press, London (1950), pp. 403–404.
9. K. Dettman and G. Leibfried, *Z. Physik*, **210**, 43 (1968).

10. R. M. Drisko, Thesis, Carnegie Institute of Technology (1955), pp. 28–68.
11. J. D. Jackson and H. Schiff, *Phys. Rev.*, **89**, 359 (1953).
12. C. P. Carpenter and T. F. Tuan, *Phys. Rev.*, **2**, 1811 (1970).
13. I. M. Cheshire, *Proc. Phys. Soc.* (*London*), **84**, 89 (1964).
14. B. H. Bransden and I. M. Cheshire, *Proc. Phys. Soc.* (*London*), **81**, 820 (1963).
15. L. H. Thomas, *Proc. Roy. Soc.* (*London*)*A*, **114**, 561 (1927).
16. T. Pradhan, *Phys. Rev.*, **105**, 1250 (1957).
17. Mott and Massey, *op. cit.*, pp. 55–57, 77–75.
18. M. L. Goldberger and K. M. Watson, *Collision Theory*, Wiley, New York (1964), pp. 203–204.
19. D. R. Bates and Anne H. Boyd, *Proc. Phys. Soc.* (*London*), **79**, 710 (1962).
20. D. R. Bates and Anne H. Boyd, *Proc. Phys. Soc.* (*London*), **80**, 1301 (1962).
21. L. Pauling and E. B. Wilson, Jr., *Introduction to Quantum Mechanics*, McGraw-Hill, New York (1935), Chapter 5.
22. J. R. Oppenheimer, *Phys. Rev.*, **31**, 349 (1928).
23. H. C. Brinkman and H. A. Kramers, *Proc. Acad. Sci. Amsterdam*, **33**, 973 (1930).
24. D. R. Bates and R. McCarroll, *Phil. Mag. Suppl.*, **11**, 39 (1962).
25. D. R. Bates, "Theoretical Treatment of Collisions between Atomic Systems," in *Atomic and Molecular Processes*, Academic Press, New York (1962), p. 578.
26. M. J. Lighthill, *Introduction to Fourier Analysis and Generalized Functions*, Cambridge University Press, London (1958), Chapter 4.
27. R. A. Mapleton, *Phys. Rev.*, **122**, 528 (1961).
28. R. A. Mapleton, *Proc. Phys. Soc.* (*London*), **83**, 35 (1964).
29. Mott and Massey, *op. cit.*, p. 621.
30. R. A. Mapleton, *J. Phys. B.*, **1**, 529 (1968).
31. R. H. Bassel and E. Gerjuoy, *Phys. Rev.*, **117**, 749 (1960).
32. R. A. Mapleton, *Proc. Phys. Soc.* (*London*), **83**, 895 (1964).
33. Mapleton, AFCRL 67-0351, app. F, pt. I.
34. R. McCarroll and A. Salin, *Proc. Phys. Soc.* (*London*), **90**, 63 (1967).
35. J. P. Coleman, *J. Phys. B*, **1**, 315 (1968).
36. R. McCarroll and A. Salin, *J. Phys. B*, **1**, 317 (1968).
37. J. P. Coleman and M. R. C. McDowell, *Proc. Phys. Soc.* (*London*), **83**, 907 (1964).
38. M. H. Mittleman and J. Quong, *Phys. Rev.*, **167**, 74 (1968).

CHAPTER 5

CLASSICAL METHODS FOR HIGH VELOCITIES

5.1 INTRODUCTION

In recent years there has been a revival of interest in the classical theory of collisions. Initially this interest in classical methods was occasioned by the success of Gryzinski in predicting cross sections for excitation and ionization using classical methods.[1] Since then many other approximate classical methods have been proposed for the calculation of cross sections for direct inelastic processes. In the field of electronic capture, however, Thomas first developed the most elaborate approximate classical theory for high impact velocities.[2] Bohr analyzed the problem less completely by defining conditions that an ejected electron must satisfy if it is to remain bound in the field of the capturing incident ion.[3] The first recent major contribution was made by Abrines and Percival, who used the Monte Carlo method to solve the classical equations for the three-body motion of two protons and an electron, and thus obtained the cross section for electronic capture in $H^+ - H(1s)$ collisions.[4] (The meaning of the classical $1s$-state is explained in a subsequent section.) Bates and Mapleton investigated the corresponding classical collision for resonant charge transfer and found that the approximate asymptotic cross section varied with impact velocity, as predicted by the first Born approximation if the two protons were treated as distinguishable particles.[5] A common feature shared by these last two problems is the classical distribution used to represent the initial conditions of the atomic target, and it therefore seems appropriate to devote the first section on classical methods to a description and explanation of some of the classical distributions used in classical scattering theory.

5.2 CLASSICAL DISTRIBUTIONS FOR THE ATTRACTIVE COULOMB FIELD

Classical distributions are used in the classical analysis of collisional theory, in particular, to represent the distributions of electrons in the atomic target. Since the effective interaction between an electron and the rest of a many-electron atom is not a single Coulomb potential, this classical distribution is severely limited in its application. Although the single Coulomb potential is of limited value in this application, the potential is very useful in other ways since classical collisional processes are commonly approximated by one or more binary interactions, and these interactions *are* single Coulomb potentials. Nevertheless, if the classical distribution is used to represent the initial velocity (or spatial) distribution of electrons in the atomic target, the application is limited to electronic capture from hydrogenic systems, because we are unable to express classical distributions in analytic form for more complicated atomic systems. This hindrance, however, is no more restrictive than the corresponding impediment of being unable to derive *exact* analytic wave functions for most quantum mechanical systems.

Assume that the atomic nucleus, a proton, is at the origin of the system of coordinates, and the position $\mathbf{r}$ and velocity $\mathbf{v}$ of the electron are measured relative to this center of force. In this frame of reference, the classical Hamiltonian, the total momentum, and the energy of the electron-proton system are denoted by H, $\mathbf{p}$, and $-\varepsilon(\varepsilon > 0)$, respectively. According to classical statistical mechanics the microcanonical, or energy-shell, ensemble for the associated six-dimensional phase space of the electron is[6]

$$\rho_c = \frac{N}{(4\pi)^2}\,\delta(H + \varepsilon), \qquad H = \frac{p^2}{2m} - \frac{e^2}{r}. \tag{56a}$$

Alternatively, in the language of classical statistical mechanics, ρ_c can be defined as the probability that all points of phase space satisfying

$$H = -\varepsilon$$

are equally probable. However, if ρ_c is a probability, the normalizing factor N is determined so that the integral of ρ_c over the allowed phase space is equal to unity. An integral over phase space, it is important to remember, is defined in terms of canonically conjugate spatial and momentum variables of integration, and other sets of variables must be obtained from the original variables by appropriate transformations.

The other designation, energy-shell ensemble, is also easy to relate to Eq. 56*a*. As explained in treatises on statistical mechanics, the energy-shell ensemble is defined in terms of a phase probability D, which is constant between the energies ε_0, and $\varepsilon_0 + \Delta\varepsilon$, and vanishes elsewhere.[7] As an aid in demonstrating the equality of the two definitions, let $\eta(x)$ represent the Heaviside unit function defined by

$$\eta(x) = \begin{cases} 1, & x > 0 \\ 0, & x < 0, \end{cases}$$

and construct the function

$$D = \frac{N}{(4\pi)^2}\left[\frac{\eta(\varepsilon_0 + \Delta\varepsilon) - \eta(\varepsilon_0)}{\Delta\varepsilon}\right].$$

This function fulfills the requirements for D if N is correctly chosen; furthermore, in the limit $\Delta\varepsilon \to 0$, we get

$$\lim_{\Delta\varepsilon\to 0} D = \frac{N}{(4\pi)^2}\frac{d\eta(\varepsilon)}{d\varepsilon} = \frac{N}{(4\pi)^2}\frac{d\eta(H+\varepsilon)}{d(H+\varepsilon)} = \frac{N}{(4\pi)^2}\,\delta(H+\varepsilon),$$

which is Eq. 1*a* for ρ_c again.[8] Since it is plausible that the two definitions are equivalent, Eq. 56*a* is used to calculate the classical distributions.

Pitaevskii,[9] Omidvar,[10] Abrines and Percival,[11] and Mapleton[12,13] all have contributed to the evaluation and the analysis of the properties of the classical distributions, but the presentation here most closely parallels the approach used by Mapleton. First the transformations and the associated Jacobian,

$$\begin{aligned} p^2 &= p_x^2 + p_y^2 + p_z^2 = p_r^2 + \frac{p_\theta^2}{r^2} + \frac{p_\phi^2}{r^2\sin^2\theta}, \\ p_r &= p\sin\theta_p\cos\phi_p, \qquad p_\phi = rp\sin\theta\cos\theta_p, \\ p_\theta &= rp\sin\theta_p\sin\phi_p, \qquad J = r^2p^2\sin\theta\sin\theta_p, \end{aligned} \tag{56b}$$

are used to transform an integration over a canonically conjugate set of variables into an integration over a noncanonically conjugate set:

$$\begin{aligned} &\frac{N}{(4\pi)^2}\int dp_r\,dp_\theta\,dp_\phi\,dr\,d\theta\,d\phi\,\delta\left(\frac{p^2}{2m} - \frac{e^2}{r} + \varepsilon\right) \\ &\qquad = N\int dpp^2\,drr^2\,\delta\left(\frac{p^2}{2m} - \frac{e^2}{r} + \varepsilon\right) = 1. \end{aligned} \tag{56c}$$

Only because ρ_c of Eq. 56*a* is spherically symmetric is it possible to express Eq. 56*c* as a two-dimensional integral over p and r. The integration is readily accomplished with the aid of the formula[14]

$$\int dx\delta[f(x)]\phi(x) = \sum_n \frac{\phi(x_n)}{|f'(x_n)|}, \tag{56d}$$

in which x_n, the zeros of $f(x)$, are in the domain of integration. This calculation yields two distributions in addition to the value of the normalizing constant N,

$$N = \frac{8}{\pi e^6}\left(\frac{2\varepsilon^5}{m^3}\right)^{1/2}, \qquad \varepsilon = \varepsilon_n = \frac{e^2}{2a_n} = \frac{mv_n^{\,2}}{2},$$
$$\frac{2}{\pi a_n^{\,3}}\int_0^{2a_n} drr^2\left(\frac{2a_n - r}{r}\right)^{1/2} = 1 \tag{56e}$$

and

$$\frac{32v_n^{\,5}}{\pi}\int_0^{\infty}\frac{dvv^2}{(v^2 + v_n^{\,2})^4} = 1, \qquad p = mv, \qquad p_n = mv_n,$$
$$\frac{32v_n^{\,5}v^2\,dv}{\pi(v^2 + v_n^{\,2})^4} = \frac{8p_n^{\,5}}{\pi^2(p^2 + p_n^{\,2})^4}(4\pi p^2\,dp). \tag{56f}$$

The distribution in velocity of Eq. 56*f* is also expressed in the corresponding momentum variables in order to facilitate a comparison with the quantum mechanical one originally obtained by Fock.[15] That relation is

$$\frac{4\pi}{n^2}\sum_{l=0}^{n-1}\sum_{m=-l}^{l}|\phi_{nlm}(\mathbf{p})|^2p^2\,dp = \frac{32p_n^{\,5}p^2\,dp}{\pi(p^2 + p_n^{\,2})^4},$$
$$v_n = \frac{v_1}{n}, \qquad p_n = \frac{p_1}{n}, \qquad v_1 = \frac{e^2}{\hbar}, \tag{56g}$$

In Eq. 56*g*, $\phi_{nlm}(\mathbf{p})$ are the wave functions in momentum space for atomic hydrogen in the energy level corresponding to the principal quantum number n. In order to get this result we must assume that all levels of the state n must be equally populated; this is an important fact to remember in connection with applications to physical systems. Furthermore, with the aid of the change of variable, $p = P/n$, it is clear that the right-hand member of Eq. 56*g*, as a function of P, is independent of n.[16] If p_n of Eq. 56*f* and Eq. 56*g* are equal, the distribution in momentum (or velocity) for the classical ensemble is formally identical with the quantum mechanical distribution in momentum (or velocity) for atomic hydrogen.[17] With this one correlation between the quantum mechanical and classical mechanical distribution, it is tempting to seek an additional analogy involving an integral over the classical angular momentum.

If lengths and velocities in Eq. 56*e* and Eq. 56*f* transform according to the relations presented in Eq. 29*h* in Section 2.1, the resulting distributions in the new variables of integration have the original form, although the subscripts are changed from 1 to n. According to the correspondence principle,[16] moreover, in the limit of large quantum numbers n, the quantum mechanical distribution of Eq. 56*g* equals the corresponding classical microcanonical distribution of Eq. 56*f*. This fact, together with the invariance in form of the classical distributions, enabled Abrines and Percival to obtain universal normalized cross sections for any given initial classical level, and justified as plausible their comparison of calculated cross sections with experiment for $n = 1$. They also noted that the complementary positional distribution does not agree with the quantum mechanical one, but they argued that the distribution of electronic momentum is the more important one for the determination of cross sections, a conjecture which will receive additional support in a subsequent section.[18] We now return to the study of the integral over the classical angular momentum.

First, we note that the angular momenta, p_θ and p_ϕ, are constants of the motion for the bound-state problem of interest,[19] but this information has not been used. A way of incorporating this knowledge is to include the classical angular momentum l as a variable of integration[20] together with the other six variables in the normalizing integral for ρ_c. This is a mathematical device. In the strict dynamical sense the specification of p_θ and p_ϕ usually dictates a particular orientation of the orbital plane; however, in the present application these orbital angular momenta are scalars, and carry no associated direction:

$$\frac{N'}{l_n^2}\int dll\, dp_r\, dp_\theta\, dp_\phi\, dr\, d\theta\, d\phi\, \delta(p_\phi)\, \delta(p_\theta - l)\, \delta(H + \varepsilon) = 1, \qquad l_n^2 = \tfrac{1}{2}(mv_n a)^2. \tag{56h}$$

The integrations over p_θ, p_ϕ, θ, and ϕ are easily effected to get

$$S = \frac{2\pi^2 N'}{l_n^2}\int dll\, dp_r\, dr\, \delta(H + \varepsilon) = 1, \tag{56i}$$

and the transformation (with the associated Jacobian J),

$$p_r^2 = p^2 - \frac{l^2}{r^2}, \qquad l = l', \qquad r = r', \qquad J = p\left(p^2 - \frac{l^2}{r^2}\right)^{-1/2}, \tag{56j}$$

is used to facilitate the integration over dl,

$$\begin{aligned} S &= \frac{2\pi^2 N'}{l_n^2}\int dllJ\, dp\, dr\, \delta\left(\frac{p^2}{2m} - \frac{e^2}{r} + \varepsilon\right) \\ &= \frac{2\pi^2 N'}{l_n^2}\int dpp^2\, drr^2\, \delta\left(\frac{p^2}{2m} - \frac{e^2}{r} + \varepsilon\right) = 1, \end{aligned} \tag{56k}$$

$$N' = \frac{v_n}{2a_n\pi^3} = \frac{l_n^2}{2\pi^2} N.$$

In the preceding integration over dl, the range of l is restricted to non-negative l such that the integrand is real, which ensures real radial momentum p_r.

All that has been demonstrated, however, is that Eq. 56h can be reduced to Eq. 56c; the desired analogy to Eq. 56g is obtained by doing the integration over dr first. The remaining integrand in dp is

$$I = \frac{v_n^2 e^4}{\pi a_n l_n^2}\, dpp \int_0^{L_3} dll \left(\frac{L_3}{e^2 p}\right)^4 (L_3^2 - l^2)^{-1/2} \tag{56l}$$

$$= \frac{32 v_n^5 v^2\, dv}{\pi (v^2 + v_n^2)^4}, \qquad L_3 = e^2 p \left(\frac{p^2}{2m} + \varepsilon\right)^{-1}.$$

Thus in analogy to the sum over the quantized angular momentum l in Eq. 56g, an integral over the classical angular momentum has been obtained.

Equation 56j has another use: it enables us to deduce distributions corresponding to *special values* of l. As a first example, the integration over dp_r is effected to get

$$S = \frac{1}{\pi a_n l_n^2} \int_0^{2a_n} drr \int_0^{L_1} \frac{dll}{[2a_n r - r^2 - (l/mv_n)^2]^{1/2}}$$

$$= \frac{1}{\pi a_n l_n^2} \int_0^{(2l_n)^{1/2}} dll \int_{r_-}^{r_+} \frac{drr}{[(r_+ - r)(r - r_-)]^{1/2}} = 1, \tag{56m}$$

$$\left.\begin{matrix} r_+ \\ r_- \end{matrix}\right\} = a_n\left[1 \pm \left(1 - \frac{l^2}{2l_n^2}\right)^{1/2}\right], \qquad L_1 = mv_n(2a_n r - r^2)^{1/2}.$$

The domain of r in the second integral is between the classical turning points $r_\pm$, and the value of this integral is πa_n. The remaining distribution in l^2 is a constant, and this same result was obtained by Abrines and Percival using a more fundamental derivation.[11] If the relation for the radial velocity used in deriving Eq. 56m from Eq. 56i is invoked again, an instructive expression can be deduced from Eq. 56m:

$$\frac{1}{\pi a_n} \frac{drr}{[(r_+ - r)(r - r_-)]^{1/2}} = \frac{v_n\, dt}{\pi a_n} = \frac{dt}{T}, \tag{56o}$$

$$T = \frac{2\pi a_n}{v_n}, \qquad dr = v_r\, dt.$$

In this equation T is the period of the classical motion; t is the time, and it is noteworthy that (dt/T) represents the classical probability for finding the electron in the range dt about t.

As a second example the integral over dr in Eq. $56i$ is evaluated first to get

$$S = \frac{2v_n}{\pi a_n m l_n^2} \int_0^\infty \frac{dv_r}{(v_r^2 + v_n^2)^2} \int_0^{L_2} \frac{dll^2(2v_c^2 - v_n^2 - v_r^2)}{|v_c^2 - v_n^2 - v_r^2|^{1/2}}$$

$$= \frac{16v_n^5}{3\pi} \int_0^\infty \frac{dv_r}{(v_r^2 + v_n^2)^3}, \tag{56p}$$

$$v_c = \frac{e^2}{l}, \qquad L_2^2 = \frac{e^4}{v_r^2 + v_n^2}.$$

In Eq. $56p$ we have a distribution in the radial velocity, and it is emphasized that the limits of integration always are consistent with conservation of energy as expressed by Eq. $56a$. Of peripheral interest here are the distributions for the degenerate line ellipse which can be deduced by first removing $\int dll/l_n^2$ from the integrals of Eq. $56m$ and Eq. $56p$ and afterwards calculating the limit, $l \to 0$. These limiting values are

$$I_1 = \frac{1}{\pi a}\left(\frac{r}{2a_n - r}\right)^{1/2} dr$$

$$I_2 = \frac{4v_n^3\, dv}{\pi(v^2 + v_n^2)^2}, \qquad l = 0, \qquad v = v_r. \tag{56q}$$

The distributions for the circular orbit, nevertheless, still remain unknown, and these results are derived next.

In this limiting case, it is first noted from Eq. $56m$ that

$$r_+ \to r_- \to a_n, \qquad l^2 \to 2l_n^2, \tag{56r}$$

and the associated velocities are

$$v(l) = \frac{p}{m} = v_n, \qquad v_r = 0.$$

The spatial distribution is determined first, and for this purpose the spatial integral of Eq. $56m$ is replaced by the equivalent contour integral (the integration $\int dll/l_n^2$ has been removed here also):

$$I(1:r_\pm) = \frac{1}{2\pi a_n} \oint_C \frac{dzz}{[(r_+ - z)(z - r_-)]^{1/2}}. \tag{56s}$$

The path of integration traverses C in a counterclockwise sense, consisting of line segments parallel to, and just above and below, the branch cut $r_- \le \operatorname{Re} z \le r_+$ (Re z denotes the real part of z), the ends of which segments are connected by small circles with centers located at (r_-, io) and (r_+, io). The phases of the square root in Eq. $56s$ on the lower and upper line segments are positive and negative, respectively. Since there is no contribution to $I(1; r_\pm)$ from the integrals over the small circles whose radii are equated to zero after the integration is performed, it is clear that Eq. 56s is equivalent to the original integral. In order to determine the form of the spatial distribution in Eq. 56s, the function $I(\phi; r_\pm)$ is studied; in this associated integral the function $\phi(z)$ is analytic on and within C. We now impose the limit defined in Eq. 56r and since the branch cut shrinks to a point at $\mathrm{Re}\, z = a_n$, the singularity of the integrand becomes a simple pole, and the standard method of residues[21] is used to find

$$I(\phi; a_n) = \frac{i\phi(a_n)}{\exp(i\pi/2)} = \phi(a_n). \tag{56t}$$

The phase of the radicand at $z = a_n$ is determined by requiring $(a - z) = (z - a)e^{\pi i}$ so that the value of $I(1; a_n) = 1$ as it must be. The result in Eq. 56t can also be expressed by the distribution

$$\delta(r - a_n) \tag{56u}$$

for the class of functions $\phi(z)$. It is in this sense that the limit of the distribution in Eq. 56m is represented by Eq. 56u. This same type of analysis is applicable to Eq. 56p and the corresponding contour integral is

$$\begin{gathered} I(\phi; v_\pm) = \frac{4v_n{}^4}{\pi} \oint_C \frac{dz z\phi(z)}{(z^2 + v_n{}^2)^2[(v_+{}^2 - z^2)(z^2 - v_-{}^2)]^{1/2}}, \\ v_\pm = v(l) \pm [v^2(l) - v_n{}^2]^{1/2}, \qquad v_n \le v(l) \le \infty, \\ 0 < l \le l_c = mv_n a_n, \\ l \to l_c, \qquad v(l) \to v_n, \qquad v_\pm \to v_n. \end{gathered} \tag{56v}$$

The branch cut in the complex velocity-plane is $v_- \le v \le v_+$, but here care is required to ensure that the other singularities at $-v_\pm$, $\pm iv_n$ are outside the region enclosed by C. The result is the same as in Eq. 56u if a_n is replaced by v_n and r by v; thus the distribution in velocity for the circular orbit is

$$\delta(v - v_n). \tag{56w}$$

The results of this section will be applied to several problems in the last section on classical methods; however, the next topic discussed is a classical method for electronic capture from neutral atoms by positively charged nuclei.

5.3 THOMAS MODEL FOR CAPTURE FROM HEAVY ATOMS

In this section a classical model originally proposed by Thomas[2] is reviewed. This model was designed for electronic capture from heavy neutral atoms by positively charged nuclei of high impact velocity, and we begin by discussing the essential ideas of this classical model.

The scattering problem confronted here is a three-body problem if the atom is approximated by an electron and an atomic core, or center of force, and without additional simplifications numerical methods must be used at the outset in order to obtain cross sections. However, much of the calculation can be effected analytically if approximating binary collisions are used, meaning that the original problem is approximated by several two-body problems. This possibility is suggested by first observing that this splitting into two-body collisions is possible if the potential of the atomic field at the electron is large compared with the potential from the incident particle at the *same distance*. Accordingly, we imagine that the incoming nucleus first passes closely to the electron in comparison with the electronic distance from the nucleus; furthermore, we suppose that sufficient energy is imparted to the electron for it to escape from the atomic nuclear field. Subject to this last condition the electron and incident particle separate with high relative velocity after this close encounter, and subsequently only the atomic field will influence the motion of the electron until after both the electron and incident particle have left this atomic field. If these two conditions—(1) the close encounter and (2) the subsequent rapid separation—are fulfilled, this three-body problem may be approximated by two binary collisions. This approximation by binary processes is called the *classical impulse approximation*.[22,23]

On the basis of this approximation, we could replace the atomic core by available atomic fields and proceed to calculate the cross sections for electronic capture from each subshell as a function of impact velocity. This proposed calculation, however, is still formidable. Instead, Thomas studied the pioneering results of Hartree[24] and found that many of these Hartree atomic potentials were closely approximated by an inverse square potential for "the region relevant to the capture of electrons by alpha particles." The point emphasized here is that the atomic model containing subshells was discarded as calculationally intractable during that era, when there were no adequate computational facilities. Before the model of Thomas is mathematically described, however, the purpose of the revival of this classical model is explained.

Since classical scattering processes are easily understood, though difficult to calculate, it was hoped that the results of these studies would suggest new approximations to the corresponding quantum mechanical problem. Furthermore, researchers desired to learn how well, or poorly, classical cross sections would predict the measured values. Finally, in order to thoroughly understand the effect of some of the assumptions, the Thomas model had to be modified and the predictions of the newly developed models had to be compared with the original ones. Thus the original model is described first.[25,26]

Let **V** represent the impact velocity of the incident nucleus P, which contains Z electronic units of positive charge, and neglect the motions of the atomic target A and its residual ion A^+. Consequently, the position of the atomic target can be chosen to be coincident with the origin of the laboratory system of coordinates. Furthermore, let $\mathbf{v}$, $\mathbf{v}'$, and $\mathbf{v}''$ represent the electronic velocities relative to this same origin before the first binary collision with P, after this collision, and after the binary collision with A^+, respectively. Moreover, denote the electronic velocities relative to P by

$$\mathbf{u} = \mathbf{v} - \mathbf{V}, \qquad \mathbf{u}' = \mathbf{v}' - \mathbf{V}, \qquad \mathbf{u}'' = \mathbf{v}'' - \mathbf{V}. \tag{57a}$$

The program of Thomas is to impart to the electron through the first elastic Coulomb scattering an intermediate velocity $\mathbf{v}'$ so that with $\mathbf{v}'$ as the initial velocity of the next binary encounter the electron will be scattered elastically by the effective atomic field into a final velocity $\mathbf{v}''$ appropriate to ultimate capture by P; this is the cross section Thomas calculates. Although momentum and kinetic energy are transferred to the electron, the incident particle is assumed to remain in uniform rectilinear motion. This is a good approximation if the impact energy is large compared with the transfer of kinetic energy to the electron. It is evident that the only important initial binary encounters are those that make

$$v' > w, \tag{57b}$$

in which w is the velocity of escape, the electronic speed required to escape from the effective field of the atom. In view of the assumption that the atomic field at the electron is large in comparison with the one of the incident particle for comparable distances, Eq. 57*b* is fulfilled only if the impact parameter p between P and e is much smaller than the electron-atomic nucleus distance r:

$$p \ll r \tag{57c}$$

(This requirement on the two potentials is satisfied if $|Ze|$ is much less than the atomic nuclear charge, and close encounters with the atomic nucleus are here implied.) Furthermore, it is assumed that r changes inappreciably during the first binary collision.

As an aid to the subsequent description the geometry of the collisional processes are depicted in Figs. 7 and 8. The angles between the vectors are

$$\begin{aligned} &\chi = \widehat{\mathbf{Vv}}, \qquad \chi' = \widehat{\mathbf{Vv}'}, \\ &\theta = \widehat{\mathbf{Vu}}, \qquad \theta' = \widehat{\mathbf{Vu}'}, \qquad \delta = \widehat{\mathbf{uu}'}, \\ &\eta' = \widehat{\mathbf{rv}'} \qquad \eta'' = \widehat{\mathbf{rv}''}, \end{aligned} \tag{57d}$$

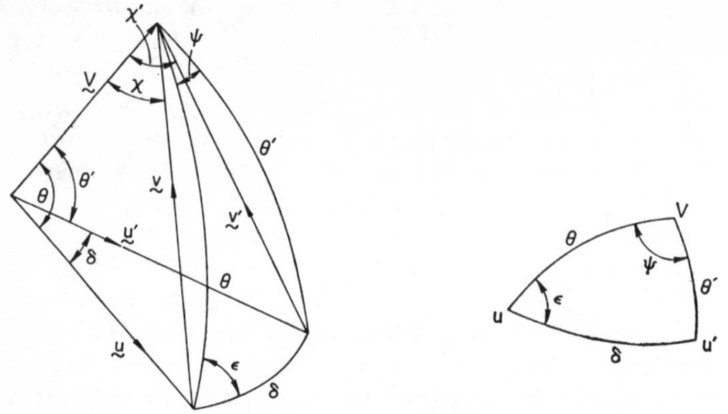

Fig. 7 Diagram of spherical triangles for first binary collision in Thomas' model for capture from heavy atoms.

and the dihedral angles between certain planes are

$$\psi = \widehat{(\mathbf{Vu})(\mathbf{Vu}')}, \qquad \phi = \widehat{(\mathbf{Vv}')(\mathbf{rv}')}. \qquad \varepsilon = \widehat{(\mathbf{Vu})(\mathbf{uu}')}, \tag{57e}$$

(Most of the angles of Eq. 57*d* are also the *face* angles of spherical trigonometry.) Although one of the planes defining ϕ is the $(\mathbf{rv}')$-plane, this plane also contains $\mathbf{v}''$, a fact explained in another part of this section. We are now prepared to finish the calculation.

If the differential scattering cross section for the $P - e$ binary collision in the domain

$$(v, r, p, \chi, \varepsilon) \rightarrow (v + dv, r + dr, p + dp, \chi + d\chi, \varepsilon + d\varepsilon) \tag{57f}$$

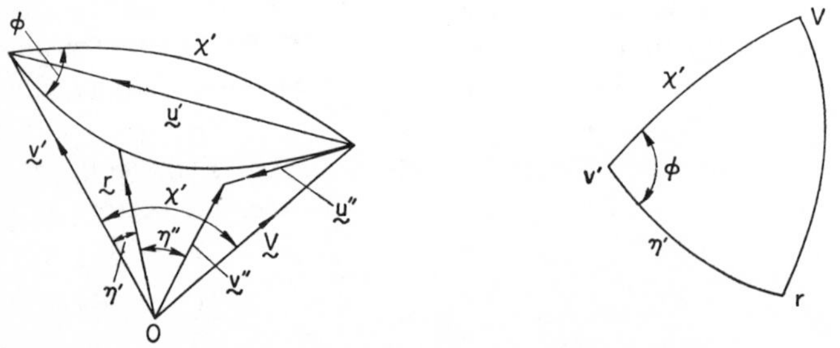

Fig. 8 Diagram of spherical triangles for second binary collision in Thomas' model for capture from heavy atoms.

is dq, and if the fraction of these collisions in which $\mathbf{r}$ is such that the electron is left in the angular domain

$$(\eta', \phi) \to (\eta' + d\eta, \phi + d\phi) \tag{57g}$$

is df, the cross section sought is

$$Q = \int_R \frac{u}{V} \, dq \, df. \tag{57h}$$

In this integral R signifies that the integration is restricted to those parts of the multidimensional space for which the conditions of capture are fulfilled. The three factors of the corresponding integrand are now explained. By definition $u\,dt$ represents the distance traveled by the electron relative to P in the time dt, and if dl denotes the distance traveled by P during the same time, it is clear that

$$u\,dt = \frac{u\,dl}{V}.$$

If we also assume that P is incident upon on a very thin plane sheet of N noninteracting atoms per unit volume with the direction of incidence parallel to the thickness of the sheet, and if we neglect multiple scattering, it follows that the probability dS of P colliding with e in traveling the distance $u\,dt$ is

$$dS = N\,d\sigma u\,dt = N d\sigma \frac{u\,dl}{V}$$

in which $d\sigma$ is the differential cross section. It is seen that the cross section $dS(N\,dl)^{-1}$ agrees in form with the integrand of Eq. 57*h*.[27]

The volume element, dq, of this integral consists of a distribution for the electrons

$$X(r, v)\,dr\,dv, \tag{57i}$$

a differential cross section

$$d\sigma = p\,dp\,d\varepsilon, \tag{57j}$$

and the classical probability that m makes an angle χ with V in the range $d\chi$

$$\frac{d(\cos\chi)}{2} \tag{57k}$$

since all directions of $\mathbf{v}$ are equally likely. Moreover, the distribution $X(r, v)$ is assumed isotropic in both $\mathbf{r}$ and $\mathbf{v}$. For a direction of $\mathbf{v}$ specified by the angle χ, the electron-atom line $\mathbf{r}$ is still equally likely to assume any orientation, and the chance that $\mathbf{r}$ has the direction specified by the angles η' and ϕ

in the ranges $d\eta'$ and $d\phi$ is

$$df = \frac{d(\cos \eta')}{4\pi}\, d\phi. \tag{57l}$$

This completes the explanation of the structure of Eq. 57*h*; however, there remains the difficult task of evaluating the integral.

Thomas noted that a change of the independent variables from (p, χ, ε) to (v', ψ, χ') is advantageous, and he accomplished this transformation in two steps as now described. Since the first binary process $(P - e)$ is elastic Coulomb scattering with scattering angle δ between the initial and final velocities of relative motion **u** and **u**′, we have the relation[28]

$$p\, dp = \frac{\mu^2\, d(\cos \delta)}{u^4(1 - \cos \delta)^2}, \qquad |\mathbf{u}| = |\mathbf{u}'|, \qquad \mu = \frac{Ze^2}{m} \tag{57m}$$

in which m is the electronic mass. Next, attention is directed to Fig. 7 for relations among the velocities and among the sides and angles of the spherical triangle; these relations are given in Eq. 57*n*:

$$\begin{aligned} \cos \delta &= \cos \theta \cos \theta' + \sin \theta \sin \theta' \cos \psi \\ \sin \delta \sin \varepsilon &= \sin \theta' \sin \psi \\ (u')^2 &= (v')^2 + V^2 - v'V \cos \chi' = u^2 = v^2 + V^2 - 2vV \cos \chi, \\ v^2 &= u^2 + V^2 - 2uV \cos \theta \\ (v')^2 &= (u')^2 + V^2 - 2u'V \cos \theta' = u^2 + V^2 - 2uV \cos \theta. \end{aligned} \tag{57n}$$

These relations are used to effect the change of variables,

$$d(\cos \delta)\, d\varepsilon\, d(\cos \chi) = J\, dv'\, d\psi\, d(\cos \chi'), \tag{57o}$$

in which the value of the Jacobian is

$$J = \frac{\partial(\cos \delta, \varepsilon, \cos \chi)}{\partial(v', \psi, \cos \chi')} = \frac{(v')^2}{vuV}. \tag{57p}$$

The purpose of this detailed explanation is to prevent confusing dependent variables with independent ones.

Next, we apply the law of cosines in Eq. 57*n* to effect the integration over $d\psi$, and with the use of this result and Eqs. 57*i* to 57*p*, Eq. 57*h* can be reduced to

$$\begin{aligned} Q &= \frac{\mu^2}{4V^2} \int_R \frac{(v')^2 X(v, r) s(u, v, v', V)}{vu^4}\, dv\, dr\, dv'\, d\phi\, d(\cos \chi')\, d(\cos \eta'), \\ s &= 2uV\left[\frac{4u^2V^2 - (V^2 + u^2 - v^2)(V^2 + u^2 - (v')^2)}{|v^2 - (v')^2|^3}\right], \\ &\int \frac{d\psi}{(1 - \cos \delta)^2} = \frac{2\pi(1 - \cos \theta \cos \theta')}{|\cos \theta - \cos \theta'|^3}. \end{aligned} \tag{57q}$$

(The integral over $d\psi$ is included for convenience.)

Thomas ingeniously simplified the integral representing Q additionally by first using the classical impulse approximation for the second binary collision

$$\frac{m}{2}(v')^2 - \frac{m\lambda}{r^2} = \frac{m}{2}(v'')^2, \tag{57r}$$

$$\frac{mw^2}{2} = \frac{m\lambda}{r^2}$$

in order to get the change of variables

$$v'\,dv' = v''\,dv'' \tag{57s}$$

holding r, or equivalently w, constant. The approximation of the *inverse square potential* for the *atomic field* first appears in Eq. 57r where it is related to the escape velocity w. (The determination of the value of λ is discussed in a subsequent part of this section.) In connection with Eq. 57r the vectors $\mathbf{r}$, $\mathbf{v}'$, and $\mathbf{v}''$ are coplanar since the second binary collision is also two-body motion about a center of force. The next part of the simplification arises from the relation between the solid angle generated by the tip of the $\mathbf{v}''$-vector as the angle ϕ between the planes ($\mathbf{V}\mathbf{v}'$) and ($\mathbf{v}'\mathbf{r}\mathbf{v}''$), and the angle χ' between $\mathbf{V}$ and $\mathbf{v}'$, are varied keeping $|\mathbf{v}'|$, η', and η'' constant. For this purpose the convention is adopted that the line element $ds = d\alpha$ generated by a unit radius and the angular change $d\alpha$, is represented by the vector of magnitude $d\alpha$ directed along the perpendicular to the plane of the rotation. The positive sense of the vector is determined according to the right-hand-screw convention. Thus the variation $d\phi$ is directed along $\mathbf{v}'$ in Fig. 8, and the component perpendicular to $\mathbf{v}''$ is

$$d\phi \sin(\eta' + \eta'').$$

The variation $d\chi'$ in the plane ($\mathbf{v}'\mathbf{V}$) is directed perpendicular to this plane (at the point 0, for reference) and is projected perpendicular to both $\mathbf{v}''$ and the preceding projected axis of rotation; its value is

$$d\chi' \cos\phi.$$

Consequently, the solid angle generated by these two perpendicular arcs is

$$d\Omega(v'') = \sin(\eta' + \eta'')\cos\phi\, d\phi\, d\chi', \tag{57t}$$

and this result combined with Eq. 57s permits the associated volume element generated by $\mathbf{v}''$ to be written as

$$d\tau = v''^2\,dv''\,d\Omega(v'') = v''v'\sin(\eta' + \eta'')\cos\phi\,dv'\,d\phi\,d\chi'. \tag{57u}$$

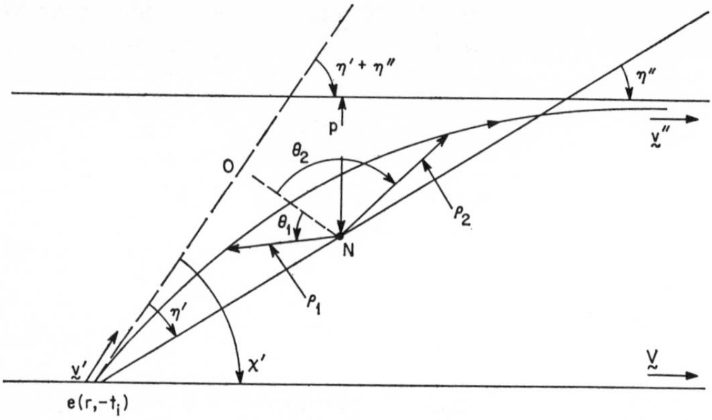

Fig. 9 Diagram of orbital parameters for second binary collision, $\eta' < \pi/2$. (**V** is not necessarily in the plane of the paper.)

Clearly the tip of the $\mathbf{u}''$-vector (Fig. 8) generates the same element of volume, and we thus get the additional relation

$$d\tau = (u'')^2\, du''\, d\omega = v''v' \sin(\eta' + \eta'') \cos\phi\, dv'\, d\phi\, d\chi', \tag{57v}$$

where $d\omega$ is the solid angle associated with $\mathbf{u}''$-space.

In order that capture can occur the vectors $\mathbf{v}''$ and $\mathbf{V}$ must be approximately equal and this means that

$$v'' \approx V, \qquad \eta' + \eta'' \approx \chi'; \qquad \phi \approx 0. \tag{57w}$$

Moreover, capture cannot occur unless

$$\frac{m(u'')^2}{2} < \frac{Ze^2}{x} \tag{57x}$$

in which x is the distance between the capturing nucleus and the electron after the electron has escaped from the influence of the atomic (now ionic) field. In order to proceed further use must be made of the orbital details for the motion of the electron in the inverse square field.

Figure 9 depicts typical orbital relations for $\eta' < \pi/2$. The electron is located at a distance r from the atomic nucleus N at some time $t = -t_i$ in its initial position of the binary collision with the atomic core. Several representative electronic positions specified by the values of the radius vectors ρ_j and corresponding polar angles θ_j are shown; positive angles are measured clockwise from the apsidal line ON; p is the perpendicular distance from the pole at N to the final asymptote to the orbit which is parallel to $\mathbf{v}''$. The velocity $\mathbf{V}$ is not necessarily coplanar with the electronic orbital plane, the plane of the paper (see Fig. 8). The dashed line is parallel to $\mathbf{v}'$.

The orbital equation of motion and the equation of conservation of angular momentum

$$\frac{m}{2}(\rho^2\dot{\theta}^2 + \dot{\rho}^2) - \frac{m\lambda}{\rho^2} = \frac{m}{2}V^2, \qquad V^2 \approx (v'')^2,$$

$$m\rho^2\dot{\theta} = mrv' \sin\eta' = mpV, \qquad \dot{\theta} = \frac{d\theta}{dt}, \qquad \dot{\rho} = \frac{d\rho}{dt} \tag{58a}$$

together yield the solution

$$\theta(\rho) = \frac{p}{\rho_0}\cos^{-1}\left(\frac{\rho_0}{\rho}\right), \ \rho_0{}^2 = p^2 - \frac{2\lambda}{V^2} \tag{58b}$$

which satisfies $\rho = \rho_0$ at $\theta = 0$ provided that the principal branch of the ($\cos^{-1}$)-function [$\cos^{-1}(1) = 0$] is used. The angle of deflection to some point (ρ, θ) for $\theta > 0$ is

$$\Delta\theta(\rho) = \left[\frac{p}{\rho_0}\cos^{-1}\left(\frac{\rho_0}{\rho}\right)\right]_{\rho_0}^{\rho} - \left[\frac{p}{\rho_0}\cos^{-1}\left(\frac{\rho_0}{\rho}\right)\right]_{r}^{\rho_0}$$

$$= \frac{p}{\rho_0}\left[\cos^{-1}\left(\frac{\rho_0}{\rho}\right) + \cos^{-1}\left(\frac{\rho_0}{r}\right)\right], \qquad \eta' < \frac{\pi}{2}, \tag{58c}$$

and the total deflection is

$$\lim_{\rho\to\infty} \Delta\theta(\rho) = \Delta = \pi + \eta'' = \frac{p}{\rho_0}\left[\frac{\pi}{2} + \cos^{-1}\left(\frac{\rho_0}{r}\right)\right]. \tag{58d}$$

With the use of conservation of angular momentum in Eq. 58*a*, and Eqs. 57*r* and 57*w*, Eq. 58*d* is transformed to

$$\pi + \eta'' = \frac{A\sin\eta'}{(A^2\sin^2\eta' - w^2)^{1/2}}\left\{\frac{\pi}{2} + \cos^{-1}\left[\frac{(A^2\sin^2\eta' - w^2)^{1/2}}{V}\right]\right\}$$

$$A^2 = w^2 + V^2, \qquad \sin^{-1}\left(\frac{w}{A}\right) \le \eta' \le \frac{\pi}{2}. \tag{58e}$$

From the structure of this equation it is obvious that η'' can approach infinity, and this largeness is a manifestation of electronic spiraling about the center of force,[29] the amount of spiraling being related to the angular momentum, the electronic mass, and the constant λ.

Before the solution to Thomas' problem can be finished, we also need the orbital results for $\eta' > \pi/2$. A typical case is illustrated in Fig. 10. In this diagram the clockwise sense is again adopted for positive angles. For ease of comparison much of Fig. 9 is reproduced in this diagram, the parenthetical designations referring to Fig. 9. The positive direction of η' in this figure, however, is counterclockwise, but this *convention* is used for *illustrative*

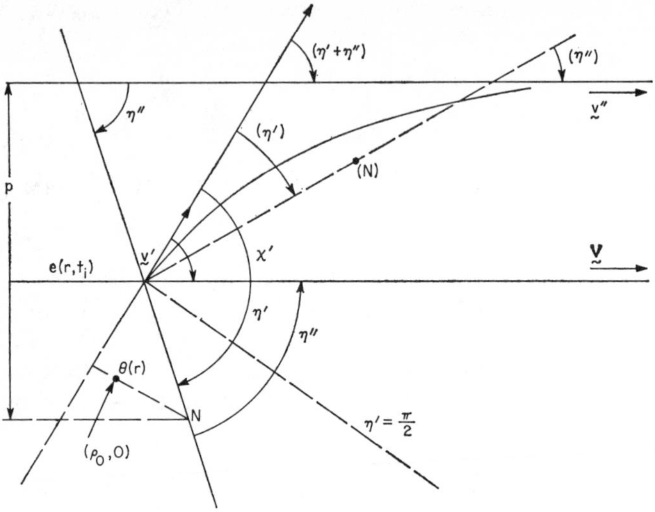

Fig. 10 Diagram of orbital parameters for second binary collision, $\eta' > \pi/2$.

convenience only. The actual values of η'' may differ by modulo $2\pi n$ ($n = 0, 1, 2 \ldots$) from the values depicted in Figs. 9 and 10. In contrast to the trajectory illustrated in Fig. 9, the initial electronic position is $e(r, t_i)$, whereas the apsidal point $(\rho_0, 0)$ is not on the physical trajectory; moreover, $\theta(r)$ is positive in Fig. 10 but negative in Fig. 9. The solutions* corresponding to Eqs. 58*c* to 58*e* are

$$\Delta\theta(\rho) = \theta(\rho) - \theta(r) = \frac{p}{\rho_0}\left\{\cos^{-1}\left(\frac{\rho_0}{\rho}\right)\right]_{\rho_0}^{\rho} - \cos^{-1}\left(\frac{\rho_0}{\rho}\right)_r^{\rho_0}\right]\right\}$$

$$= \frac{p}{\rho_0}\left[\cos^{-1}\left(\frac{\rho_0}{\rho}\right) - \cos^{-1}\left(\frac{\rho_0}{r}\right)\right], \qquad \eta' > \frac{\pi}{2};$$

$$\lim_{\rho\to\infty} \Delta\theta(\rho) = \Delta = \frac{p}{\rho_0}\left[\frac{\pi}{2} - \cos^{-1}\left(\frac{\rho_0}{r}\right)\right]; \tag{58f}$$

$$\pi + \eta'' = \frac{A \sin \eta'}{(A^2 \sin^2 \eta' - w^2)^{1/2}}\left\{\frac{\pi}{2} - \cos^{-1}\left[\frac{(A^2 \sin^2 \eta' - w^2)^{1/2}}{V}\right]\right\},$$

$$\frac{\pi}{2} \leq \eta' \leq \pi - \sin^{-1}\left(\frac{w}{A}\right).$$

It is emphasized here that the angle η'' is defined by

$$\eta'' = \Delta - \pi, \tag{58g}$$

* These results, incidentally, are identical to Thomas' results.

and it is clear from Fig. 10 that $(\Delta - \pi + \eta')$ represents the angle between $\mathbf{v}'$ and $\mathbf{V}$ if the approximation of Eq. 57*w* is used. Equations 58*e*, 58*f*, and 58*g* are used to complete the derivation of Thomas' result.

The solutions of Eq. 58*a* as a function of *t* are

$$\left.\begin{array}{ll} t < 0, & Vt = -(\rho^2 - \rho_0{}^2)^{1/2} \\ t > 0, & Vt = (\rho^2 - \rho_0{}^2)^{1/2} \\ t = -t_i, & Vt_i = (r^2 - \rho_0{}^2)^{1/2} \end{array}\right\} \quad \eta' < \frac{\pi}{2}; \tag{58h}$$

$$\left.\begin{array}{ll} t > t_i, & Vt = (\rho^2 - \rho_0{}^2)^{1/2} \\ t = t_i, & Vt_i = (r^2 - \rho_0{}^2)^{1/2} \end{array}\right\} \quad \eta' > \frac{\pi}{2} \tag{58i}$$

The fact that different times are used for these solutions is not to be interpreted that the particle collides with the electron at different physical times depending upon the orientation of the electron-nucleus line *e*N. The choice of these initial times is dictated by calculational convenience since the time is used only to relate the distances traveled by the incident particle and the electron. Although the vanishing of *t* implies that $\rho = \rho_0$ in Eq. 58*i*, the solution is written for the physical trajectory only. These two results are used to express *x* of Eq. 57*x* in terms or *r*, *u*, and *V*, and Eqs. 57*n*, 57*r*, and 57*w* are used to attain this approximate goal.

For $\eta' < \pi/2$ it is evident from Fig. 9 that the distance *x* for large values of *t* is

$$pV = rv' \sin \eta', \qquad \lim_{t\to\infty} V(t_i + t) = \rho + (r^2 - \rho_0{}^2)^{1/2},$$

$$\begin{aligned} x^2 &= (p + r \sin \eta'')^2 + [\rho + (r^2 - \rho_0{}^2)^{1/2} - (\rho + r \cos \eta'')]^2 \\ &= 2r^2 + \frac{2\lambda}{V^2} + 2rp \sin \eta'' - 2r(r^2 - \rho_0{}^2)^{1/2} \cos \eta'' \\ &= \frac{r^2}{V^2}[V^2 + (v')^2 - 2v'V \cos(\eta' + \eta'')] = \frac{r^2(u')^2}{V^2} \\ &= \frac{r^2u^2}{V^2}. \end{aligned} \tag{58j}$$

For the case that $\eta' > \pi/2$ (Fig. 10) we get

$$pV = rv' \sin(\pi - \eta'), \qquad \lim_{t\to\infty} V(t - t_i) = \rho - (r^2 - \rho_0{}^2)^{1/2},$$

$$\begin{aligned} x^2 &= (p - r \sin \eta'')^2 + [\rho - (r^2 - \rho_0{}^2)^{1/2} - (\rho + r \cos \eta'')]^2 \\ &= \frac{r^2}{V^2}[V^2 + (v')^2 + 2v'V \cos(\eta'' - \eta' + \pi)] \doteq \frac{r^2u^2}{V^2} \\ &\cos(\pi + \eta'' - \eta') = -\cos(\eta' - \eta''). \end{aligned} \tag{58k}$$

Although $(\eta' - \eta'')$ is the argument of the cosine in the last line of Eq. 58*k*, this result is not inconsistent with the argument $(\eta' + \eta'')$ appearing in Eq. 58*j*. If Δ of Eq. 58*f* is positive in the clockwise sense in Fig. 9, $(\Delta - \pi)$ is counterclockwise since $\Delta < \pi$ in that figure. Now $(\Delta - \pi) + \eta'$ is the angle between $\mathbf{v}'$ and $\mathbf{V}$, with η'' representing $(\Delta - \pi)$ in this figure; however, in Eq. 58*k* η'' and η' are scalars, and thus there is no contradiction. The important angular combination, nevertheless, is $(\Delta - \pi + \eta')$ but $(\Delta - \pi)$ equals η'' in Fig. 10, as shown, for in this case $(\Delta - \pi)$ is positive. It is appropriate at this point to comment about the assumptions used to derive Eqs. 58*j* and 58*k*; Eq. 58*j* is used for this purpose.

We see that this condition of initial coincidence of incident nucleus and electron requires a hard collision; that is, the incident particle must make practically a *direct hit* on the electron in order that the Coulomb interaction dominates the atomic field, and after this first binary collision these same two particles are again practically coincident at $t = -t_i$. However, by reason of the largeness of the incident nucleus-electron Coulomb interaction at $t = -t_i$, the relative velocity u' (Fig. 7) should be large; otherwise, the electron would not escape the influence of P and be scattered by A^+ into the final velocity $\mathbf{v}'' \approx \mathbf{V}$ appropriate for capture. It is fairly evident that these conditions are fulfilled only approximately for very fast incident particles, although through unpredictable cancellations the calculated cross sections may agree with the measured values reasonably well for unexpectedly low impact velocities.

We now return to Eqs. 57*q*, 57*r*, and 57*w* and replace v' occurring in $v's(u, v, v', V)$ by

$$(v')^2 = V^2 + \frac{2\lambda}{r^2},$$

and we then approximate the integrations

$$\int (v')^2\, dv'\, d\phi\, d(\cos\chi')$$

by

$$\int (v')^2\, dv'\, d\phi\, d(\cos\chi') = V \int dv' v'\, d(\cos\chi')\, d\phi$$

$$= \int (u'')^2\, du''\, d\omega = \frac{4\pi}{3}(u''_{\max})^3 = \frac{4\pi}{3}\left(2\frac{\mu V}{ru}\right)^{3/2}, \qquad (58l)$$

in which use was made of Eqs. 57*u*, 57*v*, and 57*x*. This inclusion of all $\mathbf{u}''$-space satisfying Eq. 57*x* ensures that our cross section includes capture into all the classically bound states of the atom (ion) being formed. Without the use of the approximation for v' in this function, (see vs in Eq. 57*q*) it would be very difficult to proceed. However, the use of this approximation has enabled us to isolate the integral of Eq. 58*l* and integrate it approximately, the accuracy of this calculation improving with decreasing values of $(u'')_{\max} = (2\mu/x)^{1/2}$ of Eq. 57*x*. Perhaps the more important result is the simplification of Eq. 57*q* to the three-dimensional integral,

$$Q = \frac{2^{3/2}\pi\mu^{7/2}}{3V^{3/2}} \int_R \frac{X(v, r)t(u, v, w, V)}{vr^{3/2}u^{1/2}}\, dr\, dv\, d(\cos\eta'),$$

$$t(u, v, w, V) = 2uV(w^2 + V^2)^{1/2} \frac{4u^2V^2 - (u^2 - w^2)(V^2 + u^2 - v^2)}{|v^2 - V^2 - w^2|^3}, \tag{58m}$$

which is numerically tractable for the distributions $X(r, v)$ used in this chapter.

Thomas chose the distribution that he used to construct the Thomas-Fermi model of the atom. In this model there are two electrons for each element h^3 of classical phase space, which is assumed spherically symmetrical in both the positional and momentum parts. The factor of 2 arises from the two spin states of the electron. Messiah gives an interesting interpretation to this distribution, and his description is paraphrased here.[30] Messiah appeals to the classical limit remarking that if the Z lowest quantum states are occupied in an atom of nuclear charge Ze, the distribution in energy of the atomic electrons is the same as that of a statistical mixture of Z classical electrons having a density in phase space equal to

$$n(\mathbf{r}, \mathbf{p}) = \begin{cases} \dfrac{2}{h^3}, & E = \dfrac{p^2}{2m} - e\Phi \le 0 \\ 0, & E > 0. \end{cases} \tag{58n}$$

The energy E here is expressed in terms of the classical Hamiltonian of the electron whose momentum is $\mathbf{p} = m\mathbf{v}$, and the highest value of E occupied by the electrons is arbitrarily equated to zero. In the spirit of the classical approximation, the atomic electrons are assumed to have the same spatial distribution as this classical statistical mixture, and it is also assumed that all accessible elements of classical phase space are fully occupied:

$$\rho(r) = \frac{2}{h^3} \int_{E \le 0} d\mathbf{p}. \tag{58o}$$

The spatial density in quantum mechanics, however, is the sum of the densities of the Z lowest quantum levels, and in the space $\mathbf{r}$ of one electron is represented by

$$\rho(\mathbf{r}) = \sum_{m=1}^{Z} |\psi_m(\mathbf{r})|^2, \tag{58p}$$

and if each ψ_m is normalized to unity for this spherically symmetric field, it is clear that

$$Z = 4\pi \int_0^\infty \rho(r) r^2 \, dr. \tag{58q}$$

Returning to the discussion of the Thomas-Fermi field, we see that although the spatial density of the Thomas-Fermi field is not used here, it is pertinent to note that it diverges at $r = 0$ and decreases at infinity as r^{-6}, both variations disagreeing with quantum mechanical boundary conditions. Nevertheless, the distribution $2/h^3$ in phase space is used despite the defects of the associated spatial distribution and the possibility of there being insufficient electrons with large principal quantum numbers for the distribution to be approximately valid.[31] Since this distribution is designed for atoms with a large number of electrons with large principal quantum numbers, and since this distribution for a degenerate electron gas is incorrect near the edge of the atom,[32] unreliable predictions from this model may be expected for capture from atoms containing a few loosely bound electrons in the outermost subshell. This unreliability should be more pronounced at low impact velocities because most of the contribution stems from capture from the outermost subshell. Moreover, the predictions also are inaccurate at very high impact velocities since the main contribution originates from capture from the innermost subshells, another region where the electronic distribution is inaccurate. This rather extensive discussion of the distribution has been presented so that the reader will be fully cognizant of the limitations of this classical model.

Spherical coordinates in positional and momentum spaces are now used for $X(r, v)$, which consist of Eq. 58n together with part of the volume element of the associated phase space,

$$X(r, v) = \frac{2}{h^3} (4\pi)^2 m^3 v^2 r^2. \tag{59a}$$

The differentials associated with Eq. 59a are already contained in Eq. 58m. In order to integrate this equation over dv we first refer to Fig. 7, which illustrates the inequalities

$$|\mathbf{V} - \mathbf{u}| \leq v \leq V + u, \qquad v' \leq V + u' = V + u, \qquad (u' = u). \tag{59b}$$

Since the electron is in the bound state

$$\frac{mv^2}{2} - \frac{m\lambda}{r^2} = \frac{mv^2}{2} - \frac{mw^2}{2} \leq 0, \tag{59c}$$

v is bound from above by w; furthermore, since v' must exceed w for the electron to escape from the atom, it is clear from Eq. 59b that the domain of

v in the integral is

$$|\mathbf{V} - \mathbf{u}| \le v \le w. \tag{59d}$$

It is instructive to note that the upper limit in Eq. 59*d* is the same as the upper bound to v in Eqs. 58*n* and 58*o* if Φ is an inverse square potential; however, the corresponding normalizing condition of Eq. 58*q* is not used in this classical model. The integration over dv is effected straightforwardly to transform Eq. 58*m* into

$$\int_{|\mathbf{V}-\mathbf{u}|}^{w} dv\, vt(u, v, w, V) = \frac{u}{2V^3}[4w^2V^2 - (w^2 + V^2 - u^2)^2](w^2 + V^2)^{1/2},$$

$$Q = \frac{(2\mu)^{7/2}m^3}{6\hbar^3 V^{9/2}} \int_R \frac{[4w^2V^2 - (w^2 + V^2 - u^2)^2](w^2 + V^2)^{1/2}}{u^{9/2}} \times r^{1/2}\, dr\, d(\cos\eta'). \tag{59e}$$

This integral is now transformed into a form that is more attractive for analysis (atomic units are used):

$$y = \frac{w}{V} = \frac{(2\lambda)^{1/2}}{rV},$$

$$e^2 = \frac{u^2}{V^2} = 2 + y^2 - 2(y^2 + 1)^{1/2}\cos(\eta' + \eta''),$$

$$\eta'' + \pi = \frac{(1 + y^2)^{1/2}\sin\eta'}{[(1 + y^2)\sin^2\eta' - y^2]^{1/2}}\left(\frac{\pi}{2} \pm \cos^{-1}\{[(1 + y^2)\sin^2\eta' - y^2]^{1/2}\}\right),$$

$$\left.\begin{matrix} \frac{\pi}{2} + \cos^{-1} & \text{if } 0 < \eta' < \frac{\pi}{2} \\ \frac{\pi}{2} - \cos^{-1} & \text{if } \frac{\pi}{2} < \eta' < \pi \end{matrix}\right\} \qquad Q = \frac{2^{21/4}\lambda^{3/4}Z^{7/2}}{3V^{11/2}}\, C,$$

$$\eta_l' = \sin^{-1}\left[\frac{y}{(1 + y^2)^{1/2}}\right],$$

$$C = \int_{y_1}^{y_2} D\, dy, \qquad D = \int_R \frac{(1 + y^2)^{1/2}[y^2 - \frac{1}{4}(1 + y^2 - e^2)^2]}{y^{5/2}e^{9/2}}\, d(\cos\eta'),$$

$$(e - 1)^2 \le y^2, \qquad \pi - \eta_l' \ge \eta' \ge \eta_l'. \tag{59f}$$

This is the integral derived by Thomas. The integral over η' includes only those regions for which the integrand is real and non-negative. It is a straightforward exercise to show that the inequality

$$(e - 1)^2 \le y^2$$

ensures the integrand to be real and positive. Thomas set $y_1 = 0$, $y_2 = \infty$ and found C to be roughly 3.4, which is close to the accurately determined[25] numerical value of 3.50. We note that this cross section $Q(T)$ varies with impact velocity as $V^{-11/2}$, and $Q(T)$ clearly become infinite as V approaches zero, but this is not a contradictory feature since the model was not designed for low impact velocities. Nevertheless, several modifications of Eq. 59*f* have been proposed; this topic is discussed next.

5.4 MODIFICATION OF THE THOMAS MODEL

The Thomas model, like most theoretical models, was modified in the attempt to predict the physical quantities of interest more accurately. We have seen from Eq. 59*c* that the electron satisfies a classical bound-state equation and that this quality is also consistent with Eq. 58*n* for the density of electrons in phase space. If we modify Eq. 59*c* to correspond to a bound-state energy of one of the subshells of the atomic target, however, we will have introduced a physical characteristic of a subshell of the atom. Accordingly, Eq. 59*c* is replaced by

$$\frac{m}{2} v^2 - \frac{mw^2}{2} = -mJ, \qquad w^2 = \frac{2\lambda}{r^2}, \tag{60a}$$

in which mJ is the first ionization potential of the atomic target.[25] From this equation it is clear that the smallest real value of w and the associated value of y_1 are

$$w_1 = 2J, \qquad y_1 = \left(\frac{2J}{V^2}\right)^{1/2}, \tag{60b}$$

y_1 being the new lower limit to C of Eq. 59*f*. As already mentioned in connection with analysis of $X(r, v)$ in Eq. 59*a*, this distribution overestimates the density of the electrons near the atomic nucleus at $r = 0$; thus, according to the definition of y in Eq. 59*f*, the upper limit to C also should be revised downward to some finite value. However, the value of the integral is relatively insensitive to this alteration except at extremely high values of V for which the theory is inadequate; consequently, the upper limit is not changed. The modified integral is $C(y_1)$, and this model is named the first modification of Thomas, the corresponding cross section being denoted by $Q(T1)$.

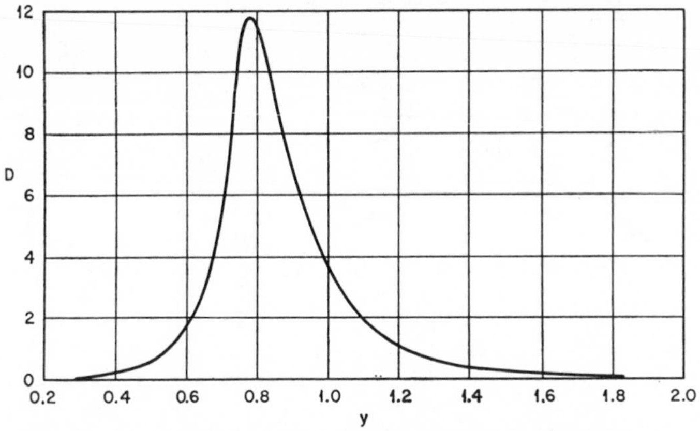

Fig. 11 A plot of the function $D(y)$ occurring in the Thomas cross section integral (see Eq. 59*f*). [This plot is Fig. 3 of Bates and Mapleton (1966)].

If the impact energy is used in place of the impact velocity, an instructive expression for $Q(T1)$ is obtained:

$$Q(T1) = 2.48 \times 10^{-12} \lambda^{3/4} Z^{7/2} M_p^{11/4} E^{-11/4} C(y_1)\,\text{cm}^2,$$
$$\frac{1.836}{y_1^2} = \frac{E(keV)}{M_p[mJ(eV)]}. \qquad (60c)$$

M_p in this equation represents the mass of the incident nucleus in units of the mass of the proton. According to this equation, $C(y_1)$ is a function of the ratio, y_1; therefore the quantity P defined by

$$P = \frac{E^{11/4}(\text{keV})}{\lambda^{3/4} Z^{7/2} M_p^{11/4}} Q(\text{cm}^2) = 2.48 \times 10^{-12} C(y_1), \qquad (60d)$$

also is a function of y_1 only. We return to this point shortly.

Another item of interest concerning the structure of $Q(T1)$ in terms of impact velocity is that

$$Q_a = Z^{-7/2} Q(T1) \qquad (60e)$$

is a function of the impact velocity and the structure of the atomic target; thus the impact-velocity dependence of Q_a is the same for a given atom and all positively charged incident nuclei. Nevertheless, according to Eq. 57*x*, the larger Z is, the larger are the associated escape velocities of the captured electrons, and according to Eqs. 57*a* and 57*w*, this means that V also must increase with Z to maintain the same relative smallness of u'' in comparison to V. This comparison exemplifies the caution required to avoid violating the restrictive assumptions of this classical model. The integral of D of Eq. 59*f* is shown in Fig. 11 in order to illustrate the variation of C as a function of its lower limit y_1.

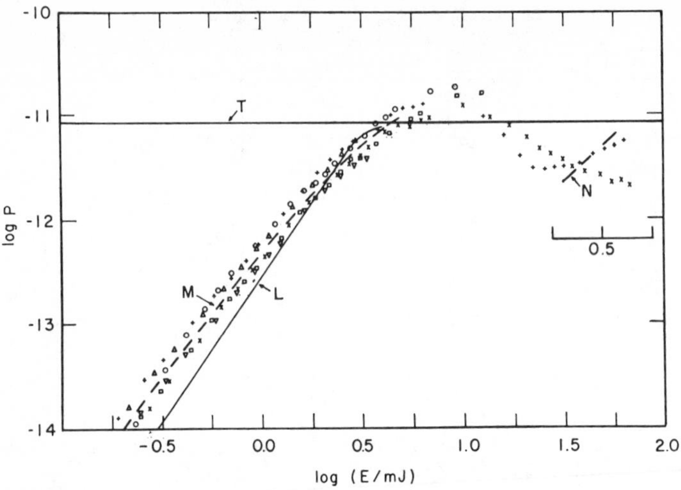

Fig. 12 The universal curve. Electron capture parameter P with the cross section Q in cm^2, λ in atomic units, the impact energy E of the proton in keV, and mJ in eV. (Common logarithms are used.) The curve T is the prediction of the original theory of Thomas (1927),[2] and the full curve L is the prediction of the modified theory with the cross section dependent upon the ionization potential of the gaseous target. The broken curve M is an empirical curve drawn through the points based on the experimental data. The points are obtained from measurements in various gases: O, Ne; +, Ar; Δ, Kr; ∇, Xe; X, N_2 (per atom); $\square$, O_2 (per atom). The raised horizontal scale on the right refers to the argon points + only; for this scale mJ is the ionization potential of the $2p$-subshell; the slope of the curve N is identical to the slope of the curve M. [This curve is Fig. 4 of Bates and Mapleton (1966)].

We now return to Eqs. 60*c* and 60*d* to consider comparisons of calculated and measured cross sections. In Fig. 12 $\log_{10} P$ is plotted as a function of $\log_{10} (E/mJ)$ with P defined by Eq. 60*d*. The full curve L represents

$$P = 2.48 \times 10^{-12} C(y_1)$$

of Eq. 60*d* and the full curve T is the prediction given by the original theory of Thomas ($y_1 = 0$). The experimental points for this comparison were obtained by using the measured values for $Q(cm^2)$ in Eq. 60*d* and as explained in the caption, the incident nucleus is a proton so that M_p and Z are both equal to one. The values of λ, however, were derived from the effective interactions given in tables on atomic structure.[33] Since these published interactions are expressed in Rydbergs (ε_0), and atomic units are being used, the values of λ appropriate to the different gases are given by

$$\lambda = \frac{r^2 V_t(r)}{2}, \tag{60f}$$

with V_t taken from the tables. In practice the arithmetic average of λ is used for that domain of electronic distances r over which V_t is a good approximation to the inverse-square potential. The broken curve M in Fig. 12 is an empirical curve drawn through the points based on the experimental data.

It is observed that the points corresponding to these atomic gases, and molecular gases per gas atom, are predicted fairly well on the average, the error in prediction increasing as (E/mJ) decreases. Thus P of Eq. 60d calculated with the experimental values of Q is closely a function of y_1 for a restricted domain of y_1. However, as (E/mJ) increases, the points representing Ar finally deviate markedly from those representing N_2 per gas atom, and it is noteworthy that the broken curve N, drawn parallel to M, is displaced to the right by an amount in close agreement with

$$\log_{10}\left[\frac{mJ(3p)}{mJ(2p)}\right] = \log_{10}(15.7), \tag{60g}$$

which relates the ionization potentials of the inner $2p$-subshell to the outermost $3p$-subshell. (Capture from the $2s$-subshell also contributes, but we ignore these finer details here.) The classical model, $T1$, fails in this region since it contains no provision for capture from different subshells of the atom. In an effort to accommodate and differentiate between capture from different subshells, Bates and Mapleton[34] proposed the use of an electronic distribution different from the one of Eq. 59a used by Thomas; moreover, Mapleton[35,36] interpreted this model so that it was also applicable to atoms containing unfilled subshells of equivalent p-electrons.*

The point of departure for designing this second modification of Thomas, designated $T2$, is Eq. 58m. A study of the assumptions used to derive Eq. 58m reveals that a numerically tractable model requires the retention of all the approximations that were used. This leaves only the electronic distribution $X(r, v)$, which can be altered. For the spatial part of this distribution we incorporate Eqs. 58p and 58q with the assumed spherically symmetrical density $\rho(r)$. However, $\rho(r)$ is defined as in Eq. 58p by a sum of quantum mechanical probability densities, one for each subshell of the atomic target. Finally, we retain the classical feature described by Eq. 60a, but in this application there is a different mJ for each subshell. There is a further difference here; the interaction used in this bound-state equation is not the inverse square potential, but is a potential calculated using quantum mechanics. Consequently, the inverse square potential used to deduce Eq. 58m can be interpreted as an approximation to this approximate quantum mechanical interaction. The distribution chosen is

$$X = \sum \frac{\delta(v - v_k)}{4\pi v_k^2}\rho(r), \qquad v_k^2 = 2|V(r)| - 2J_k \tag{61a}$$

* This is also valid for unfilled subshells of equivalent d-electrons (unpublished work).

in atomic units. In this equation k denotes one of the j-subshells of the atom.

A few clarifying remarks about the units are pertinent here. If $X(r, v)\,dr\,dv$ is removed from Eq. 58*m*, it is easily established by a *transformation* to atomic units that the dimensional factor a_0^2 belongs to the remaining part of Eq. 58*m*. Moreover, an analysis of the quantity $X(r, v)\,dr\,dv$, with $X(r, v)$ defined by Eq. 59*a*, reveals that this function is dimensionless, and if it is expressed in atomic units, the numerical factor $(4/\pi)$ appears since $e = \hbar = m = 1$ in this system of units. Similarly, X of Eq. 61*a* multiplied by $dr\,dv$ is dimensionless, but if we wish to relate this distribution to phase space in the cgs system, our function is

$$(4\pi)^2\,dr r^2 \rho_k(r)\,\frac{\delta(p - p_k)}{4\pi p_k^2}\,p^2\,dp\,\frac{m^3}{\hbar^3}.$$

If atomic units are now used, the original form is regained and the point of this last comparison is to show that $\hbar^{-3}$ must be used here instead of h^{-3}, which appears in the Thomas's distribution.

Equation 61*a* is substituted into Eq. 58*m*, and the integration over dv amounts to replacing v by the v_k of Eq. 61*a*. However, we have to determine the spatial function, $\phi(r)$. First, the application is restricted to atoms consisting of closed subshells. The quantity calculated is the probability density that an electron is in the volume element $4\pi r^2$ irrespective of the location of the other $(N - 1)$-electrons of this N-electron atom. The domain of the r-variable is *not* $(0, \infty)$, as in quantum mechanics, but is restricted to the classical radial spaces which are defined by positive values of v_k^2 for each subshell k. The wave functions of the atom are approximated by the appropriate antisymmetrized products of one-electron orbitals; the LS coupling scheme is assumed for our purposes, and the LSM_LM_S-representation is used.

The antisymmetrical wave function for the atoms of interest can be represented by a sum of N_a terms,

$$N_a = N!\,(n_1!\,n_2!\cdots n_j!)^{-1}, \tag{61b}$$

each term consisting of a product of j groups of orthonormal antisymmetrical functions, with each group denoting a different subshell in its classically restricted space. Now in the wave function that is antisymmetrical in all electronic coordinates, the coordinate $\mathbf{x}_1(=\mathbf{r})$ occurs

$$N_{k-1} = (N - 1)!\,[n_1!\,n_2!\cdots(n_k - 1)!\cdots nj!]^{-1}$$

times in the kth subshell that contains n_k-electrons. Thus for calculational convenience the appropriately normalized wave function is rewritten as

$$\begin{aligned}\Psi &= \sum_{k=1}^{j} N_{(k-1)}^{1/2}\Psi_{n_k}(\mathbf{x}_1\cdots\mathbf{x}_k)\Psi_{N-n_k}(\mathbf{x}_{k+1}\cdots\mathbf{x}_N),\\ N &= \sum_{k=1}^{j} n_k, \qquad N_a = \sum_{k=1}^{j} N_{(k-1)}.\end{aligned} \tag{61c}$$

(The usual minus signs are omitted in Eq. 61*c* since they are not needed here.) In this formula $\mathbf{x}_1$ is isolated in one of the terms that represents the *k*th subshell, and since there are $N_{(k-1)}$ of these terms in the *N*-electron wave function, the factor $N_{(k-1)}^{1/2}$ remains after normalizing with $N_{(k-1)}^{-1/2}$. The function ψ_{N-n_k} represents the product for the other subshells. Next we calculate the probability density in $\mathbf{x}_1$-space,[37] which is normalized to N:

$$\begin{aligned} P &= \int d\mathbf{x}_2\, d\mathbf{x}_3 \cdots d\mathbf{x}_N |\Psi|^2 (N_a N^{-1})^{-1} \\ &= \sum_{k=1}^{j} n_k |\phi_k(\mathbf{x}_1)|^2, \qquad \mathbf{x}_1 = \mathbf{r}. \end{aligned} \tag{61d}$$

[The calculational principle used is that the sum of the integrals over each of the $N_{(k-1)}$ orthonormal products is equal to $N_{(k-1)}$ times the integral over one of the products.] Since Eq. 61*d* represents the situation for closed subshells, the one-electron function appearing in Eq. 61*d* is (only *s*- and *p*-orbitals)

$$|\phi_k(m_1)|^2 = \frac{R_k{}^2(r)}{4\pi}. \tag{61e}$$

[This relation, which is obvious for subshells $(ns)^2$, also is correct for $(np)^6$-subshells; this can be validated easily by integrating the square of the orthonormal $(np)^6$-single determinal function over $d\mathbf{x}_2 \cdots d\mathbf{x}_6$.] Equation 61*d* is rewritten as

$$P = (4\pi)^{-1} \sum_{k=1}^{j} n_k R_k{}^2(r) = \rho(r), \tag{61f}$$

and if this equation is multiplied by the volume element in **r**-space, there results

$$4\pi r^2\, dr P = \sum_{k=1}^{j} n_k (rR_k)^2\, dr = 4\pi\rho(r) r^2\, dr. \tag{61g}$$

If the delta functions in v are reintroduced and Eq. 61*g* is used, Eq. 61*a* can be modified to

$$X\, d\mathbf{r}\, d\mathbf{v} = \sum_{k=1}^{j} \delta(v - v_k) n_k (rR_k)^2\, dr\, dv, \tag{61h}$$

which is the final form. However, before this result is used in Eq. 58*m*, let us see how to derive a result similar to Eq. 61*f* for atoms containing unfilled subshells of equivalent *p*-electrons.

In these cases of unfilled subshells of equivalent *np*-electrons there also exists a degeneracy of the ground-state term. [Although a ^1S-term occurs in the configurations $(np)^2$ and $(np)^4$, these terms are *not* the ground-state terms.] Therefore the purpose of this development is to define the associated $\rho(r)$ so that this degeneracy is removed by an appropriately defined averaging process. Attention is first directed to Eq. 44*f* in Section 3.4, and the normalized antisymmetrical function for a subshell of $n(n \leq 6)$ equivalent *p*-electrons is represented by the expansion containing only one fpc. The restrictions implicit in the vector-coupling coefficients of that sum are recorded for reference:

$$\begin{aligned} M_l'' + m_l'' &= M_l, \qquad & M_s'' + m_s'' &= M_s, \\ L' + 1 \geq L &\geq |L' - 1|, \qquad & S' + \tfrac{1}{2} \geq S &\geq |S' - \tfrac{1}{2}|. \end{aligned} \tag{61i}$$

The value of the probability density associated with this subshell is readily obtained with the aid of the orthonormal property of the spin functions and a formula for summing vector-coupling coefficients in the spin:[38]

$$\begin{aligned} I = \int |\psi(p^n)|^2 \, d\mathbf{x}_2 \cdots d\mathbf{x}_n &= \sum_{(S'L')} \sum_{M_l'} \sum_{m_l'} |(L'M_l'lm_l'|L'1LM_l)|^2 \\ &\quad \times |p(m_l'; \mathbf{x}_1)|^2 |(p^{n-1}(S'L')pSL|\}p^nSL)|^2. \end{aligned} \tag{61j}$$

Next the arithmetic average of I for the degenerate states of the ground-state term $p^n(LS)$ is calculated:

$$\begin{aligned} P_n &= [(2L + 1)(2S + 1)]^{-1} \sum_{M_l} \sum_{M_S} I(M_l) \\ &= (2L + 1)^{-1} \sum_{(S'L')} \sum_{M_l} \{|p(-1; \mathbf{x}_1)|^2 |(L'(M_l + 1), 1 - 1|L'1LM_l)|^2 \\ &\quad + |p(0; \mathbf{x}_1)|^2 |(L'M_l1, 0|L'1LM_l)|^2 + |p(1; \mathbf{x}_1)|^2 \\ &\quad \times |(L', (M_l - 1), 1, 1|L'1LM_l)|^2\} |(p^{(n-1)}(S'L')pSL|\}p^nSL)|^2. \end{aligned} \tag{61k}$$

This expression is essentially the same as Eq. 44*k* (Section 3.4), and the value of the sum over M_l is $(2L + 1)/3$. Since the sum of the $(\text{fpcs})^2$ over the allowed terms is unity, there is left

$$\begin{aligned} P_n &= \frac{1}{4\pi} \left\{ \frac{\sin^2 \theta}{2} + \cos^2 \theta + \frac{\sin^2 \theta}{2} \right\} R_n^2(r) \\ &= \frac{1}{4\pi} R_n^2(r). \end{aligned} \tag{61l}$$

The normalizing factor,

$$N_{(k-1)} N_a^{-1} N,$$

of Eqs. 61*c* and 61*d* is now inserted, and the result for this subshell agrees with those for Eq. 61*f*, the equality asserted following Eq. 61*h*. We return to Eq. 58*m* and determine the cross section $Q(T2)$.

For this goal we substitute Eq. 61*f* into Eq. 61*a* and multiply the distribution by the remaining part of the volume element of phase space that is not contained in Eq. 58*m* (atomic units are used). This product,

$$(4\pi)^2 v^2 r^2 X(r, v),$$

is inserted into Eq. 58*m*, and the integration over dv is effected to get

$$Q(T2) = \frac{4\pi\sqrt{2}Z^{7/2}}{3V^{1/2}} a_0^2 \sum_k \int_R \frac{\phi_k(r)(w^2 + V^2)^{1/2}}{v_k r^{3/2} u^{9/2}} \times \frac{[4u^2V^2 - (u^2 - w^2)(V^2 + u^2 - v^2)]}{|v_k^2 - v^2 - w^2|^3} \, dr \, d(\cos\eta'),$$

$$v_k^2 = 2|V(r)| - 2J_k, \qquad \phi_k(r) = \phi_{nl}(r) \qquad (61m)$$

$$= \frac{N_{nl}[rR_{nl}(r)]^2}{A_{nl}}, \qquad N_{nl} = n_k,$$

$$A_{nl} = \int_0^{r_{nl}} dr (rR_{nl})^2.$$

The velocities occurring in Eq. 61*m* are expressed in units of V, as before, and the definitions of e and η'' in Eq. 59*f* are thus applicable; in terms of these dimensionless variables Eq. 61*m* for each (nl) becomes

$$Q_{nl}(T2) = \frac{4\sqrt{2}\,Z^{7/2}}{3V^7} (\pi a_0^2) N_{nl} \int_0^{r_{nl}} \frac{dr\phi_{nl}(r)(1 + y^2)^{1/2}}{r^{3/2}(1 + a^2)^3 (y^2 - a^2)^{1/2}} \int_{-1}^{1} \frac{d(\cos\eta')F}{e^{9/2}},$$

$$F = [4e^2 - (e^2 - y^2)(1 + a^2 + e^2 - y^2)], \; |e - 1| \le (y^2 - a^2)^{1/2}, \qquad (61n)$$

$$\frac{v_k^2}{V^2} = y^2 - a^2 = w^2 - a^2, \qquad y^2 = \frac{2|V(r)|}{V^2},$$

$$a^2 = \frac{2J_k}{V^2}, \qquad y^2(r_{nl}) - a^2 = 0.$$

To repeat, the radial functions in Eq. 61*m* are normalized to the domain $(0, r_{nl})$ as contrasted to the domain $(0, \infty)$ usually occurring in quantum mechanics. Attention is also directed to the apparent singularity at the end point r_{nl} defined by the zero of $(y^2 - a^2)^{1/2}$, and to the supplementary condition imposed on $|e - 1|$. This condition, together with the restriction of positive real values of the integrand, determines the contributing subdomains of η' in the total allowable domain, $-1 \le \cos\eta' \le 1$.

The restrictions, repeated for convenient reference and clarity,

$$\pi - \eta_l' \ge \eta' \ge \eta_l', \qquad \eta_l' = \sin^{-1}\left[\frac{y}{(1 + y^2)^{1/2}}\right],$$

$$|e - 1| \le (y^2 - a^2)^{1/2}, \qquad (61o)$$

are used to analyze the contribution to the integrand of Eq. 61*n* in the neighborhood of $r = r_{nl}$. A study of e^2 (defined in Eqs. 59*f* and 61*o*) at $r = r_{nl}$, reveals that contributions to the integrand only occur if e, a^2, and $(\eta' + \eta'')$ also satisfy

$$e = 1, \qquad a^2 \leq 3, \qquad \cos(\eta' + \eta'') \leq \frac{(1 + a^2)^{1/2}}{2}. \tag{61p}$$

However, according to Eq. 61*n* only one value of a^2 occurs for a given impact velocity V and a given subshell $k(=nl)$; therefore, if a^2 satisfies Eq. 61*p*, only a countably infinite number of values of η' fulfill this equation, and η'' of Eq. 59*f* also is determined by η' and a^2. Now the Riemann integral over $d(\cos\eta')$ of this countable set is not defined, but the associated Lebesgue integral is defined for a bounded integrand, and its value on this set vanishes.[39] Unless this vanishing of the η'-integral can be represented as a power series in $(r - r_{nl})$ so that the variation of the integrand near r_{nl} is known exactly, we can say only that the integrand as a function of r is indeterminate at $r = r_{nl}$. [The Riemann integral over dr exists since the singularity, exclusive of the η' integral, is $(r_{nl} - r)^{-1/2}$. If the η' integral, exclusive of $(y^2 - a^2)^{-1/2}$ can be represented by $(r_{nl} - r)^p f(r)$, $(p > 0)$ near r_{nl}, the total integrand is less singular, and if $p \geq \frac{1}{2}$, the singularity is removed.] Nevertheless, calculational experience suggests that the integrand vanishes at the endpoint r_{nl}. This completes the description of these modifications of the original model, and attention is directed next to some classical models originally proposed by Gryzinski.

5.5 GRYZINSKI MODEL

In two papers Gryzinski effected a detailed review and study of classical collisional processes in the laboratory and center-of-mass systems,[40] devoting particular attention to Coulomb collisions;[41] based on this research he proposed a classical theory of atomic collisions, including a classical model for charge transfer.[42] This theory for electronic capture is based on a single binary encounter between the incident particle and the electron to be captured, with the interaction between these two particles being a single Coulomb potential. From an analysis of the associated transfer of kinetic energy in the laboratory frame of reference, Gryzinski suggested how to calculate the cross section for electronic capture by a positively charged ion from a neutral atom. This formula is

$$Q(v_1, v_2) = \int_{\Delta E_l}^{\Delta E_u} \sigma_{\Delta E}{}^e \, d(\Delta E), \tag{62a}$$
$$\Delta E_l = \tfrac{1}{2}mv_1{}^2 + U_A - U_B, \qquad \Delta E_u = \tfrac{1}{2}mv_1{}^2 + U_A + U_B.$$

In Eq. 62*a*, v_1 and v_2 are the initial velocities of the incident ion and the atomic electron, respectively, m is the mass of the electron, and U_A and U_B are the initial and final binding energies of the electron, respectively. The symbol $\sigma_{\Delta E}{}^e$ denotes the effective cross section for the transfer of kinetic energy from the incident ion to the electron, and the upper and lower limits to the integral are postulated. The major part of this section is devoted to the explanation of the meaning of Eq. 62*a*; for this purpose, a recent paper by Gerjuoy is studied first.[43]

In this paper the exact expression for $\sigma_{\Delta E}{}^e$ is calculated, whereas in the papers by Gryzinski only approximations to $\sigma_{\Delta E}{}^e$ are derived. (The notation $\sigma_{\Delta E}{}^e$ is used here instead of $\sigma_{\Delta E}^{\text{eff}}$ used by Gerjuoy.) In order to derive $\sigma_{\Delta E}{}^e$, the cross section $\sigma_{\Delta E}$ is needed, and to calculate $\sigma_{\Delta E}$ we must define the collision system first. Initially two particles of masses m_1 and m_2 have velocities $\mathbf{v}_1$ and $\mathbf{v}_2$ in a frame of reference at rest with respect to the observer, the laboratory system, and after the single collision they have final velocities $\mathbf{v}_1'$ and $\mathbf{v}_2'$ with respect to the same system:

$$\begin{aligned}
\mathbf{v}_i &= v_i\hat{n}_i = \mathbf{u}_i + \mathbf{V} = \mathbf{u}_i + \hat{n}_V V,\\
\mathbf{v}_i' &= v_i'\hat{n}_i' = \mathbf{u}_i' + \mathbf{V} = \mathbf{u}_i' + \hat{n}_V V, \qquad i = 1, 2,\\
\mathbf{V} &= (m_1\mathbf{v}_1 + m_2\mathbf{v}_2)M^{-1}, \qquad M = m_1 + m_2, \\
\mathbf{u}_1 &= m_2M^{-1}\mathbf{v}, \qquad \mathbf{u}_2 = -m_1M^{-1}\mathbf{v}, \qquad \mathbf{v} = \mathbf{v}_1 - \mathbf{v}_2 = \hat{n}v,\\
\mathbf{u}_1' &= m_2M^{-1}\mathbf{v}', \qquad \mathbf{u}_2' = -m_1M^{-1}\mathbf{v}', \qquad \mathbf{v}' = \mathbf{v}_1' - \mathbf{v}_2' = \hat{n}'v'.
\end{aligned} \tag{62b}$$

The relations among velocities in the center-of-mass system $\mathbf{u}_i$ and $\mathbf{u}_i'$, the relative velocities $\mathbf{v}$ and $\mathbf{v}'$, the velocity of the center of mass $\mathbf{V}$, and the corresponding quantities in the laboratory system are all displayed in Eq. 62*b*. In this treatment of the collisional process the coordinate axis of the laboratory and center-of-mass frames of reference are presumed parallel so that the relations between the vectors, $\mathbf{u}_i + \mathbf{V}$ and $\mathbf{u}_i' + \mathbf{V}$ are valid also for the components of these sums. If $\mathbf{v}_1$ and $\mathbf{v}_2$ are specified, $\mathbf{v}$ and $\mathbf{V}$ are determined by definition; consequently, the polar axis of a fixed system of spherical coordinates can be chosen with the polar axis (Z-axis) coincident with $\mathbf{V}$. With respect to this system the unit vectors $\hat{n}$ and $\hat{n}'$ have polar and azimuthal angles $(\bar{\theta}, \bar{\phi})$ and $(\bar{\theta}', \bar{\phi}')$. respectively. (The barred symbols signify the center-of-mass quantities.)

Next the gain of kinetic energy ΔE by particle 2 from the incident particle 1 is calculated:

$$\begin{aligned}
\Delta E &= \tfrac{1}{2}m_2v_2'^2 - \tfrac{1}{2}m_2v_2{}^2 = \tfrac{1}{2}m_1v_1{}^2 - \tfrac{1}{2}m_1v_1'^2\\
&= m_2(\mathbf{u}_2' - \mathbf{u}_2)\cdot\mathbf{V} + \frac{m_2}{2}(\mathbf{u}_2' - \mathbf{u}_2)\cdot(\mathbf{u}_2' + \mathbf{u}_2)\\
&= \mu\mathbf{V}\cdot(\mathbf{v} - \mathbf{v}') = \mu Vv(\cos\bar{\theta} - \cos\bar{\theta}'), \qquad \mu = m_1m_2M^{-1}.
\end{aligned} \tag{62c}$$

These relations follow directly from the definitions in Eq. 62*b* and from the fact that the scattering is elastic, $v = v'$. Since $\mathbf{v}_1$ and $\mathbf{v}_2$ are given, v and V are determined, and

$$d(\Delta E) = \mu v V \sin \bar{\theta}' \, d\bar{\theta}' \tag{62d}$$

relates $d(\Delta E)$ to $d\bar{\theta}'$. The cross section for transfer of kinetic energy ΔE is defined by

$$\sigma(\mathbf{v}_1\mathbf{v}_2) = \int d(\Delta E)\sigma_{\Delta E}(\mathbf{v}_1, \mathbf{v}_2), \tag{62e}$$

in which σ is the total cross section for scattering.

For the determination of $\sigma_{\Delta E}$, let $\bar{\sigma}(v; \hat{n} \to \hat{n}')$ be the differential cross section for elastic scattering in the center-of-mass system, and denote the solid angle in this system by $d\hat{n}'$. Next express σ of Eq. 62*e* in terms of $\bar{\sigma}$ and use Eq. 62*d*; this yields

$$\begin{aligned} \sigma(\mathbf{v}_1, \mathbf{v}_2) &= \int d\hat{n}' \sigma(v; \hat{n} \to \hat{n}') \\ &= \frac{1}{\mu v V} \int d(\Delta E) \, d\bar{\phi} \bar{\sigma}(v; \hat{n} \to \hat{n}'), \end{aligned} \tag{62f}$$

and with the use of Eq. 62*e* we deduce

$$\sigma_{\Delta E}(\mathbf{v}_1, \mathbf{v}_2) = \frac{1}{\mu v V} \int_0^{2\pi} d\bar{\phi}' \bar{\sigma}(v; \hat{n} \to \hat{n}'). \tag{62g}$$

For fixed $\mathbf{v}_1$ and $\mathbf{v}_2$, or, correspondingly, fixed $\bar{\theta}$ and $\bar{\phi}$, the right side of Eq. 62*g* is a function of $\bar{\theta}'$, or, equivalently, a function of ΔE. This result is now specialized to represent scattering by a single Coulomb potential.

The corresponding differential cross section and angular relations are

$$\begin{aligned} \bar{\sigma} &= \left(\frac{Z_1 Z_2 e^2}{2\mu v^2}\right)^2 \csc^4\left(\frac{\chi}{2}\right), \\ \cos\chi &= \cos\bar{\theta}\cos\bar{\theta}' + \sin\bar{\theta}\sin\bar{\theta}'\cos(\bar{\phi} - \bar{\phi}'). \end{aligned} \tag{62h}$$

The integral of Eq. 62*g* corresponding to $\bar{\sigma}$ of Eq. 62*h* is easily integrated, and is

$$\begin{aligned} \sigma_{\Delta E}(\mathbf{v}_1, \mathbf{v}_2) &= \frac{1}{\mu v V}\left(\frac{Z_1 Z_2 e^2}{\mu v^2}\right)^2 \int_0^{2\pi} \frac{d\bar{\phi}'}{(a - b\cos\bar{\phi}')^2} \\ &= \frac{2\pi}{\mu v V}\left(\frac{Z_1 Z_2 e^2}{\mu v^2}\right)^2 \frac{(1 - \cos\bar{\theta}\cos\bar{\theta}')}{(\cos\bar{\theta} - \cos\bar{\theta}')^3}, \\ a &= 1 - \cos\bar{\theta}\cos\bar{\theta}', \qquad b = \sin\bar{\theta}\sin\bar{\theta}'. \end{aligned} \tag{62i}$$

We digress here to compare this result with a result in Section 5.3. The angle χ of Eq. 62*h* corresponds to the angle δ in Fig. 7 for the case that $\mathbf{v}_1 = \mathbf{v}_1'$ are equal to $\mathbf{V}$ in the figure, and $\mathbf{v}_2$, $\mathbf{v}_2'$, $\mathbf{v}$, and $\mathbf{v}'$ correspond, respectively, to $\mathbf{v}$, $\mathbf{v}'$, $\mathbf{u}$, and $\mathbf{u}'$ of Fig. 7. The integral, moreover, corresponds to the integral over $d\psi$ in Eq. 57*q* of Section 5.3, and apart from the factor $(uvV)^{-1}$ of Eq. 62*i*, we get the same result by integrating Eq. 57*m* of Section 5.3, $d(\cos\delta)$ excluded, over $d\psi$. The point of this comparison has been to relate $d\bar{\phi}'$ of Eq. 62*i* to the angular variation between the orbital planes in Fig. 7. The calculation is now resumed.

First, several useful relations are deduced from Eqs. 62*b* and 62*c*, and $\sigma_{\Delta E}$ is expressed in a calculationally more convenient form:

$$v = (v_1^2 + v_2^2 - 2v_1v_2\hat{n}_1\cdot\hat{n}_2)^{1/2}$$

$$V = M^{-1}(m_1^2v_1^2 + m_2^2v_2^2 + 2m_1m_2v_1v_2\hat{n}_1\cdot\hat{n}_2)^{1/2}$$

$$\cos\bar{\theta} = \frac{\mathbf{v}\cdot\mathbf{V}}{vV} = \frac{1}{MvV}\left[m_1v_1^2 - m_2v_2^2 + (m_2 - m_1)v_1v_2\hat{n}_1\cdot\hat{n}_2\right] \qquad (62j)$$

$$\sigma_{\Delta E} = \frac{2\pi(Z_1Z_2e^2)^2}{|\Delta E|^3}\left(\frac{V}{v}\right)^2\left(1 - \cos^2\bar{\theta} + \frac{\Delta E}{\mu vV}\cos\bar{\theta}\right).$$

If the target, particle two, is distributed isotropically in velocity in the laboratory system, the effective cross section is defined by

$$\sigma_{\Delta E}{}^e = \frac{1}{v_1}\int\frac{d\hat{n}_2 v}{4\pi}\,\sigma_{\Delta E}(\mathbf{v}_1, \mathbf{v}_2),$$
$$V = |\mathbf{v}_1 - \hat{n}_2v_2|, \qquad d\hat{n}_2 = d\phi_2\,d\theta_2\sin\theta_2. \qquad (62k)$$

(It is instructive to realize that $d\hat{n}_2v/4\pi$ represents the classical probability that the line segment $\mathbf{v}_1 - \hat{n}_2v_2$ has the direction corresponding to $\hat{n}_2$ in the range $d\hat{n}_2$.) Since the distribution of $\mathbf{v}_2$ is isotropic, $\sigma_{\Delta E}{}^e$ must be independent of the direction of $\mathbf{v}_1$; consequently, $\sigma_{\Delta E}{}^e$ can depend only on the scalar values of v_1 and v_2, and as a result, $\sigma_{\Delta E}{}^e$ can be represented by an integral over the solid angles of both $d\hat{n}_2$ and $d\hat{n}_1$; thus with the use of Eqs. 62*j* and 62*k* we get

$$\sigma_{\Delta E}{}^e = \frac{(Z_1Z_2e^2)^2}{8\pi|\Delta E|^3v_1}\int d\hat{n}_1\,d\hat{n}_2\,\frac{V^2}{v}\left(1 - \cos^2\bar{\theta} + \frac{\Delta E}{\mu vV}\cos\bar{\theta}\right),$$

$$-1 + \frac{\Delta E}{\mu vV} \le \cos\bar{\theta} \le 1, \qquad \Delta E > 0, \qquad (62l)$$

$$-1 \le \cos\bar{\theta} \le 1 + \frac{\Delta E}{\mu vV}, \qquad \Delta E < 0.$$

It is important to note the allowable domain of $\cos\bar{\theta}$ in Eq. 62*l*; see the paper by Gerjuoy for a detailed study of the domains of the variables of integration.[43] The restrictions needed for the intelligent application of these results will be recorded in less detail later in this section. The integral in Eq. 62*l* is now converted to a six-dimensional integral by using delta functions,

$$d\hat{n}_1\, d\hat{n}_2 = \int_{v_1 v_2} \frac{d\dot{\mathbf{v}}_1\, d\dot{\mathbf{v}}_2}{\dot{v}_1{}^2\dot{v}_2{}^2}\, \delta(v_1 - \dot{v}_1)\delta(v_2 - \dot{v}_2),$$

in which the variables of integration $\dot{v}_1$ and $\dot{v}_2$ are discriminated from v_1 and v_2 by using dots. The transformations

$$\dot{\mathbf{v}}_1 = \mathbf{V} + \frac{m_2}{M}\mathbf{v} \qquad \text{and} \qquad \dot{\mathbf{v}}_2 = \mathbf{V} - \frac{m_1\mathbf{v}}{M}$$

implicit in Eq. 62*b* and the fact that the associated Jacobian is unity are used to transform Eq. 62*l* into the more suitable form

$$\begin{aligned}
\sigma_{\Delta E}{}^e &= \frac{(Z_1Z_2e^2)^2}{8\pi|\Delta E|^3 v_1{}^3 v_2{}^2} \int d\mathbf{v}\, d\mathbf{V}\, \delta(\dot{v}_1 - v_1)\delta(\dot{v}_2 - v_2) \\
&\quad \times \frac{V^2}{v}\left(1 - \cos^2\bar{\theta} + \frac{\Delta E}{\mu v V}\cos\bar{\theta}\right), \qquad \cos\bar{\theta} = \hat{n}\cdot\hat{n}_V, \\
\dot{v}_1 &= \left(V^2 + \frac{m_2{}^2}{M^2}v^2 + \frac{2m_2}{M}vV\cos\bar{\theta}\right)^{1/2}, \qquad (62m) \\
\dot{v}_2 &= \left(V^2 + \frac{m_1{}^2 v^2}{M^2} - \frac{2m_1}{M}vV\cos\bar{\theta}\right)^{1/2}.
\end{aligned}$$

The simple expressions for $\dot{v}_1$ and $\dot{v}_2$ result from the choice of V as the polar axis. Since the integrand of Eq. 62*m* is independent of the solid angle $d\Omega_V$ and the angular variable $\bar{\phi}$, and since the domains of $\cos\bar{\theta}$ (defined in Eq. 62*l*) are also free of these variables, the associated integrations ($d\Omega_V$ and $d\bar{\phi}$) can be effected and produce the factor $8\pi^2$. In order to integrate over $\bar{\theta}$ (the polar angle of $\mathbf{v}$) $\cos\bar{\theta}$ is expressed in terms of $\dot{v}_1(=v_1)$, v_2, and V from Eq. 62*m*; this result is substituted for $\cos\bar{\theta}$ in the expression defining $\dot{v}_2$ in the same equation. The formula given in Eq. 56*d* of Section 5.2 is used next to evaluate the contribution (integration over $d\bar{\theta}$) stemming from $\delta(\dot{v}_1 - v_1)$; the resulting integral to be evaluated is

$$\begin{aligned}
\sigma_{\Delta E}{}^e &= \frac{\pi(Z_1Z_2e^2)^2}{|\Delta E|^3 v_1{}^2 v_2{}^2}\int dv \int dV \frac{M}{m_2} V^3 \delta\left[\left(\frac{MV^2}{m_2} + \frac{m_1v^2}{M} - \frac{m_1v_1{}^2}{m_2}\right)^{1/2} - v_2\right] \\
&\quad \times \left(1 - \cos^2\bar{\theta}_i + \frac{\Delta E}{\mu v V}\cos\bar{\theta}_i\right), \qquad (62n) \\
\cos\bar{\theta}_i &= \frac{M}{2m_2 vV}\left(v_1{}^2 - V^2 - \frac{m_2{}^2v^2}{M^2}\right), \quad \bar{\theta}_i = \bar{\theta}(v_1 = \dot{v}_1).
\end{aligned}$$

The expression for $\cos \bar{\theta}_i$ is included in the last equation to emphasize its dependence upon the variables of integration. Observe that the remaining delta function expresses conservation of energy in the laboratory system, and this means that the allowable values of v and V must be consistent with conservation of energy. Equation 56d of Section 5.2 is again invoked to evaluate the integral over dV, and this result is

$$\sigma_{\Delta E}{}^{e} = \frac{\pi(Z_1Z_2e^2)^2}{v_1{}^2v_2|\Delta E|^3}\int_{v_l}^{v_u} dv\Bigg(\frac{m_1}{M}v_1{}^2 + \frac{m_2v_2{}^2}{M} - \frac{\mu}{M}v^2 - \frac{\{v_1{}^2 - v_2{}^2 + [(m_1 - m_2)/M]v^2\}^2}{4v^2} + \frac{\Delta E}{2\mu v^2}\left(v_1{}^2 - v_2{}^2 + \frac{m_1 - m_2}{M}v^2\right)\Bigg). \tag{62o}$$

This integral is straightforwardly evaluated to yield the final results,

$$\sigma_{\Delta E}{}^{e} = \frac{\pi(Z_1Z_2e^2)^2}{4v_1{}^2v_2|\Delta E|^3}\Big[(v_1{}^2 - v_2{}^2)(v_2'^2 - v_1'^2)(v_l{}^{-1} - v_u{}^{-1}) + (v_1{}^2 + v_2{}^2 + v_1'^2 + v_2'^2)(v_u - v_l) - \tfrac{1}{3}(v_u{}^3 - v_l{}^3)\Big],$$

$$v_1'^2 = v_1{}^2 - \frac{2\Delta E}{m_1}, \qquad v_2'^2 = v_2{}^2 + \frac{2\Delta E}{m_2}. \tag{62p}$$

The limits v_u and v_l, which conform to the application of interest, are recorded in Eq. 63a, and for this application, $\Delta E > 0$ and $m_1 \gg m_2 = m$.

<u>α (Alpha)</u>

$$v_l = v_1 - v_2, \qquad v_u = v_1 + v_2, \qquad 0 < \Delta E < b,$$
$$b = \frac{4\mu}{M}\left[E_1 - E_2 - \frac{v_1v_2}{2}(m_1 - m)\right],$$
$$E_j = \frac{m_jv_j{}^2}{2}, \qquad j = 1, 2.$$

<u>β (Beta)</u>

$$\Delta E > 0, \qquad b < \Delta E < a, \qquad v_l = v_2' - v_1', \qquad v_u = v_1 + v_2,$$
$$a = \frac{4\mu}{M}\left[E_1 - E_2 + \frac{v_1v_2}{2}(m_1 - m)\right]. \tag{63a}$$

<u>γ (Gamma)</u>

$$\Delta E > 0, \qquad \Delta E > a, \qquad 2m_2v_2 > (m_1 - m)v_1,$$
$$v_l = v_2' - v_1', \qquad v_u = v_1' + v_2'.$$

<u>δ (Delta)</u>

$$\Delta E > 0, \qquad \Delta E > a, \qquad 2m_2v_2 < (m_1 - m_2)v_1,$$
$$\sigma_{\Delta E}{}^{e} = 0.$$

Equations 62p and 63a are used to calculate the integral in Eq. 62a. This integration is routine, and only the results are recorded. In view of the subsequent application, Z_1 and Z_2 each is equated to unity:

$$Q(v_1, v_2) = \frac{\pi e^4}{3v_1{}^2 v_2}\left[-\frac{2v_2{}^3}{(\Delta E)^2} - \frac{6v_2/m}{\Delta E}\right], \qquad 0 < \Delta E < b;$$

$$Q(v_1 v_2) = \frac{\pi e^4}{3v_1{}^2 v_2}\left[\frac{3(v_1/m - v_2/m)}{\Delta E} + \frac{v_2'^3 - v_2{}^3 - (v_1'^3 + v_1{}^3)}{(\Delta E)^2}\right],$$

$$b < \Delta E < a; \qquad (63b)$$

$$Q(v_1, v_2) = \frac{\pi e^4}{3v_1{}^2 v_2}\left[-\frac{2v_1'^3}{(\Delta E)^2}\right], \qquad \Delta E > a, \qquad m_2 v_2 > (m_1 - m_2)v_1;$$

$$Q(v_1 v_2) = 0, \qquad \Delta E > a > 0, \qquad 2mv_2 < (m_1 - m_2)v_1.$$

In order to evaluate the cross section the limits of Eq. 62a must be used for the ΔE in Eq. 63b.

Garcia and Gerjuoy[44] used this classical procedure to calculate cross sections for the reactions

$$H^+ + H(1s) \rightarrow H(2s) + H^+$$
$$H^+ + H_2 \rightarrow H(2s) + H_2{}^+$$
$$H^+ + He(1s^2) \rightarrow H(1s) + He^+(1s).$$

In the application of Eq. 63b to capture from an atomic subshell containing n equivalent electrons, Q of Eq. 63b is multiplied by n. Furthermore, the velocity v_2 is assumed to satisfy

$$\frac{m}{2} v_2{}^2 = U_A, \qquad (63c)$$

and U_A is the ionization potential of the subshell from which capture occurs, whereas U_B denotes the ionization potential of the atomic level into which capture takes place. Garcia and Gerjuoy have some criticism for this model. As they explain, the primary assumptions upon which this classical model is based are: (1) the Coulomb interaction between the incident proton and the atomic electron is the primary one in determining the cross section; (2) the trajectory of the incident proton is unaffected by the transfer of kinetic energy; (3) the magnitude of ΔE is the primary criterion for deciding whether or not capture occurs. Assumption (3) appears to be the weakest since the model contains no provision for the transfer of momentum to the electron. Thus not all electrons having energies in the correct range, $\Delta E_l < \Delta E < \Delta E_u$, would be captured since capture is not equally probable for all magnitudes and directions of the electronic momentum relative to that of the proton. The authors conjecture that the major correction would occur at high impact velocities, the omission of momentum transfer being of less consequence at low impact energies. (Recall that approximate quantum mechanical cross sections depend similarly upon transfer of momentum.) It is impossible to separate entirely assumption (1) from assumption (3) since the Coulomb interaction provides the mechanism to bind the captured electron to the incident ion.

Garcia, Gerjuoy, and Welker[45] applied Eq. 63*b* to the processes

$$H^+ + A \rightarrow H(nl) + A^+$$

for atoms A consisting of the noble gases and the alkali atoms. Equation 63*c* is again used to represent v_2 for the atomic target; only capture from outer subshells was calculated since they found that capture from inner subshells was unimportant for the range of impact energies they investigated, this energy seldom exceeding 100 keV. The exponential velocity distributions for the bound electron used by Gryzinski were not adopted by Garcia et al. since they were unable to justify this assumption. Garcia et al. then reiterate, that the binary approximation is untenable at low impact velocities. Several features of Eq. 63*b* are discussed by Garcia et al., the first pertaining to scaling.

It is assumed that $U_A > U_B$, $v_1 \geq 2v_2$, and $m_2/m_1 = \lambda \ll 1$. In this instance it is readily established that ΔE_u is in the range under α of Eq. 63*a*; thus the corresponding cross section of Eq. 63*b* is applicable, and the resulting cross section per equivalent electron is

$$Q = \frac{\pi e^4}{3v_1{}^2v_2{}^2}\left[-\frac{2v_2{}^4}{(\Delta E)^2} - \frac{6v_2{}^2}{m_2\Delta E}\right]_{\Delta E_l}^{\Delta E_u}$$
$$= \frac{\pi e^4}{3\lambda U_A E_1}\left[-\frac{2U_A{}^2}{(\Delta E)^2} - \frac{3U_A}{\Delta E}\right]_{\Delta E_l}^{\Delta E_u}. \tag{63d}$$

If $U_B \ll \lambda E_1 + U_A$, the dominant part of Eq. 63*d* is

$$Q = \frac{2\pi e^4}{3\lambda U_A{}^3}\frac{U_B}{(E_1/U_A)}\left[\frac{4}{(\lambda E_1/U_A + 1)^3} + \frac{3}{(\lambda E_1/U_A + 1)^2}\right]. \tag{63e}$$

Now imagine the capturing process from atoms A and A′ by incident ions with the same ionization energy of the capturing level U_B and with impact energies E_1 and E_1', respectively. The cross sections per electron are related by

$$Q(E_1; A)U_A{}^3 = Q(E_1'; A')U_A'^3,$$
$$U_B' = U_B, \qquad E_1'U_A = E_1U_A'. \tag{63f}$$

This is the scaling relation for high impact velocities.

Also of interest is the convergence of the sum of the cross sections into all final states of the atom formed by capture. Numerical results indicate that the sum does remain finite provided that $U_B < U_A$; this condition, according to Eqs. 62*a* and 63*b*, ensures that ΔE does not vanish and causes Q to become infinite. In case the incident ions are protons and the atom is H(1*s*), it has been stated that the sum indeed converges. As Garcia et al. remark,[45] the cross section $\sigma_{\Delta E}{}^e$ has been integrated over all final angular momenta consistent with the energy transfer ΔE. For this reason, they do not multiply each cross section for capture into the level n_B by the degeneracy $n_B{}^2$ in order to accommodate capture into each substate m_l of that level. (As Gerjuoy explains, Eq. 63*b* is the total cross section for capture with given change in energy and thus includes the integral over all final angular momenta. Therefore it would be incorrect to include the statistics and weights of the final capturing energy levels.*) According to Eq. 63*e* the sum to be evaluated for this example is

$$Q(\Sigma) = \sum_{n_B=1}^{\infty} \frac{2\pi e^4}{3\lambda E_1 U_A{}^2} \frac{U_A}{n_B{}^2} \frac{4\lambda E_1/U_A + 7}{[(\lambda E_1/U_A) + 1]^3} = \frac{\pi^3 e^4}{9\lambda E_1 U_A{}^2} \frac{4\lambda E_1/U_A + 7}{[(\lambda E_1/U_A) + 1]^3},$$

$$U_B = U_A n_B{}^{-2}, \qquad \sum_{n=1}^{\infty} \frac{1}{n^2} = \frac{\pi^2}{6}. \tag{63g}$$

It is recalled from Section 4.4 that in first Born approximations the cross section into the level n_B and its variation with E_1 vary as $n_B{}^{-3}$ and $E_1{}^{-6}$, respectively. In contrast, the respective variations given by Eq. 63*g* are $n_B{}^{-2}$ and $E_1{}^{-3}$. Transfer of momentum to the electron, however, is not the only cause of this marked difference, since Thomas' cross sections only vary as $E_i{}^{-11/4}$, notwithstanding accommodation for momentum transfer. It was mentioned previously that ΔE_l of Eq. 62*a* vanishes for some v_1 if $U_A < U_B$.

Moreover, the cross sections are very large for U_A only slightly greater than U_B, and several modifications have been proposed to remove these troublesome features. A discussion of these semiempirical approaches, however, is more profitably accomplished in sections on comparison of calculated and experimentally determined cross sections.

Another defect mentioned in connection with this classical procedure is its failure to satisfy detailed balancing. For example; if $U_A < U_B$ for the reaction

$$B^+ + A \rightarrow B + A^+,$$

we have seen that ΔE_l vanishes for sufficiently small v_1; however, for the same velocity v_1, the cross section for the reverse process

$$A^+ + B \rightarrow A + B^+$$

does not diverge since U_A and U_B are now interchanged and the resulting ΔE_l does not vanish. Some remarks by Garcia, Gerjuoy, and Welker in this

* Private communication.

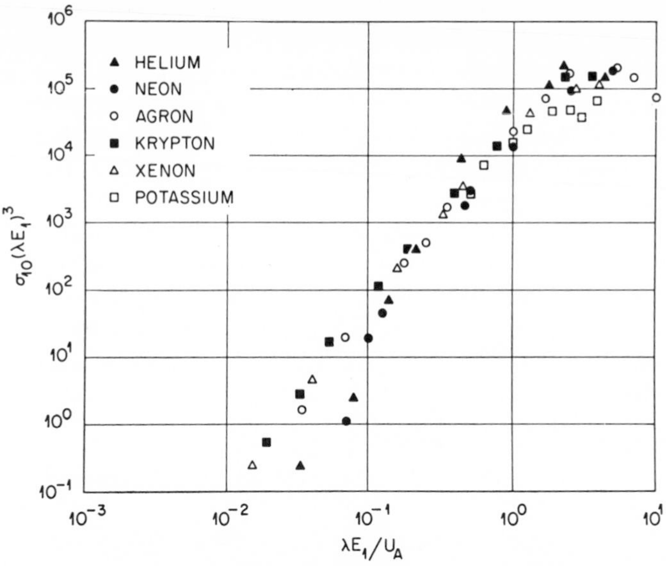

Fig. 13 Scaled experimental charge-transfer cross sections. [Data are from Figs. 1–6 of Garcia, Gerjuoy, and Welker (1968)].

connection are paraphrased here. Within the confines of the classical binary approximation, Gryzinski may have proposed nearly the correct bounds on ΔE as given by Eq. 62*a* provided that U_A exceeds U_B. It is not necessarily a defect that his prescription violates detailed balancing and causes the cross section to diverge if U_B exceed U_A. In this case, it is suggested, the cross section for the reverse process should be calculated, and detailed balancing should be used to obtain the desired cross section.

Returning now to scaling, we rewrite Eq. 63*e*, the form valid at high impact velocities,

$$\frac{E_1^3\lambda^3 Q}{U_B} = \frac{2\pi e^4}{3}\frac{\lambda^2 E_1^2}{U_A^2}\left[\frac{4\lambda E_1/U_A + 7}{(\lambda E_1/U_A + 1)^3}\right] = F\left(\frac{\lambda E_1}{U_A}\right). \tag{63h}$$

We again depart from the format of this chapter and compare the predictions of Eq. 63*h* with the corresponding experimental values, as we did in Fig. 12 of Section 5.4. The experimental values of $(E_1\lambda)^3 U_B^{-1} Q$ are plotted against $(\lambda E_1)U_A^{-1}$ in Fig. 13 for He, four noble gases, and K. [In this figure (E_1/U_A) is dimensionless, and U_B is omitted from the ordinates, an omission that suggests that all electrons are captured into the same final state. Therefore this omission is to be interpreted as an additional approximation. Furthermore, the scaling law of Eq. 63*h* is deduced for capture by protons into H(1*s*).]

In Fig. 13 the cross sections for He and K were multiplied by 3 and 6, respectively, in order to obtain 6 equivalent electrons in each case. These scaled points are plotted only for impact energies that are greater than the energy where the observed maximum occurs, for it would be inconsistent to use Eq. 63*h* at lower impact energies. It is observed that the heavy noble gases scale exceptionally well. In a comparison of the methods of this section with Thomas' methods (Section 5.4) we should not overlook the advantage of the procedure of Gryzinski, which provides cross sections for capture into the separate final states; in contrast, Thomas' method only yields the sum of the cross sections into all classically allowed final states.

5.6 ASYMPTOTIC VARIATION OF CLASSICAL CROSS SECTIONS AT HIGH NONRELATIVISTIC IMPACT VELOCITIES

This section is the classical counterpart to Section 4.4. The phrases *high impact velocity* and *asymptotic variation* retain the significance given to them in Section 4.4; the last classical method studied is treated first here.

The classical cross section proposed by Gryzinski and calculated by Gerjuoy et al.[43,45] for capture from neutral atoms by protons varies as E_i^{-3} according to Eq. 63*e*. This variation remains valid for capture into any level n of the hydrogen atom; furthermore, according to Eq. 63*g*, this same variation is obtained for the sum of the cross sections. This formula, however, was not designed for these high velocities, as Gryzinski clearly states.[42] Next, the cross sections derived from the models of Thomas are scrutinized.

According to Eqs. 59*f* and 60*d*, both the original and first modifications of the cross section of Thomas vary as $E_i^{-11/4}$ for capture from neutral atoms by positively charged nuclei. As discussed in Section 5.3, however, these predictions are not reliable since the electronic distribution of Eq. 59*a* is not valid near the nucleus, which is the region that presumably provides the major contribution to the cross section. Support for this presumption originates from the analyses of the approximate quantum mechanical cross sections for the same reactions. This same $E_i^{-11/4}$ variation also occurs for cross sections defined by the second modification of Thomas for the same processes, and again this prediction is unreliable asymptotically. In this instance, it is the inverse square potential that is one source of the failure. Near the atomic nucleus this interaction is not a good approximation to the actual potential field; consequently, the use of this inverse square law to deduce Eqs. 58*j* and 58*k* of Section 5.3 (the same results being used in Section 5.4) is not a valid procedure. Thus the second modification of Thomas is not

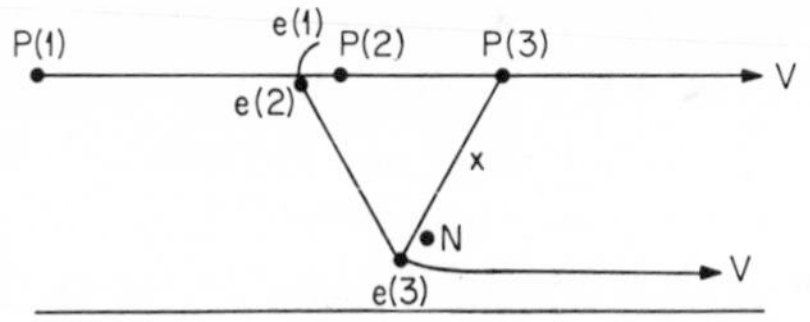

Fig. 14 Classical trajectory with several positions of the proton P, electron e, relative to the nucleus N, in Thomas' model for capture from light atoms at high impact velocities.

applicable to capture from the $(1s)^2$-subshell, the dominant capturing process at sufficiently high impact velocities.

There is one other case treated by Thomas.[2] This cross section obtained for high relative velocities V of the two nuclei has been an object of interest in recent years by reason of the close agreement between it and quantum mechanical cross sections for the same process:

$$H^+ + H(1s) \rightarrow H(1s) + H^+. \tag{64a}$$

All three quantum mechanical calculations, the second Born approximation of Drisko,[46] the impulse approximation of Bransden and Cheshire,[47] and the continuum distorted wave approximation of Cheshire,[48] yield the V^{-11} dependence on impact velocity and nearly the same numerical coefficient as the classical result. In order to understand clearly what classical cross section is being compared, Thomas' model is reviewed.[25,49]

This classical cross section is for capture into all classically allowed bound states of the H atom instead of the reactions in the right-hand number of Eq. 64a. According to this classical model, the required final translational velocity for capture is imparted to the electron by two binary collisions, and the sequence of scattering events is portrayed in Fig. 14 by designating with the same numerals the positions of the incident ion and the electron at the same times. The electron initially at point 1 at a distance r from its atomic nucleus N receives a vectorial impulse from the ion P so that the electron has the correct initial velocity and impact parameter to be scattered by N with the appropriate velocity and direction for subsequent capture by P. The condition for classical capture requires a final relative velocity that is less than the escape velocity w of the electron e with respect to P as its nucleus; this requirement for a distance x between P and e is

$$\frac{m}{2} w^2 - \frac{e^2}{x} \leq 0. \tag{64b}$$

We next examine the assumptions implicit in this representation of the capturing process; for ease of description the distance between a particle X and e is designated (Xe).

For the first binary collision to be a good representation of the scattering, the minimum value of (Pe) must be very much less than the mean value of

(Ne) in order for the field between P and e to be dominant; similarly, the corresponding conditions on the distances must be interchanged for the second binary process to be equally representative of the scattering. (As mentioned in the preceding sections, this is the *classical impulse approximation.*) Moreover, the velocity of the electron is assumed negligible in comparison to V just before the first binary collision. In this model the deflection of P and the corresponding recoil of N are omitted as negligible so that the velocity of P is the same in the laboratory system and in the frame of reference with origin of coordinates at N. Thus, if the initial velocity of the electron in its classical orbit is denoted by v, it is necessary that

$$v \ll V \tag{64c}$$

for the first condition $(\mathrm{P}e)_{\mathrm{min}} \ll (\mathrm{N}e)_{\mathrm{mean}}$ to be satisfied. One reason why Eq. 64c enters is seen by scrutinizing a bound-state equation for the atomic electron,

$$\frac{m}{2}v^2 - \frac{e^2}{r} = -\frac{e^2}{2a}, \qquad \frac{m}{2}v^2 = \frac{p^2}{2m} = \frac{p_r^{\,2}}{2m} + \frac{l^2}{2mr^2}, \tag{64d}$$

and it is clear from this equation that larger values of r correspond to smaller values of v. The relation for the kinetic energy in terms of momentum variables in spherical coordinates is included to emphasize that only conservation of energy restricts the value of the classical angular momentum unless a particular orbit is specified, and this important fact enters subsequently. It is pertinent to note that these conditions on distances and velocities are increasingly difficult to fulfill as the P-N distance decreases. We proceed now to calculate the cross section for classical analogues of Eq. 64a.

The differential cross section for the first binary collision is the Rutherford scattering cross section,[50]

$$q = 2\pi p\,dp = \frac{\pi e^4}{2m^2V^4}\csc^4\left(\frac{\delta}{2}\right)\sin\delta\,d\delta \tag{64e}$$

in which p, e, V, and δ are the impact parameter, the electronic charge, the impact velocity, and the center-of-mass scattering angle, respectively; m, the electronic mass, is closely equal to the reduced mass $mM(M + m)^{-1}$. If U is the velocity of the electron in the laboratory system after the scattering, and noting Eq. 64c, it is evident that the relation among U, V, and δ is

$$\begin{gathered} U = 2V\sin\left(\frac{\delta}{2}\right), \qquad dU = V\cos\left(\frac{\delta}{2}\right)d\delta, \\ q = \frac{\pi e^4}{m^2V^5}\csc^3\left(\frac{\delta}{2}\right)dU, \end{gathered} \tag{64f}$$

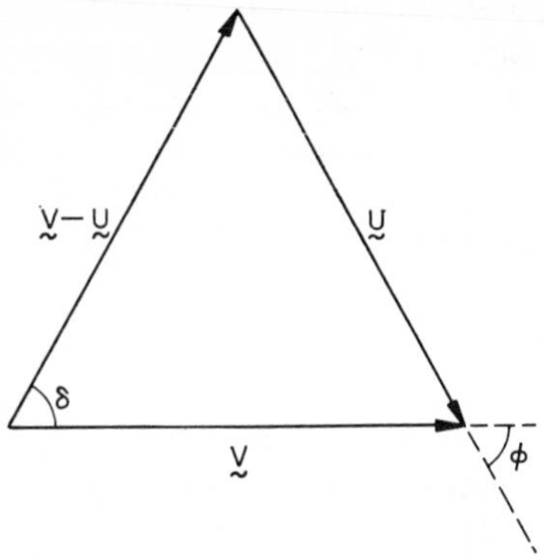

Fig. 15 Vector diagram for the velocities in Thomas' model for capture at high impact velocities corresponding to Fig. 14.

and the associated vectors are illustrated in Fig. 15. The angle ϕ in this figure denotes the angle of recoil of the electron in the laboratory system.[55] * The chance that this scattered electron, initially at a distance r from its nucleus before the scattering, has impact parameter p' with respect to the same nucleus and is deflected through an angle δ', is

$$P = \frac{p'\,dp'\,d\phi'}{4\pi r^2} = \frac{\text{target area}}{\text{area of sphere}}, \tag{64g}$$

where ϕ' is the usual azimuthal angle. [We may also interpret this relation as the probability that the line (N*e*) falls within a certain solid angle.] The Coulomb scattering law is used again with Eq. 64*g* to get

$$P = \frac{e^4}{16\pi m^2 U^4 r^2} \csc^4\left(\frac{\delta'}{2}\right) d\omega, \qquad d\omega = 2\pi \sin\delta'\, d\delta'\, d\phi'. \tag{64h}$$

In this second binary collision there is negligible difference between the laboratory and center-of-mass systems since the electronic mass is negligible relative to the nuclear mass; moreover, since the transfer of translational energy is likewise negligible, the initial and final speeds, U and U', respectively, are approximately equal. In order that capture can occur, the final relative velocity $\mathbf{w} = \mathbf{V} - \mathbf{U}'$ must satisfy Eq. 64*b*, and Thomas ingeniously used this condition for capture to eliminate $d\omega$ from Eq. 64*h* by noting that the tip of the $\mathbf{U}'$ vector must fall within the sphere whose volume is $(4\pi/3)w^3$ in velocity-space:

$$(U')^2\,dU'\,d\omega = \frac{4\pi}{3}w^3, \qquad w^2 = \frac{2e^2}{mx}, \qquad V - w \le U' \le V + w. \tag{64i}$$

* The diagrams in reference 55 are for momenta, not velocities.

This relation between the two elements of volume is an approximation, and the smaller w is in comparison to U', the more accurate is the approximation. If δ and δ' are both equal to 60°, $V \approx U \approx U'$, $dU \approx dU'$, $r \approx x$, and if the additional approximation $dU' \approx w$ is used, the cross section originally derived by Thomas is obtained by combining Eqs. 64*f*, 64*h*, 64*i*, and the approximations just mentioned:[51]

$$Q(r) = Pq, \qquad q = \frac{8\pi e^4}{m^2 V^5}\left(\frac{2e^2}{mr}\right)^{1/2}, \qquad P = \frac{8e^6}{3m^3 V^6 r^3}. \tag{64j}$$

Some of the assumptions used to deduce Eq. 64*j* are discussed next.

The value of P given by Eq. 64*j* tends to infinity as r tends to zero; yet, as a probability that the electron-nucleus line falls within a certain solid angle, P should not exceed unity. From the bound-state condition in Eq. 64*d*, moreover, the velocity v increases with decreasing r in contradiction to the requirement that $v \ll V$ for the first collision to be approximately binary. We cannot maintain $(eP) \ll (eN)$ as r decreases unless the impact velocity is increased and the impact parameter decreased so that the Coulomb scattering relation,[52]

$$\cot\left(\frac{\delta}{2}\right) = \cot 30° = \frac{mV^2 p}{e^2},$$

is not violated. Unless the classical angular momentum is fixed in Eq. 64*d*, the one-dimensional barrier imposed by the centrifugal potential $(l^2/2mr^2)$ can vanish, and thus r can vanish with v simultaneously becoming infinite if $l = 0$—the case of the degenerate line ellipse discussed in Section 5.2 in connection with Eq. 56*q*. We thus see that Eq. 64*c* may be invalidated in addition to P exceeding unity as r decreases. If this equation is not satisfied, the assumed geometry of the scattering and the assumed binary collisions are not good approximations to the scattering mechanism. It is evident that some additional restrictions must be imposed to prevent violation of some of these assumptions.

Perhaps the most obvious solution is to assume an isotropic distribution of circular orbits with a common center at the nucleus N (Eq. 56*u*). If the radius is a_0, the first Bohr radius, the original result of Thomas (atomic units) is obtained from Eq. 64*j*:

$$Q = \int \delta(r - a_0) Q(r)\, dr = \frac{64\sqrt{2}\,\pi}{3V^{11}}. \tag{64k}$$

Note that this result is $(\sqrt{2}/3\pi)$ times the result of Eq. 52*k* of Section 4.3, which represents the cross section in the second Born approximation for nonresonance. Another approach is to represent the electron by one of the classical distributions derived in Section 5.2, and since the distribution of

Eq. 56*f* is the correct quantum mechanical distribution for H(1*s*) of Eq. 64*a* if $v_n = v_0 = (e^2/\hbar)$, the associated spatial distribution of Eq. 56*e* is suggested. However, a straightforward combination of this distribution with Eq. 64*j* leads to a divergent integral. An arbitrary procedure adopted by Bates and Mapleton[25] to avoid this divergence is to equate P to unity for distances less than the distance at which P becomes unity (atomic units still being used):

$$\begin{aligned} P(r_0) &= \frac{8}{3V^6 r_0^{\,3}} = 1, \\ Q(r > r_0) &= \frac{64\sqrt{2}\,\pi}{3V^{11}} \int_{r_0}^{2} \frac{2}{\pi} \frac{(2-r)^{1/2}}{r^2}\, dr = \frac{128}{3^{2/3} V^9}, \\ Q(r < r_0) &= \frac{8\sqrt{2}\,\pi}{V^5} \int_{0}^{r_0} \frac{2}{\pi} (2-r)^{1/2} r\, dr = \frac{64}{3^{2/3} V^9}, \\ Q &= \frac{192}{3^{2/3} V^9}. \end{aligned} \tag{64l}$$

These integrals were evaluated approximately by using the fact that $r_0 \ll 1$. The distinguishing feature of this result is the V^{-9}-dependence instead of the V^{-11}-variation of Eq. 64*k*.

Another distribution is suggested by the fact that the 1*s*-state of H has quantized angular momentum equal to zero, and we thus choose the isotropic distribution of degenerate line ellipses since the associated classical angular momentum likewise is zero. [This type of distribution, as well as the one used in Eq. 64*k*, refers to orbits with appropriate squared angular momentum l^2 in the range $\frac{1}{2}d(l/l_n)^2$, a fact implicit in the analysis of Section 5.2.] If this spatial distribution of Eq. 56*q* is used with Eq. 64*j*, the resulting cross section is

$$Q = \int_0^2 dr \frac{Q(r)}{\pi} \left(\frac{r}{2-r}\right)^{1/2} = \frac{24}{3^{1/3} V^7}, \qquad P(r) = 1, \quad 0 < r < r_0; \qquad P(r) = P, \quad r_0 < r. \tag{64m}$$

The distinguishing feature of this result is the V^{-7}-variation as contrasted to Eq. 64*k* and 64*l*. Although the use of the distribution for the circular orbit does not require the imposition of additional restrictions, this distribution disagrees markedly with the corresponding quantum mechanical distribution for H(1*s*). It is interesting, nevertheless, that the dependence upon impact velocity is the same as the quantum mechanical ones discussed in Section 4.4. However, the value of a_0, used in Eq. 64*k*, should be greater than r_0, given by Eq. 64*l*, to ensure that P of Eq. 64*j* never exceeds unity.

The classical model used for the calculation of the cross section for capture into all classically allowed states for the initial reactants on the left of Eq. 64*a*

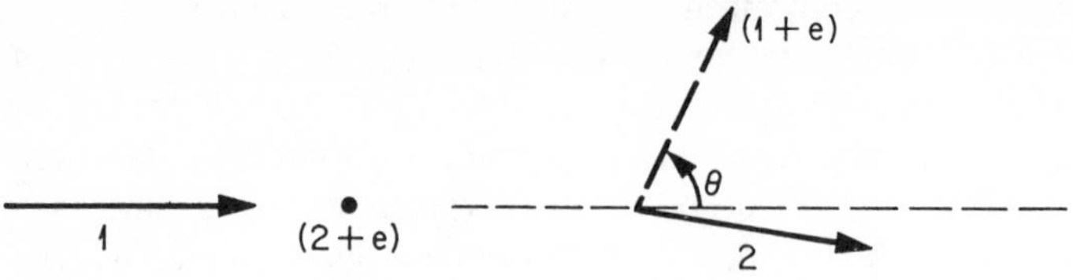

Fig. 16 Diagrammatic illustration of initial and final velocities in the laboratory system for the classical model of Bates and Mapleton. In this model for high velocities the angle θ is roughly 90° for the scattered atom $(1 + e)$ formed by capture.

is also interesting. As we learned in Section 4.4, if the protons are treated as distinguishable particles, the cross section calculated in the first Born approximation using the prior (or post) interaction yields Eq. 55*g* for Eq. 64*a*; moreover, the dominant contribution to this cross section originates from the center-of-mass angular region near $\theta = \pi$. This capturing at large angles of scattering suggests a classical description for the two protons since their motion appears to conform to the laws of classical mechanics for these large deflections at small impact parameters. Motivated by these considerations, Bates and Mapleton proposed a classical model to describe this scattering as shown in Fig. 16.[5,53]

Proton 1 is incident upon the hydrogen atom whose nucleus, proton 2, is initially at rest in the laboratory system at the origin of the observer's frame of reference. If the presence of the electron is ignored, it is seen that this figure portrays, in particular, the repulsive motion of proton 1, which is deflected through nearly 90° corresponding to a very small impact parameter; that is, if V_i and V_f represent the impact velocity and final velocity, respectively, of proton 1, and if ρ is the associated impact parameter, according to classical mechanics for the repulsive Coulomb interaction, we have[54]

$$V_f = V_i \cos\theta \ll V_i, \qquad \rho = \frac{2e^2 \cot\theta}{MV_i^2} \ll 1 \tag{65a}$$

for θ near 90°. This obviously corresponds to backward scattering in the center-of-mass system. We proceed to construct a model consistent with this assumption of large angular deflection. First, it is stressed that only for identical nuclear masses is it possible for V_f of Eq. 65*a* to vanish. The minimum value of the kinetic energy of the incident particle after scattering is equal to

$$T_m = \frac{M_1}{2}\left(\frac{M_2 - M_1}{M_2 + M_1}\right)^2 V_i^2 = \frac{M_1}{2} V_f^2$$

from classical mechanics, and the vanishing of V_f for large V_i implies that $M_2 = M_1$.[55] Now prior to the collision the electron is classically bound to proton 2, and the electronic motion satisfies

$$\tfrac{1}{2}mv^2 - \frac{e^2}{r} = -\varepsilon_1 = -\frac{e^2}{2a_0} = -\frac{me^4}{2\hbar^2} = -\frac{mv_1{}^2}{2}. \tag{65b}$$

It is also assumed that the distance of closest approach, $R_{\text{c.a.}}$, satisfies

$$R_{\text{c.a.}} = \frac{4e^2}{MV_i^2} \ll r; \tag{65c}$$

furthermore, the velocity and position of the electron are postulated to change inappreciably during the effective duration of the capturing process. In this model the interaction between proton 2 and the electron only determines the distribution in velocity and position of the electron in accord with Eq. 65*b*; consequently, the position of the electron changes very little during the effective duration of the capturing process provided that the *important* electronic velocities satisfy

$$v \ll V_i. \tag{65d}$$

Important electronic velocities are those electronic velocities of the distribution

$$f(v)\,dv = \frac{32v_1{}^5v^2}{\pi(v^2 + v_1{}^2)^4}\,dv, \qquad v_1 = \frac{e^2}{\hbar} \tag{65e}$$

that contribute most to the cross section. We have chosen the distribution in Eq. 65*e* from Eq. 56*f* (Section 5.2) since this function represents both the quantum mechanical and classical hydrogenic atom with binding energy ε_1. Since this distribution attains its maximum value at $v^2 = v_n{}^2/3$ and decreases strongly with increasing v, the condition expressed by Eq. 65*d* is satisfied for the domain of v that provides the major contribution to the cross section. We proceed to investigate the correctness of the assumptions, and for this purpose we estimate the effective duration of the capturing process to be

$$\Delta t = \frac{R_{\text{c.a.}}}{V_i}; \tag{65f}$$

hence the fractional change in position during this time is

$$\frac{\Delta r}{r} \approx \frac{v\Delta t}{r} = \frac{v}{V_i}\frac{R_{\text{c.a.}}}{r} \ll 1. \tag{65g}$$

Consequently, in view of Eq. 65*c*, it is clear that Eq. 65*g* is satisfied if Eq. 65*d* is fulfilled. In this estimation there is also the tacit assumption that Δv is negligible during the time Δt.

Consistent with the foregoing postulates, the position of the electron can be referred during the collision, with small error, to proton 1 instead of the origin. Furthermore, by reason of Eq. 65*d*, the first relation of Eq. 65*a* and the postulate $\Delta v \approx 0$, the relative velocity during the encounter is approximately $(\mathbf{V}_f - \mathbf{v})$ so that the condition for electronic capture by proton 1 can be expressed by

$$\frac{m}{2}|\mathbf{V}_f - \mathbf{v}|^2 - \frac{e^2}{r} \leq 0. \qquad (65h)$$

This capturing condition is operative only during the collision since only then are the distances from the electron to the origin and to proton 1 nearly the same; furthermore, proton 1 is assumed to achieve its final velocity $\mathbf{V}_f$ while these preceding two distances are approximately equal since $\mathbf{V}_f$ is so small that the distance proton 1 travels during the encounter $\approx V_f \Delta t$ is negligible in comparison to r. It is clear by now that $V_f \Delta t \ll r$ if the previously mentioned restrictions are fulfilled. In view of these conditions it is a good approximation to use Eq. 65*b* to eliminate the distance r from Eq. 65*h*:

$$\frac{m}{2}(V_f^2 - 2vV_f \cos\chi) - \varepsilon_1 \leq 0, \qquad \cos\chi = \frac{\mathbf{v}\cdot\mathbf{V}_f}{vV_f}. \qquad (65i)$$

The equality sign in Eq. 65*i* provides a bound on the condition for classical capture, and the physical (non-negative) solution to this equation for V_f gives the maximum allowable value of V_f corresponding to the values of v, $\cos\chi$, and v_1 of Eq. 65*b* and 65*i*,

$$V_f \leq V_c = v\cos\chi + (v^2\cos^2\chi + v_1^2)^{1/2}. \qquad (65j)$$

With the use of this equation and the first relation in Eq. 65*a*, we get the range of the scattering angles

$$0 \leq \cos\theta \leq \frac{V_c}{V_i}, \qquad (65k)$$

and we now have all ingredients needed to calculate the cross section.

Equation 64*e* corresponding to the laboratory system is used, and this is written here for convenience:

$$d\sigma = 2\pi\left(\frac{2e^2}{MV_i^2}\right)^2 \frac{\cos\theta}{\sin^3\theta}\, d\theta. \qquad (65l)$$

Since θ depends upon v and χ through Eqs. 65*j* and 65*k*, the chance that the vector $\mathbf{v}$ of the electron is in the range $d\chi$ about χ,

$$\frac{1}{2}\sin\chi\, d\chi, \qquad (65m)$$

is needed together with Eq. 65*e*, which gives the chance that the velocity $\mathbf{v}$ is in the range dv about v. Consequently, the differential cross section for proton 1 to emerge with the velocity $V_f \approx V_c$ at the angle θ and the electron to be classically bound to proton 1 with relative velocity $(\mathbf{v} - \mathbf{V}_f)$, is

$$dQ = f(v)\,dv\,\frac{\sin\chi}{2}\,d\chi\,d\sigma. \tag{65n}$$

The integration of $d\sigma$ over $d\theta$, consistent with Eqs. 65*k* and 65*l*, is effected first to give

$$\int_0^{\cos^{-1}(V_c/V_i)} d\sigma = 2\pi\left(\frac{2e^2}{MV_i^2}\right)^2 \frac{V_c^2}{2V_i^2(1 - V_c^2/V_i^2)}$$

$$\approx \pi\left(\frac{2e^2V_c}{MV_i^3}\right)^2. \tag{65o}$$

We recall the restriction of Eq. 65*d*, but the domain of v is $(0, \infty)$; nevertheless, these conflicting requirements are only apparent since $f(v)$ decreases very strongly with increasing v. Equations 65*n* and 65*o* are combined next, and the value of V_c given by Eq. 65*j* is used; the resulting expression is easily integrated, and the answer (in units of 100 keV) is

$$Q = \pi\left(\frac{2e^2}{MV_i^3}\right)^2 \int_0^\infty dv f(v) \int_0^\pi d\chi\,\frac{\sin\chi}{2}\,V_c^2$$

$$= \frac{5}{4}\left[\frac{1}{12}\left(\frac{m}{M}\right)^2 \frac{\pi a_0^2}{E_i^3}\right]. \tag{65p}$$

This cross section is 1.25 times the quantum mechanical one of Eq. 55*g* of Section 4.4 for capture into H(1*s*) only; however, it can be proven that Eq. 55*g* actually represents capture into all excited states; thus there is remarkably good agreement. (In this connection the reader is reminded that the classical and quantum mechanical first Born approximation also yield the same differential cross section for elastic scattering in the Coulomb field.)

Next, a summary of the asymptotic cross sections for Eq. 64*a* is given; however, the atomic target is an imaginary hydrogen atom whose mass exceeds M of the proton for the cross sections whose chief contributions stem from small scattering angles. For those cases in which the chief contribution originates from the angular region near $\theta = \pi$, the mass is equated to M so that this cross section for distinguishable protons likewise is defined. Consequently, the cross sections are defined for each of the cases presented. The notation b, cl, t, d, i, and cdw denote, respectively, first Born approximation with prior interaction, classical of last Eq. 65*p*, Thomas, second Born approximation of Drisko, quantum mechanical impulse approximation, and continuum distorted wave approximation. Moreover, V designates the impact velocity and v_0 denotes $(e^2/\hbar)$. These formulas are expressed in units of πa_0^2:

$$\begin{aligned}
Q(\mathrm{OBK}) &= \frac{2^{18}}{5}\left(\frac{v_0}{V}\right)^{12},\\
Q_{\mathrm{b}}(E_i^{-6}) &= 0.661\,Q(\mathrm{OBK}),\\
Q_{\mathrm{d}} &= \left[\frac{5\pi}{2^{12}}\frac{V}{v_0} + 0.295\right]Q(\mathrm{OBK}) = Q_{\mathrm{cdw}},\\
Q_{\mathrm{i}} &= \left[\frac{5\pi}{2^{11}}\frac{V}{v_0} + 0.295\right]Q(\mathrm{OBK}),\\
Q_{\mathrm{t}} &= \frac{64\sqrt{2}}{3}\left(\frac{v_0}{V}\right)^{11}, \qquad Q_{\mathrm{b}}(E_i^{-3}) = \frac{16}{3}\left(\frac{m}{M}\right)^2\left(\frac{v_0}{V}\right)^6,\\
Q_{\mathrm{cl}} &= 1.25\,Q_{\mathrm{b}}(E_i^{-3}).
\end{aligned} \tag{65q}$$

Finally, the impact energies at which certain pairs of cross sections become equal are listed, and the results are displayed so that the cross sections on the left dominate at higher values of E_i:

$$\begin{aligned}
\frac{5\pi}{2^{12}}\frac{V}{v_0}\,Q(\mathrm{OBK}) &= 0.295\,Q(\mathrm{OBK}), && E_i = 147.9\ \mathrm{MeV};\\
Q_{\mathrm{d}} &= Q_{\mathrm{b}}(E^{-6}), && E_i = 227.7\ \mathrm{MeV};\\
Q_{\mathrm{b}}(E_i^{-3}) &= Q_{\mathrm{d}}, && E_i = 49.5\ \mathrm{MeV};\\
Q_{\mathrm{b}}(E_i^{-3}) &= Q_{\mathrm{b}}(E_i^{-6}), && E_i = 70\ \mathrm{MeV};\\
Q_{\mathrm{b}}(E_i^{-3}) &= Q(\mathrm{OBK}), && E_i = 80.3\ \mathrm{MeV}.
\end{aligned} \tag{65r}$$

It is instructive to note that only three of these results occur at nonrelativistic velocities of the proton.

REFERENCES

1. M. Gryzinski, *Phys. Rev.*, **115**, 374 (1959).
2. L. H. Thomas, *Proc. Roy. Soc. (London)A*, **114**, 561 (1927).
3. N. Bohr, *Kgl. Danske Videnskab. Selskab, Mat Fys. Skrifter*, **18**, no. 8 (1948).
4. R. Abrines and I. C. Percival, *Phys. Letters*, **13**, 216 (1964).
5. D. R. Bates and R. A. Mapleton, *Proc. Phys. Soc. (London)*, **85**, 605 (1965).
6. L. D. Landau and E. M. Lifshitz, *Statistical Physics*, Addison-Wesley, Reading, Mass. (1958), pp. 10–12.
7. R. B. Lindsay, *Introduction to Physical Statistics*, Wiley, New York (1948), Chapter 6, Section 7.
8. I. M. Gel'fand and G. E. Shilov, *Generalized Functions: Properties and Operations*, Vol. 1, Academic Press, New York (1964), Chapter 1, Section 2.2.

9. L. P. Pitaevskii, *Zh. Eksperim. i Theor. Fiz.*, **42**, 1326 (1962); [*Soviet Phys.-JETP*, **15**, 919 (1962)].
10. K. Omidvar, "Electron Capture by Protons in Hydrogen in an Electric Field," Goddard Space Flight Center Report X-641-65-306, July 1965.
11. R. Abrines and I. C. Percival, *Proc. Phys. Soc.* (*London*), **88**, 861 (1966).
12. R. A. Mapleton, *Proc. Phys. Soc.* (*London*), **87**, 219 (1966).
13. R. A. Mapleton, "Classical and Quantal Calculations on Electron Capture," Air Force Cambridge Research Laboratories Report AFCRL 67-0351, Physical Science Research Papers, No. 328, 1967, Part II, Sections 1.2, 1.3, 1.5, App.A.
14. Gel'fand and Shilov, *op. cit.*, p. 184.
15. V. Fock, *Z. Phys.*, **98**, 145 (1936).
16. A. Burgess and I. C. Percival, "Theoretical Treatment of Classical Theory of Atomic Scattering," in *Advances in Atomic and Molecular Physics*, Vol. 4 (D. R. Bates and I. Estermann, eds.), Academic Press, New York (1968), p. 135.
17. R. Abrines and I. C. Percival, *Proc. Phys. Soc.* (*London*), **88**, 873 (1966).
18. Burgess and Percival, *op. cit.*, pp. 109–114.
19. H. Goldstein, *Classical Mechanics*, Addison-Wesley, Reading, Mass. (1950), Chapter 3.
20. Landau and Lifshitz, *Statistical Physics*, p. 12.
21. E. C. Titchmarsh, *The Theory of Functions*, Oxford University Press, London (1950), Chapter 3.
22. R. C. Stabler, *Phys. Rev.*, **133**, A1268 (1964).
23. I. C. Percival and N. A. Valentine, *Proc. Phys. Soc.* (*London*), **88**, 885 (1966).
24. D. R. Hartree, *Proc. Cambridge Phil. Soc.*, **21**, 625 (1923).
25. D. R. Bates and R. A. Mapleton, *Proc. Phys. Soc.* (*London*), **87**, 657 (1966).
26. Mapleton, AFCRL 67-0351, Part II, Section 3.1, Apps. B and C.
27. D. Bohm, *Quantum Theory*, Prentice-Hall, Englewood Cliffs, N.J. (1951), Chapter 12.
28. Goldstein, *op. cit.*, p. 84.
29. L. D. Landau and E. M. Lifshitz, *Mechanics*, Addison-Wesley, Reading, Mass. (1960), Problem 2, p. 40.
30. A. Messiah, *Quantum Mechanics*, Vol. 2, (J. Potter, trans.) (1962); reprint ed., North-Holland, Amsterdam, and Interscience, New York (1963), pp. 615–616.
31. L. I. Schiff, *Quantum Mechanics*, McGraw-Hill, New York (1949), pp. 271–273.
32. Lindsay, *op. cit.*, pp. 226–227.
33. F. Herman and S. Skillman, *Atomic Structure Calculations*, Prentice-Hall, Englewood Cliffs, N.J. (1963).
34. D. R. Bates and R. A. Mapleton, *Proc. Phys. Soc.* (*London*), **90**, 909 (1967).
35. R. A. Mapleton, *Phys. Rev.*, **164**, 51 (1967).
36. Mapleton, AFCRL 67-0351, Part II, Section 3.6.
37. A. Messiah, *Quantum Mechanics*, Vol. 1, (G. M. Temmer, trans.) (1961); reprint ed., North-Holland, Amsterdam, and Interscience, New York (1963), p. 126.
38. A. R. Edmonds, *Angular Momentum in Quantum Mechanics*, Princeton University Press, Princeton, N.J. (1957), p. 38.
39. Titchmarsh, *op. cit.*, Chapter 10, pp. 324, 335.
40. M. Gryzinski, *Phys. Rev.*, **138**, A305 (1965).
41. M. Gryzinski, *Phys. Rev.*, **138**, A322 (1965).
42. M. Gryzinski, *Phys. Rev.*, **138**, A336 (1965).
43. E. Gerjuoy, *Phys. Rev.*, **148**, 54 (1966).
44. J. D. Garcia and E. Gerjuoy, *Phys. Rev. Letters*, **18**, 944 (1967).
45. J. D. Garcia, E. Gerjuoy, and J. E. Welker, *Phys. Rev.*, **165**, 72 (1968).

46. R. M. Drisko, Thesis, Carnegie Institute of Technology, 1955.
47. B. H. Bransden and I. M. Cheshire, *Proc. Phys. Soc.* (*London*), **81**, 820 (1963).
48. I. M. Cheshire, *Proc. Phys. Soc.* (*London*), **84**, 89 (1964).
49. Mapleton, AFCRL 67-0351, Part II, Sections 2.1 to 2.4.
50. Goldstein, *op. cit.*, p. 84.
51. D. R. Bates, C. J. Cook, and F. J. Smith, *Proc. Phys. Soc.* (*London*), **83**, 49 (1964).
52. Goldstein, *op. cit.*, p. 83.
53. R. A. Mapleton, AFCRL 67-0351, Part II, Sections 1.1, 1.3, 1.6.
54. Goldstein, *op. cit.*, pp. 83 and 89.
55. Landau and Lifshitz, *Mechanics*, Section 17, Chapter 4.

CHAPTER 6

COMPARISONS OF THEORETICAL AND EXPERIMENTAL CROSS SECTIONS

6.1 INTRODUCTION

In this chapter comparisons are made between theoretically predicted cross sections and the corresponding measured values when they exist. For many atoms, however, only measured *total* cross sections are available, whereas the calculated values represent a *single* process, often the dominant one. In other cases the only cross sections available for comparison are the total cross sections per gas atom,

$$\tfrac{1}{2}Q(A_2),$$

for capture from the associated homonuclear diatomic molecule. These experimental facts are always emphasized here. The purpose of these comparisons is to test some of the approximate calculational procedures explained in this book; thus there is the implication that the experimental values are the more accurate cross sections, in accord with common belief. This judgment however, is left to others. No attempt is made here to assess the accuracy of the measured values; rather we simply present typical results. The reader is cautioned about the deceptive appearance of good agreement because logarithmic scales are often used. In all cases the cross sections, given in units of cm^2, are plotted as a function of the impact energy E_i expressed in units of eV, keV, or MeV. To a very good approximation the impact velocity, in atomic units, is

$$v_i(\text{a.u.}) = 0.2\left(\frac{E(\text{keV})}{M_p}\right)^{1/2}, \qquad E_i = E,$$

with M_p the mass of the incident particle expressed in units of the mass of the proton. Thus, incident protons of 25 keV correspond to an impact velocity of 1 a.u. (see Appendix 1).

6.2 COMPARISONS

The cross sections calculated by Smith using the PSS method (Section 1.3) for resonant charge transfer between H^+ and H(1s) are presented first.[1] Although this two-state approximation is similar to the one used by Dalgarno and Yadav,[2] allowance is also made for the indistinguishability of the protons. Apart from this alteration, however, as Smith explains, more accurate numerical methods constitute the main improvement over the previously calculated cross sections of Dalgarno and Yadav. In Fig. 17 these results are compared with measured values.[3,4] Above $E = 3$ eV, the cross sections calculated for indistinguishable protons $Q(\pi/2)$ differ negligibly from the cross sections calculated for distinguishable protons $Q(x)$. The reader is referred to Smith's paper for a detailed discussion of nuclear interference effects for $E < 3$ eV and for a detailed analysis of the significance of the measured cross sections. The prediction of this two-state approximation is good if the plausible assumption that capture into excited states would not increase the calculated cross sections in excess of 20% in the given range of E. Also note the good prediction of the slope of a smooth curve drawn through

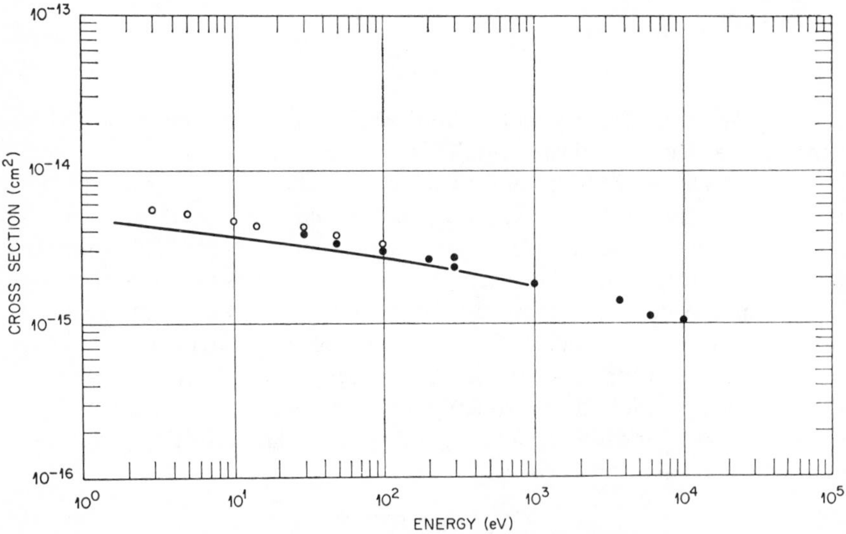

Fig. 17 Cross sections for charge transfer between protons and atomic hydrogen. Theoretical results using the PSS approximation are for the collision of $H^+ + H(1s) \rightarrow H(1s) + H^+$: —— Smith.[1] Measured results are for $H^+ + H(1s) \rightarrow \sum_n H(nl) + H^+$: ● Fite, Smith, and Stebbings;[3] ○ Belyaev, Brezhnev, and Erastov.[4] Impact energy of protons is in eV.

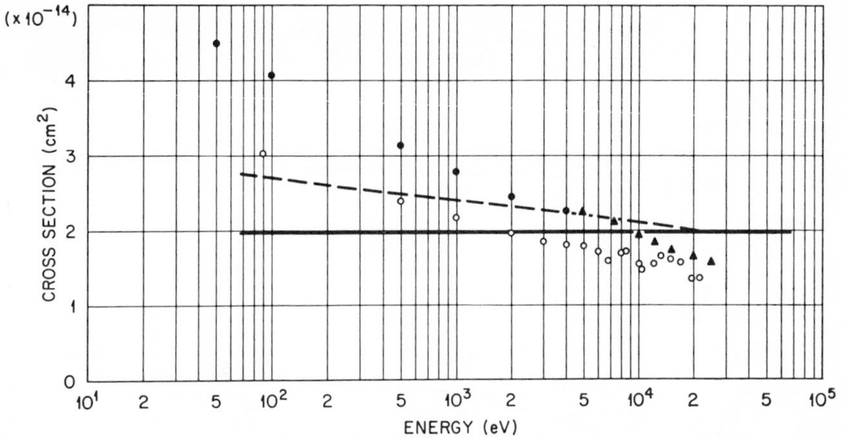

Fig. 18 Cesium charge transfer cross sections as a function of the incident ion energy. Theoretical results are for $Cs^+(^1S) + Cs(^2S) \rightarrow Cs(^2S) + Cs^+(^1S)$: —— Bates and Mapleton (C_1);[5] - - - Bates and Mapleton (C_2).[5] Experimental results are for $Cs^+(^1S) + Cs(^2S) \rightarrow \sum_n Cs(nl) + \sum_n Cs^+(n'l')$: ● Marino, Smith, and Caplinger;[6] ○ Perel, Vernon, and Daley;[8] △ Chkuaceli, Nikoleychvili, and Gouldamachvili.[7]

the experimental points for $E < 1$ keV. Although agreement between predicted and measured slopes is an important test of a theory, this feature is seldom mentioned in subsequent comparisons (see Section 2.2).

Next the semiclassical cross sections of Bates and Mapleton[5] for Cs^+-Cs resonant charge transfer is compared in Fig. 18 with the corresponding measured total cross sections[6–8] (the phrase "corresponding total cross sections" is used to indicate that the initial reactants are the same for theory and experiments). Bates and Mapleton's classical cross section, C_1,[5] also is displayed, predicting the correct order of magnitude for the measured values (see Section 2.1). The semiclassical result, C_2, predicts more successfully, of course, although the dependence on E is incorrect. In view of the many approximations used to derive C_2 (Section 2.1), the agreement is unexpectedly good. Note that a *linear* scale is used for the cross section in this figure.

A linear scale is used in Fig. 19 as well; this figure displays calculated cross sections derived from the pseudocrossing of potential energy curves discussed in Section 2.4. All experimental cross sections of Fig. 19 were measured by Hasted and Smith,[9] and the theoretical predictions, based on the Landau-Zener formula, were calculated by Boyd and Moiseiwitsch.[10] The agreement is good; nevertheless, the interpretation of this agreement requires careful study, as Boyd and Moiseiwitsch caution. The real test is a comparison of the

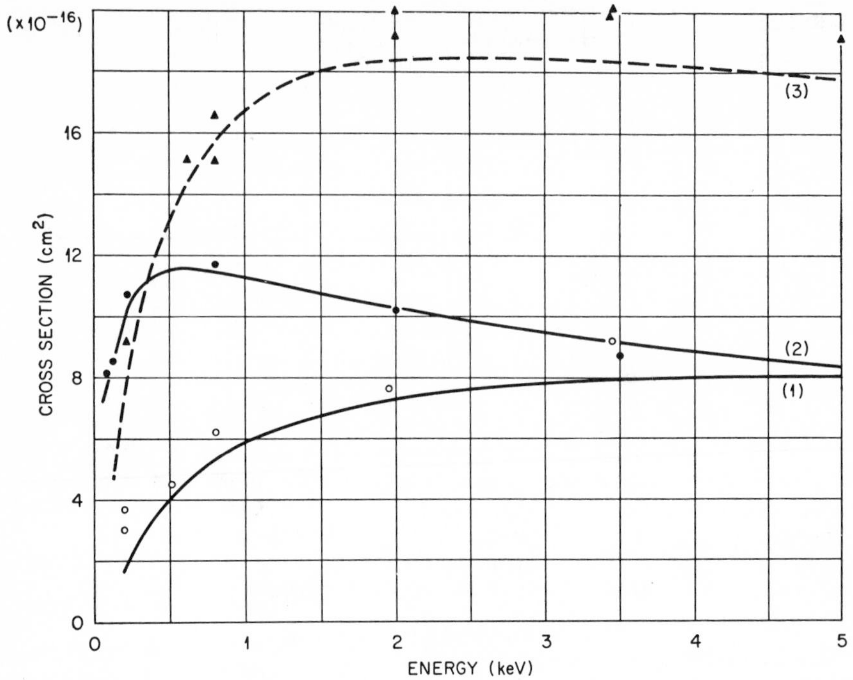

Fig. 19 Single-electron capture cross sections by the doubly charged ions of Ar^{2+}, N^{2+}, N^{2+} from Ne, He, Ne, respectively. Theoretical values calculated using the pseudo-crossing of potential energy curves (Section 2.4) are for the collisions (1) $Ar^{2+}(3p^4:{}^3P) + Ne(2p^6:{}^1S) \rightarrow Ar^+(3p^5:{}^2P) + Ne^+(2p^5:{}^2P)$; (2) $N^{2+}(2p:{}^2P) + He(1s^2:{}^1S) \rightarrow N^+(2p^2:{}^3P) + He^+(1s:{}^2S)$; and (3) $N^{2+}(2p:{}^2P) + Ne(2p^6:{}^1S) \rightarrow N^+(2p^2:{}^1S) + Ne^+(2p^5:{}^2P)$; Boyd and Moiseiwitsch.[10] Experimental results are from ground-state configurations into all allowable final reactant states: ○ Ar^{2+} + Ne, Hasted and Smith;[9] ● N^{2+} + He, Hasted and Smith;[9] △ N^{2+} + Ne, Hasted and Smith.[9] Impact energy of A^{2+} is in keV.

same measured and calculated cross sections. A few remarks about these comparisons are thus pertinent. The three processes measured are

$$Ar^{2+}(3p^4:{}^3P) + Ne(2p^6:{}^1S) \rightarrow A^+ + Ne^+, \qquad (66M)$$

$$N^{2+}(2p:{}^2P) + He(1s^2:{}^1S) \rightarrow N^+ + He^+, \qquad (67M)$$

$$N^{2+}(2p:{}^2P) + Ne(2p^6:{}^1S) \rightarrow N^+ + Ne^+. \qquad (68M)$$

In all three cases each of the initial states is postulated to be in the ground-state term of the ground-state configuration defined by LSM_LM_S- (or $LSJM_J$-) representation, LS-coupling being assumed. Equation 68M is used to illustrate some points about the possible final states. The total spin initially is $S_i = \frac{1}{2}$, and since spin-dependent forces are omitted, the final spin, S_f, has the same value (see Section 3.4 for a discussion of Wigner's spin-conservation rule). Since the change of integral energy of the first excited state of Ne^+ is larger than several of the low-lying states of N^+, only the ground state of $Ne^+(2p^5\!:{}^2P)$ is used for this discussion. The reason for this choice is that the corresponding cross sections for the excited Ne^+ states are probably much smaller than those for the excited N^+ states in the range of E shown in Fig. 19. These states of N^+ are $N^+(2p^2\!:{}^1D, {}^1S)$ and $N^+(2s, 2p^3\!:{}^5S)$. This last state is forbidden according to Wigner's rule, but the other two states are allowed since $S_f = \frac{1}{2}$. This point, and other factors relating to the distances of the crossing points, led the authors to approximate the measured cross sections by the following calculated cross sections:[8]

$$Ar^{2+}({}^3P) + Ne({}^1S) \rightarrow Ar^+(3p^5\!:{}^2P) + Ne^+(2p^5\!:{}^2P), \tag{66C}$$

$$N^{2+}({}^2P) + He({}^1S) \rightarrow N^+(2p^2\!:{}^3P) + He^+({}^2S), \tag{67C}$$

$$N^{2+}({}^2P) + Ne({}^1S) \rightarrow N^+(2p^2\!:{}^1S) + Ne^+(2p^5\!:{}^2P). \tag{68C}$$

In the case of the process

$$Cs^+({}^1S) + Cs({}^2S) \rightarrow Cs + Cs^+, \tag{69}$$

treated in connection with Fig. 18, not only is the first ionization potential small ($\approx$3.89 eV), but there is a large number of available excited states of $Cs({}^2L)$. Again it is assumed that the residual ion is in its ground state, $Cs^+(5p^6\!:{}^1S)$, for the reason previously discussed. The important feature is that many excited states, consistent with Wigner's rule, may contribute to the measured cross section, a fact mentioned by Perel et al.[8] A final point is that the impact velocity is small for the last four processes; for example, in the case of Eq. 69, $M_p \approx 110$ and $v_i \approx 0.06$ a.u. at an impact energy of 10 keV. Based on this fact, arguments can be developed about the relative efficiencies of charge transfer and simultaneous excitation; this, however, is not attempted here. Instead the Bates two-state method as applied to resonant charge transfer between H^+ and $H(1s)$ is studied next (see Section 3.1).

In Fig. 20 the calculated values of McCarroll[11] are compared with the measured values of McClure,[12] and Fite, Smith, and Stebbings[3] for the processes

$$H^+ + H(1s) \rightarrow \sum_n H(nl) + H^+. \tag{70}$$

The IP version with allowance for *momentum transfer* (transfer of translational motion to the electron) is used in this calculation. If it is assumed that the theoretical predictions would not be increased more than 20% by contributions from the omitted excited states, the agreement is excellent for

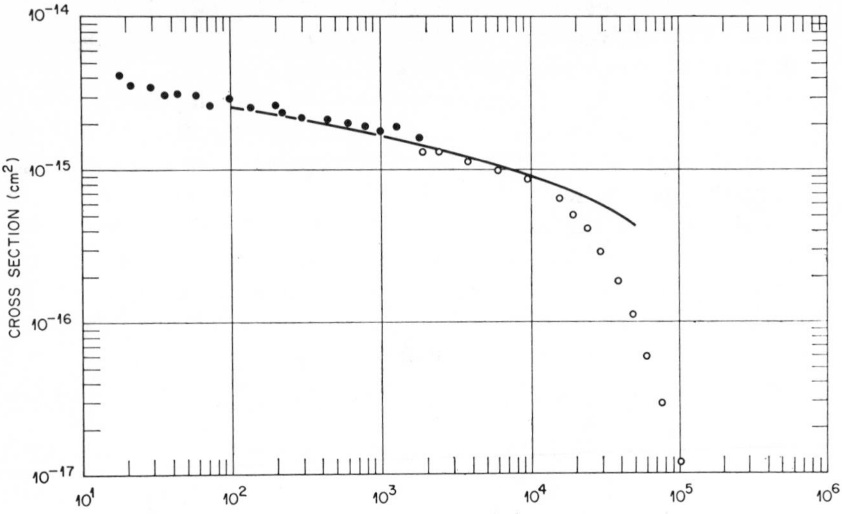

Fig. 20 Cross sections for electron capture from atomic hydrogen by protons. Theoretical results using the Bates two-state method (Section 3.1) are for $H^+ + H(1s) \rightarrow H(1s) + H^+$: —— McCarroll.[11] Experimental cross sections are for $H^+ + H(1s) \rightarrow \sum_n H(nl) + H^+$: ○ McClure;[12] ● Fite, Smith, and Stebbings.[2] Impact energy of H^+ is in eV.

0.1 keV $\leq E \leq$ 15 keV. For larger values of E the calculated values overestimate the experimental ones. McCarroll's solution of this two-state approximation is exact; another interesting aspect is that the solution agrees with the results of Dalgarno and Yadav[2] as E decreases; as E increases, the solution tends to the OBK result (Section 3.4), which notoriously overestimates the experimental predictions. This same approximate method was applied by Green, Stanley, and Chiang[13] to the process

$$H^+ + He(1s^2) \rightarrow H(1s) + He^+(1s), \tag{71C}$$

and these results are compared to the experimentally determined total cross section[14,15] for

$$H^+ + He(1s^2) \rightarrow \sum_n H(nl) + \sum He^+(n'l'). \tag{71M}$$

The results used in Fig. 21 were calculated using the IP version allowing for distortion, back-coupling, and momentum transfer to the electron (see the original paper for an assessment of the sensitivity of the value of the cross section to these corrections). The measured values are predicted most successfully in the vicinity of the maximum of the curve that is drawn through the points recorded by Stier and Barnett. In such calculations involving many electron atoms, some of the atomic wave functions are also approximate,

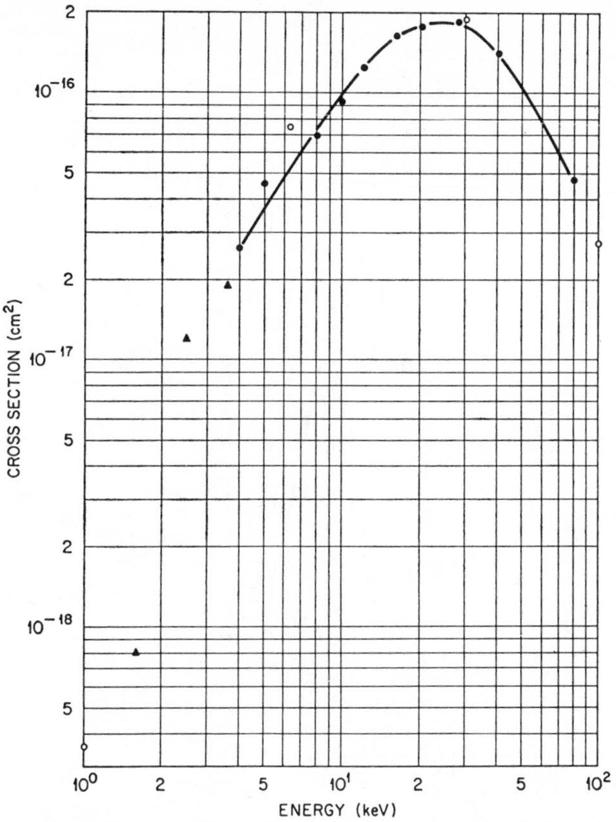

Fig. 21 Electron capture cross sections for protons passing through helium. Theoretical cross sections are calculated by the Bates two-state method (Section 3.1) for $H^+ + He(1s^2) \rightarrow H(1s) + He(1s)$: ○ Green, Stanley, and Chiang.[13] Experimental cross sections are for $H^+ + He(1s^2) \rightarrow \sum_n H(nl) + \sum He(n'l')$: ● Stier and Barnett;[14] ▲ Hasted.[15] Impact energy of H^+ is in keV.

although experience with calculations on capture from $He(1s^2)$ has revealed small improvement of cross sections with large improvement of the He-wave function. As Green et al. have remarked, inclusion of contributions from excited states worsens their predictions. (They calculated only four cross sections for the case compared here.)

Lovell and McElroy[16] used the IP version of the Bates three-state approximation, with allowance for momentum transfer, to calculate cross sections for the process

$$H^+ + H(1s) \rightarrow H(2s) + H^+. \tag{72C}$$

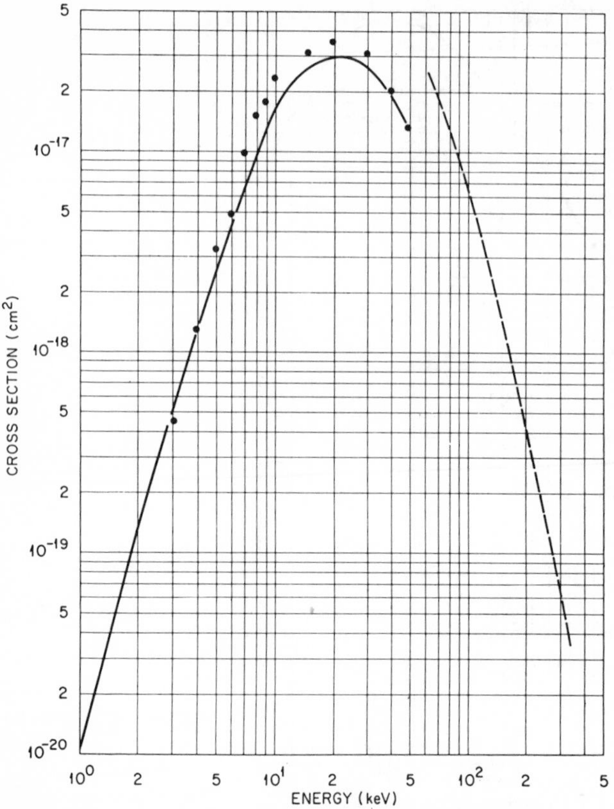

Fig. 22 Cross sections for electron capture from H into the 2*s* state of atomic hydrogen by protons. Both theoretical (the Bates three-state method) and the experimental results are for the same reaction $H^+ + H(1s) \rightarrow H(2s) + H^+$: —— Lovell and McElroy;[16] - - - - Mapleton;[18] ● Bayfield.[17] Impact energy of H^+ is in keV.

If the incident proton is denoted by B, and the other by A, the three states used to derive the curve in Fig. 22 are labeled 1*s*A, 2*s*A, 2*s*B; the three associated coupled differential equations were solved numerically. A physically meaningful test of the numerical accuracy is conservation of charge, and this means that the sum of the squares of the time-dependent coefficients at $t = \infty$ should equal unity ideally. Indeed this sum was nearly equal to unity, as we saw in Section 3.1. These predictions are compared with Bayfield's measured cross sections[17] for the process of Eq. 72*C*. The agreement is very good since measurement and calculation are for the *same* process in this instance. Also included in Fig. 22 are the cross sections calculated using the first Born approximation with prior interaction.[18] These values

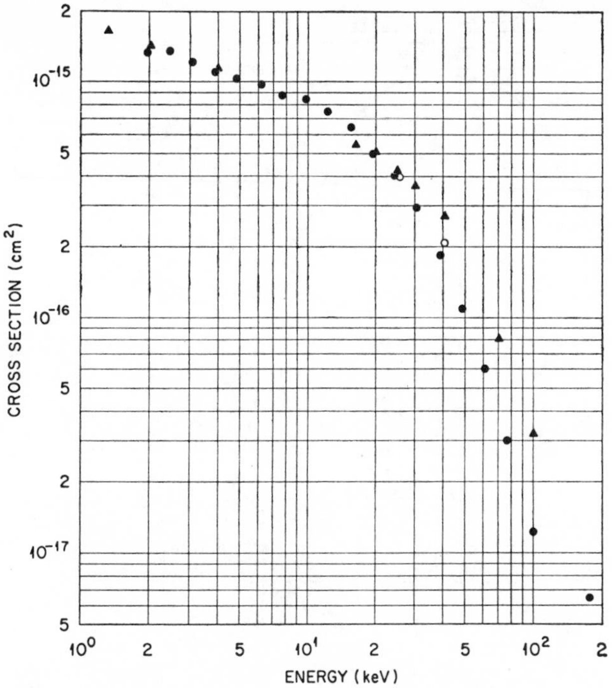

Fig. 23 Total electron capture cross sections from atomic hydrogen by protons. Theoretical calculations are by Gallaher and Wilets:[19] ▲ the four-state approximation employs the Sturmian functions $S(1s)$, $S(2s)$, $S(2p_0)$, and $S(2_{p+1})$; ○ the six-state approximation uses the additional Sturmian functions $S(3s)$ and $S(4s)$. Impact energy of H^+ is in keV. Experimental results are for the reaction $H^+ + H(1s) \rightarrow \sum H(nl) + H^+$: ● McClure.[12]

overestimate the measured values significantly. Next, a four-state calculation for Eq. 70 is described.

In contrast to customary expansions, Gallaher and Wilets expanded the wave function representing Eq. 70 in a set of Sturmian functions.[19,20] This is a discrete set containing only bound-state types of functions. Thus this type of representation is free of the convergence problems associated with atomic wave functions in the continuum. The Sturmian functions are referred to a rotating frame of coordinates with the Z-axis parallel to the internuclear line. The IP version is used with allowance for momentum transfer to the electron. The wave function is approximated by four linear combinations of the Sturmians designated by the states $1s$, $2s$, $2p_0$, and $2p_{+1}$. In the approximate wave function there are eight terms because of the following conditions: (1) each linear combination is constructed from two Sturmians with separate origins on the two protons; (2) the spherical harmonic associated with

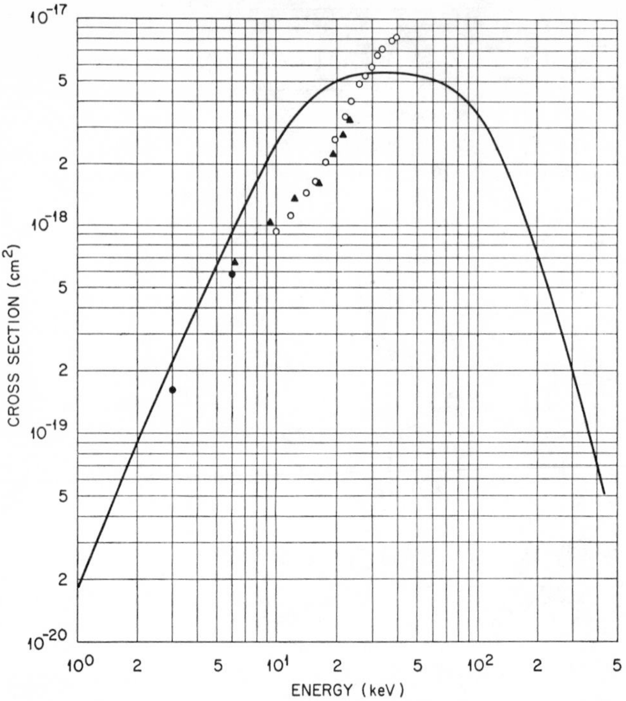

Fig. 24 Electron capture cross sections for protons passing through helium. Theoretical calculations using the Bates five-state approximation are for the reaction $H^+ + He(1s^2) \rightarrow H(2s) + He^+(1s)$: —— Sin Fai Lam.[21] Experimental results are for $H^+ + He(1s^2) \rightarrow H(2s) + \sum He^+(n'l')$: ● D^+ and ○ H^+, Jaecks, Van Zyl, and Geballe;[23] ▲ Andreev, Ankudinov, and Bobashev.[22] Impact energy of H^+ is in keV.

$2p_{+1}(Y_{1m})$ contains both $m = \pm 1$. Results of this calculation are compared with the measured values of McClure[12] in Fig. 23. The predictions are excellent. Although the calculated values overestimate the measured ones for 30 keV $\leq E \leq$ 100 keV, inclusion of the $3s$- and $4s$-terms (6-state) in the expansion reduces the disagreement in this range of E.[19] (Nevertheless, there are convergence difficulties associated with this type of expansion, a study of which is discussed in the article by Cheshire, Gallaher, and Taylor.[54]) Two other approximate calculations tested represent the processes $nl = 2s$ and $nl = 2p_x$, $2p_z$, both for $n'l' = 1s$, in Eq. 71*M*. Approximation of the system's wave function is by a linear combination of five states for the calculation of these cross sections.[21] The IP version, with allowance for momentum transfer, is used by Sin Fai Lam, but a nonrotating frame of reference and ordinary atomic wave functions are employed. These wave functions represent the initial atomic state $He(1s^2)$ and the product states

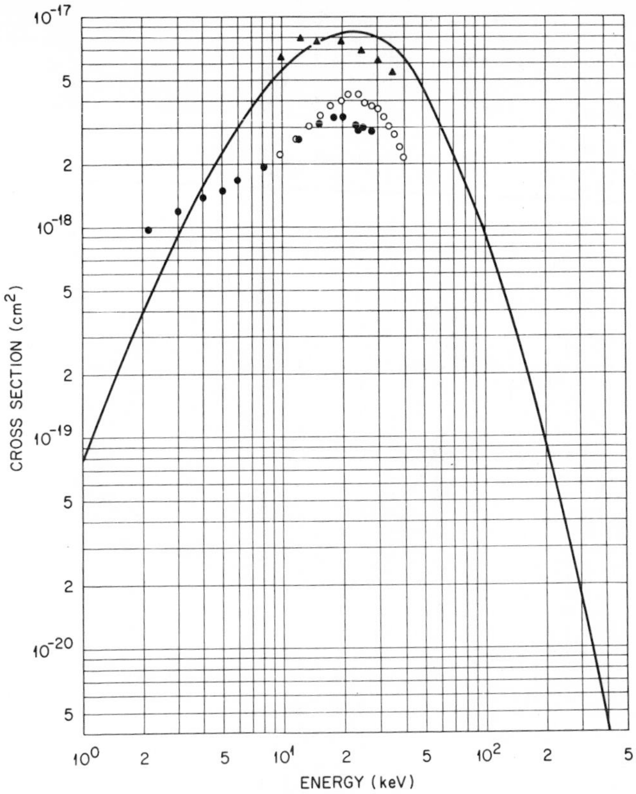

Fig. 25 Cross sections for electron capture into $\sum_{m=-1}^{1}$ H($2p_m$) by protons passing through helium gas. The Bates five-state approximation was used by Sin Fai Lam[21] (——) to calculate the cross sections for $H^+ + He(1s^2) \rightarrow H(2p_x; 2p_z) + He^+(1s)$ using the impact-parameter version. Experimental points are for the reactions $H^+ + He(1s^2) \rightarrow \sum_{m=-1}^{1} H(2p_m) + \sum He^+(n'l')$: ● Pretzer, Van Zyl, and Geballe;[24] ▲ de Heer, van Eck, and Kistemaker;[25] ○ Andreev, Ankudinov, and Bobashev.[22]

$He^+(1s)H(nl)$ with $nl = 1s, 2s, 2p_x, 2p_z$. Since the xz-plane is the plane of the collisions, H($2p_y$) vanishes for the usual representation of the p-states of H. The comparisons for capture into H($2s$) are displayed in Fig. 24.[22,23] Although the calculated predictions are good, the curvature of the calculated curve differs markedly from the curvature of a smooth curve drawn through the experimental points. In Fig. 25 the predictions for the sum of the cross sections, $Q(2p_x) + Q(2p_z)$, are compared with three sets of experimentally determined cross sections.[22,24,25] We see that the values of de Heer et al. are predicted most closely and that the variation with impact energy is predicted well for all three sets of the measured cross sections. Finally, in relation to

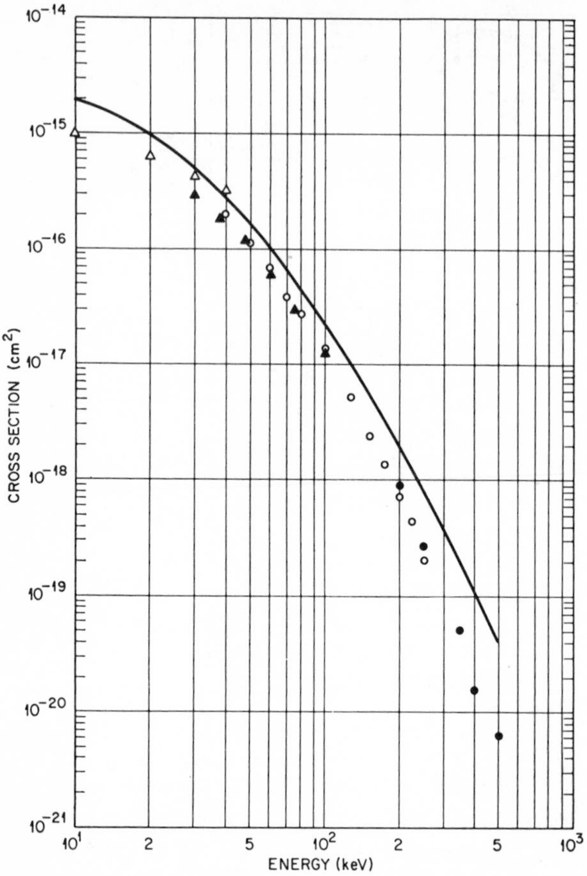

Fig. 26 Total electron capture cross sections for protons in atomic hydrogen. Cross sections for $H^+ + H(1s) \rightarrow H(1s) + H^+$ are calculated using the Bassel-Gerjuoy method from which total cross sections are estimated using Eq. (73): —— Grant and Shapiro.[26] Experimental results are for the reaction $H^+ + H(1s) \rightarrow \sum H(nl) + H^+$: △ Fite, Stebbings, Hummer, and Brackmann;[29] ▲ McClure;[12] ○ Wittkower, Ryding, and Gilbody.[30] For collisions in molecular hydrogen results are presented for one-half the cross section for $H^+ + H_2$ (ground) $\rightarrow \sum H(nl) + H_2^+$; ● Barnett and Reynolds.[31] Impact energy of H^+ is in keV.

Figs. 24 and 25, the measured cross sections represent all values of $n'l'$ in Eq. 71*M*.

Grant and Shapiro,[26] using the formulation of Bassel and Gerjuoy[27] (see Section 3.2) expressed in terms of the post interactions, calculated the cross sections for resonant charge transfer between H^+ and $H(1s)$ to the third order in the interactions. Grant and Shapiro allowed for capture into all excited states by expressing the total cross section for Eq. 70 as

$$\sigma = K\sigma_{1s}, \qquad K = 1 + \frac{\sum_l \sigma_{2l}(V_f)}{\sigma_{1s}(V_f)}\left[1 + \frac{\sum_{n=3}^{\infty}\sum_l \sigma_{nl}(\text{OBK})}{\sum \sigma_{2l}(\text{OBK})}\right], \tag{73}$$

and they evaluated K approximately using the results of Jackson and Schiff[28] for $\sigma_{2l}(V_f)$ and $\sigma_{1s}(V_f)$ (see Sections 3.4 and 3.5). The notation V_f and OBK signify the interaction used to evaluate the corresponding cross section. In Fig. 26 these calculated results for Eq. 70 are compared with the corresponding experimentally determined ones.[12,29,30] Also included are the measurements of Barnett and Reynolds[31] per gas atom for the process

$$H^+ + H_2(\text{ground}) \rightarrow \sum H(nl) + \sum H_2^+. \tag{74}$$

As noted before, $\frac{1}{2}Q(H_2)$ is the quantity plotted. Although the agreement is fair, these predictions overestimate as E increases, an expected result. The overestimation, as the authors[26] explain, is because this higher-order distorted wave calculation tends to the OBK result as does the first-order result of Bassel and Gerjuoy. Another point of interest is the agreement between the results of Barnett and Reynolds for Eq. 74 per gas atom and those of Wittkower et al. for Eq. 70. Next, several predictions of the first Born approximation are viewed, and the wave version is used in all cases (see Sections 3.3, 3.4, 3.5).

The approximate total cross section for the processes

$$H^+ + He(1s^2) \rightarrow \sum H(nl) + \sum He^+(n'l') \tag{75}$$

was calculated by Mapleton[32] using both the prior and post interactions. "Approximate total cross section" here connotes that cross sections were obtained for only eleven of the final states. In Fig. 27 the arithmetic average of the corresponding prior and post cross sections are compared with the experimentally determined cross sections.[14,31] The agreement is good for 50 keV $\leq E \leq$ 1 MeV. This agreement may be fortuitous, however, for in the next example the agreement is not good. The same approximation was used by Mapleton[33] to obtain cross sections for the process

$$H^+ + O(2p^4{:}\,^3P) \rightarrow H(1s) + O^+(2p^3{:}\,^4S), \tag{76}$$

a nearly accidentally resonant reaction (see Section 1.4).

Since the prior and post cross sections differ negligibly in this example, the prior values are used and compared with the total cross sections per gas atom, $\frac{1}{2}Q(O_2)$ for capture by protons. These results are displayed in Fig. 28.[14,34] From a study of OBK cross sections for Eq. 76 and for capture leaving the residual ions $O^+(2p^3{:}\,^2D)$ and $O^+(2p^3{:}\,^2P)$, it is speculated that the sum of the last two cross sections should exceed the $O^+(^4S)$-case by roughly a factor of two.[35] Moreover, these estimates only refer to $2p$-orbital

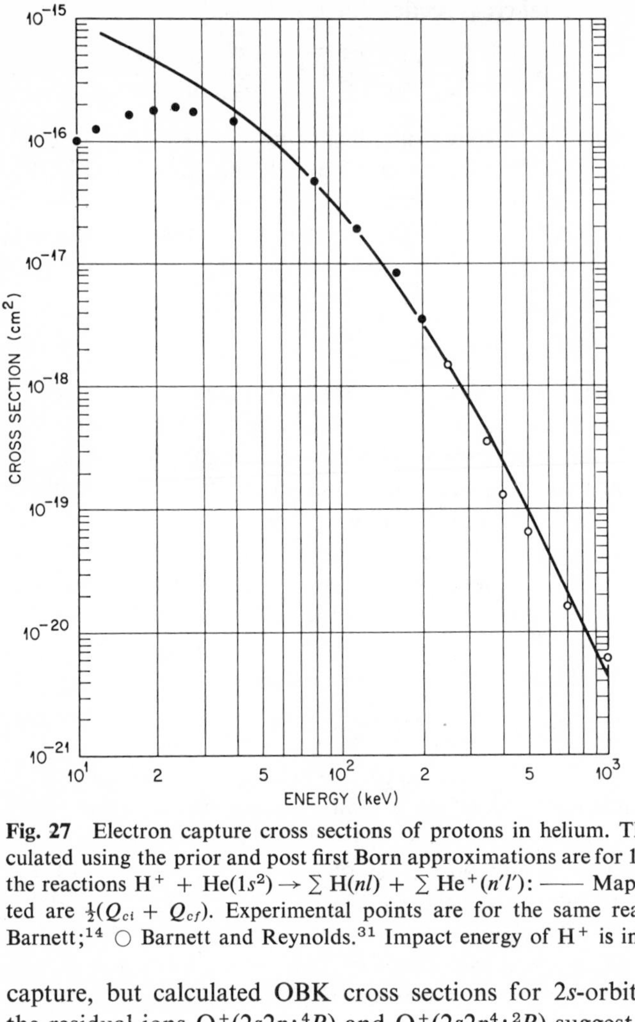

Fig. 27 Electron capture cross sections of protons in helium. Theoretical results calculated using the prior and post first Born approximations are for 11 of the final states of the reactions $H^+ + He(1s^2) \rightarrow \sum H(nl) + \sum He^+(n'l')$: —— Mapleton's[32] results plotted are $\frac{1}{2}(Q_{ci} + Q_{cf})$. Experimental points are for the same reaction: ● Stier and Barnett;[14] ○ Barnett and Reynolds.[31] Impact energy of H^+ is in keV.

capture, but calculated OBK cross sections for $2s$-orbital capture leaving the residual ions $O^+(2s2p:{}^4P)$ and $O^+(2s2p^4:{}^2P)$ suggest the importance of these contributions in the range of impact energies of Fig. 28.[33] The evidence is that the present theoretical curve would be revised upward appreciably by these omitted contributions. It is therefore clear that these cross sections for capture from O_2 per gas atom are predicted poorly by the prior and post first Born approximations. One reason for this failure stems from inadequate cancellation of the contribution from the proton-nucleus interaction, a prediction made earlier by Bassel and Gerjuoy.[27] The constituent scattering amplitude originating from this interaction has a so-called *tail*; that is, this amplitude decreases with scattering angle less strongly than the rest of the

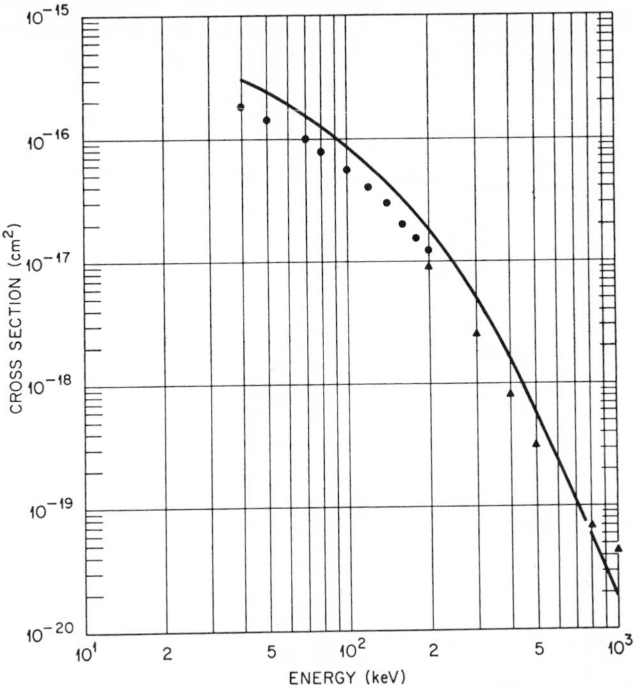

Fig. 28 Electron capture cross sections for protons in atomic oxygen. Theoretical results using the prior first Born approximation are for the reaction $H^+ + O(2p^4 : {}^3P) \to H(1s) + O^+(2p^3 : {}^4S)$: —— Mapleton.[33] Experimental results are taken as one-half the total cross section of the reaction $H^+ + O_2(\text{ground}) \to \sum H(nl) + \sum O_2^+$; ● Stier and Barnett;[14] ▲ Toburen, Nakai, and Langley.[34] Impact energy of H^+ is in keV.

amplitude. In the case of capture from He, cancellation at small angles, together with the tail at large angles, fortuitously produces cross sections in good agreement with measurements as displayed in Fig. 27. Since the nuclear charge on O is 8, the contribution from the tail is so pronounced that the prior and post cross sections are nearly equal in spite of the inexact atomic wave functions used (see Section 3.3). Consequently, it is not surprising that these theoretical predictions are poor for capture from O. (It is conjectured that the total cross section for capture from O is closely approximated by $\frac{1}{2}Q(O_2)$ for the range of E in Fig. 28.)

The last example of the first Born approximation using prior and post interactions is an atom-atom process,

$$H(1s) + H(1s) \to H^-(1s^2) + H^+. \tag{77}$$

Mapleton originally calculated these cross sections using the post interaction[36] and later recalculated the cross sections using both prior and post interactions.[37] Allowance is made for the identity of the protons and electrons in

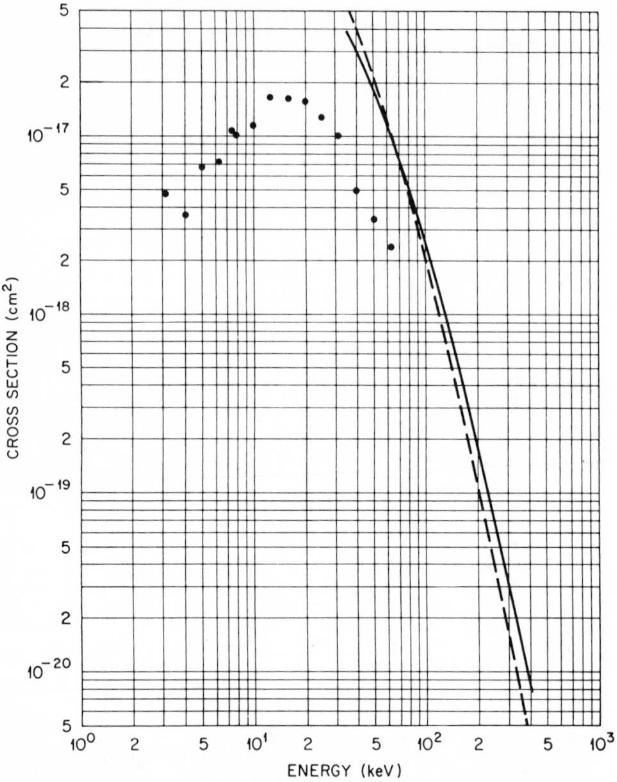

Fig. 29 Total production of H^- ions in collisions of hydrogen atoms. Theoretical results using the prior and post first Born approximation are for the reaction $H(1s) + H(1s) \rightarrow H^-(1s^2) + H^+$: —— Q_i, Mapleton;[37] - - - Q_f, Mapleton.[37] Experimental results include all energetically allowed H^- ions: ● McClure.[38] Impact energy of H is in keV.

these calculations, and the plane wave approximation is used, a feature explained next. It is evident from Eq. 77 that a Coulomb force exists at infinite separation of the final reactants; therefore the final state contains a Coulomb wave function (attractive force) as a factor. This feature greatly complicates the evaluation of the cross section; in order to reduce the calculation to a feasible level, the Coulomb function is replaced by a plane wave—the limiting value of this function at high impact velocities. This is the *plane wave approximation.* In the table of cross sections in the second paper by Mapleton[37] there are tabulated values of a factor $F(\alpha)$, which is a partial correction to the plane wave amplitude. This factor, which would increase the calculated values, is *not* used in the tabulated cross sections. In Fig. 29 these predictions are compared with McClure's measurements, which also represent capture into all excited states of H^-, if excited states exist.[38] The predictions overestimate the measurements; moreover, the prior and post cross sections differ significantly, and it is speculated that the inexact atomic wave functions are the chief source of this difference at these impact energies. Nevertheless, as was explained in Section 4.4, the exact prior and post plane wave amplitudes, derived from interactions containing the $R_{i,f}^{-1}$ terms, have the same asymptotic limits only if $Z_p = Z_a$. Those results, however, were deduced for a different capturing process, and it is conjectured that the present plane wave amplitudes are equal in the asymptotic limit. Before a simpler calculational method is applied, the calculated cross sections of Eq. 75 are compared with measured values for 0.1 MeV $\leq E \leq$ 100 MeV since much attention has been devoted to the variation of cross sections with E at high values of E.[39]

Actually only the process for final 1s-states was needed since previous studies of this atom showed that a very good estimate of the total cross section for this range of E is

$$Q = 1.202 \frac{(Q_i + Q_f)}{2} \tag{78}$$

(see discussion following Eq. 54g, Section 4.4). The use of this formula implies that contributions from excited He^+-states are negligible. In Fig. 30 these cross sections are compared with corresponding experimentally determined total cross sections.[34,40,55] The agreement is very good.

The wave version of the simpler OBK method is tested next. As already discussed in Section 3.4, Eq. 44p has been used with qualitative success to predict capture from complex atoms by protons. This formula, which utilizes the OBK cross sections and the ratios $R(H)$ for atomic hydrogen,* is reproduced here for convenience:

$$Q(\mathrm{H}) = \sum_{m,n} Q_{mn}(\mathrm{OBK}; \mathrm{A}; E)R(\mathrm{H}; E) \tag{79A}$$

The OBK cross sections are calculated for capture from each subshell into H(1s), and the sum over n is replaced by the factor 1.202. Only capture into the ground states of the residual ions are used, it being assumed that contributions from the excited ionic states are negligible for this estimation. In the case of atomic oxygen, cross sections were calculated for[39]

$$\mathrm{H}^+ + \mathrm{O}(2p^4\!:{}^3P) \rightarrow \mathrm{H}(1s) + \begin{cases} \mathrm{O}^+(2p^3\!:{}^4S;\, {}^2D;\, {}^2P), \\ \mathrm{O}^+(ms2p^4\!:{}^4P;\, {}^2P), \\ m = 1, 2. \end{cases} \tag{79B}$$

* The original concept of this estimating procedure stemmed from the author's private discussions with Bates and Dalgarno.

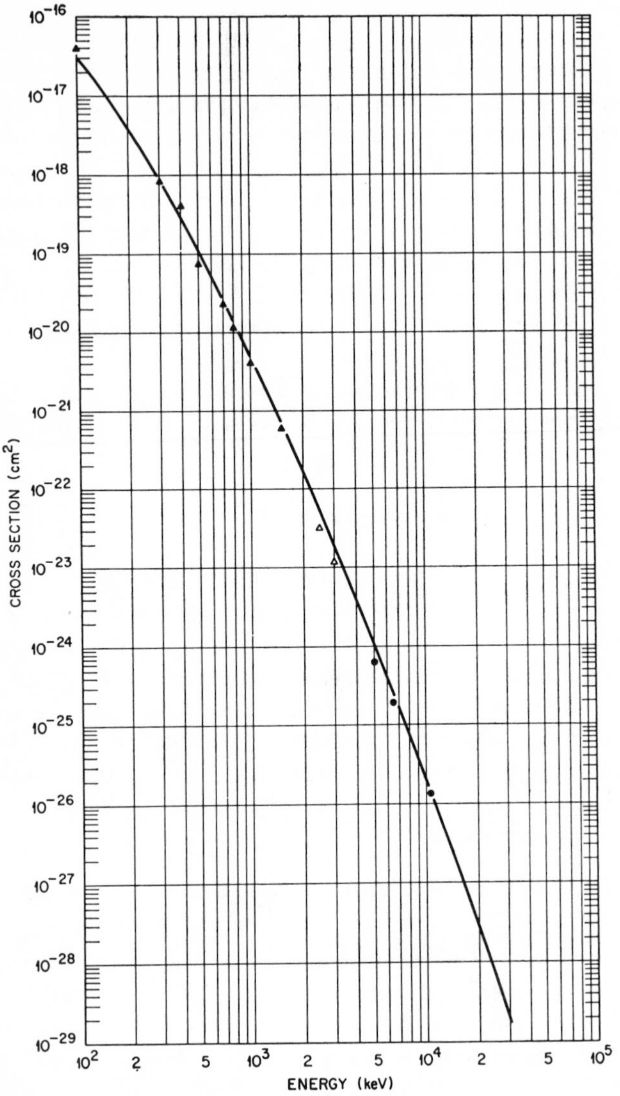

Fig. 30 Electron capture cross sections for protons in helium at high impact energies. Calculated Born total cross sections are estimated by Eq. (78): —— Mapleton.[39] Both theoretical and experimental results are for $H^+ + He(1s^2) \rightarrow \sum H(nl) + \sum He^+(n'l')$: ▲ Toburen, Nakai, and Langley;[34] △ Welsh, Berkner, Kaplan, and Pyle;[40] ● Berkner, Kaplan, Paulikas, Pyle.[55] Impact energy of H^+ is in keV.

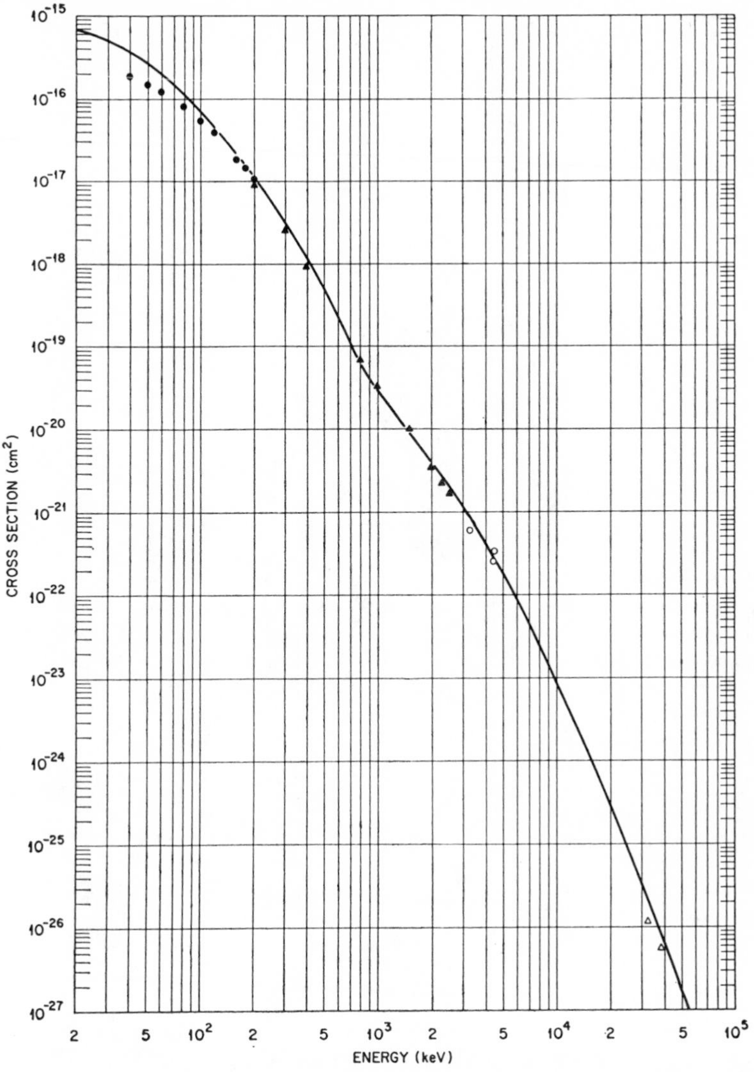

Fig. 31 Electron capture cross sections for protons in atomic oxygen. Theoretical estimated total cross sections obtained from $1.202 \sum_m Q_m(\mathrm{OBK}; O; E)\, R(\mathrm{H}; E)$ (approximation to Eq. 79*a*) and calculated OBK cross sections for $H^+ + O(2p^4{:}\,^3P) \rightarrow H(1s) + [O^+(2p^3{:}\,^4S;\ ^2D;\ ^2P);\ O^+(ms2p^4{:}\,^4P,\ ^2P)\ m = 1,2]$: —— Mapleton.[39] Experimental results for one-half the cross section for $H^+ + O_2$ (ground) $\rightarrow \sum H(nl) + \sum O_2^+$: ● Stier and Barnett;[14] ▲ Toburen, Nakai, and Langley;[34] ○ Schryber;[41] △ Acerbi, Castiglioni, Dutto, Resmini, Succi, and Togliaferri.[42] Impact energy of H^+ is in keV.

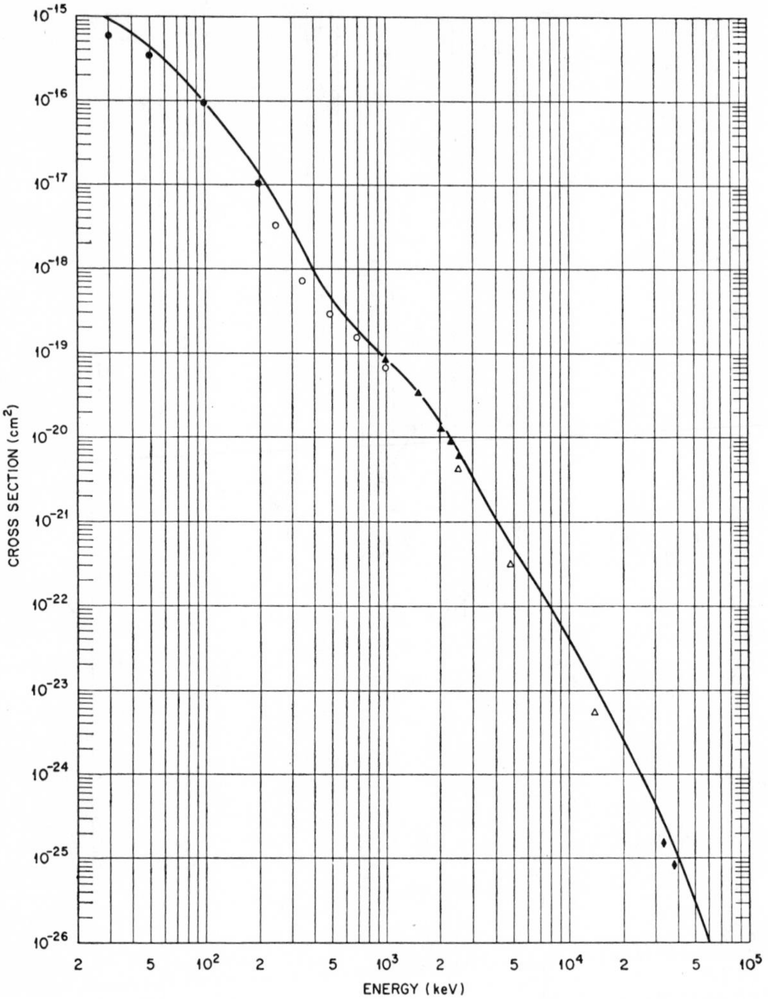

Fig. 32 Electron capture cross sections for protons in argon. Theoretical estimates obtained as in Fig. 31 using calculated OBK results for $H^+ + Ar(3p^6{:}\,^1S) \rightarrow H(1s) + [Ar^+(mp^5{:}\,^2P);\ Ar^+(ns{:}\,^2S);\ m = 2, 3;\ n = 1, 2, 3]$: —— Mapleton.[39] Experimental results for $H^+ + Ar(3p^6{:}\,^1S) \rightarrow \sum H(nl) + \sum Ar^+$: ● Stier and Barnett;[14] ○ Barnett and Reynolds;[31] ▲ Toburen, Nakai, and Langley;[34] △ Welsh, Berkner, Kaplan, and Pyle;[40] ◆ Acerbi, Castiglioni, Dutto, Resmini, Succi, and Togliaferri.[42] Impact energy of H^+ is in keV.

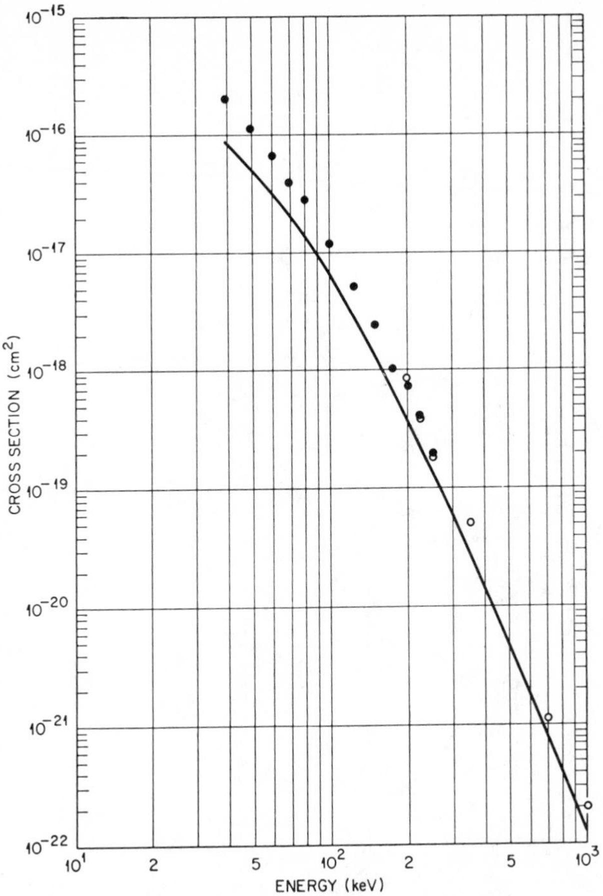

Fig. 33 Electron capture cross sections for protons in atomic hydrogen. Total cross sections estimated from calculated results obtained using the impulse approximation (consult this chapter) for $H^+ + H(1s) \rightarrow H(1s) + H^+$: —— Cheshire.[43] Experimental results for $H^+ + H(1s) \rightarrow \sum H(nl) + H$: ● Wittkower, Ryding, and Gilbody.[30] Experimental results for one-half the cross section of $H^+ + H_2$ (ground) $\rightarrow \sum H(nl) + \sum H_2{}^+$: ○ Stier and Barnett,[14] and Barnett and Reynolds.[31] Impact energy of H^+ is in keV.

In Fig. 31 the associated predictions are compared with the measured cross sections per gas atom, $\frac{1}{2}Q(O_2)$.[14,34,41,42] In view of the crude calculational approximation, the prediction is unexpectedly good. Equation 79*A*, modified as before, also was used to calculate cross sections for[39]

$$H^+ + Ar(3p^6\!:{}^1S) \rightarrow H(1s) + \begin{cases} Ar^+(mp^5\!:{}^2P), \\ Ar^+(ns\!:{}^2S), \\ m = 2, 3;\ n = 1, 2, 3. \end{cases} \tag{80}$$

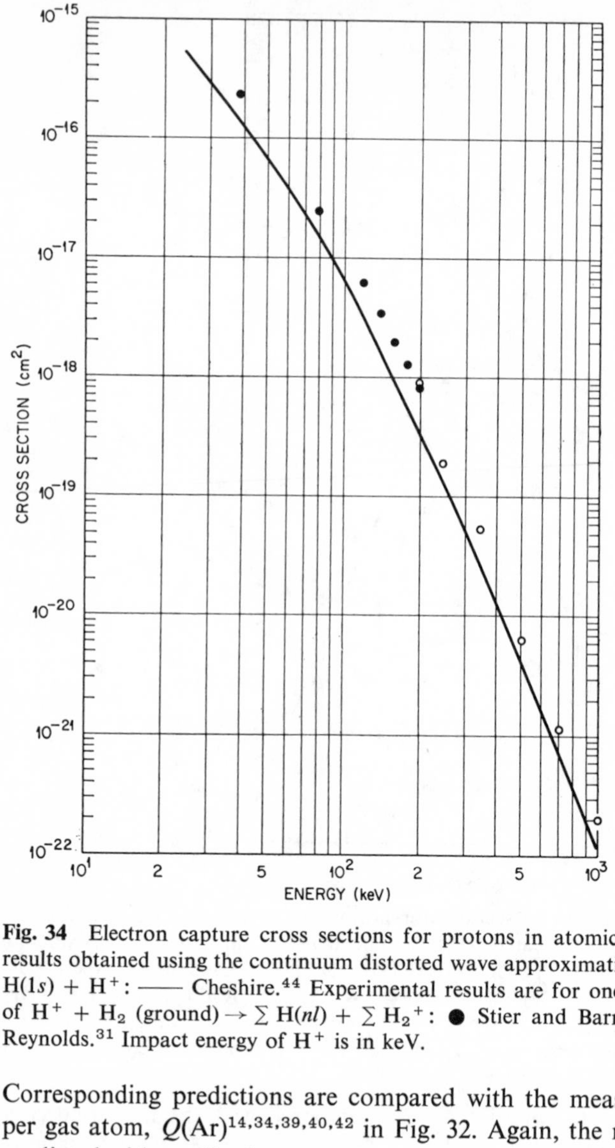

Fig. 34 Electron capture cross sections for protons in atomic hydrogen. Calculated results obtained using the continuum distorted wave approximation for $H^+ + H(1s) \rightarrow H(1s) + H^+$: —— Cheshire.[44] Experimental results are for one-half the cross section of $H^+ + H_2$ (ground) $\rightarrow \sum H(nl) + \sum H_2{}^+$: ● Stier and Barnett;[14] ○ Barnett and Reynolds.[31] Impact energy of H^+ is in keV.

Corresponding predictions are compared with the measured cross sections per gas atom, $Q(\text{Ar})$[14,34,39,40,42] in Fig. 32. Again, the measured values are predicted with qualitative success.

Next, the impulse approximation procedure is tested. Cheshire[43] calculated the associated cross sections for resonant charge transfer between $H(1s)$ and H^+, and allowed for capture into all excited states by using the results of Jackson and Schiff[28] and Mapleton.[18] In Fig. 33 these adjusted cross sections are compared with the measured cross sections per gas atom of Eqs. 70 and

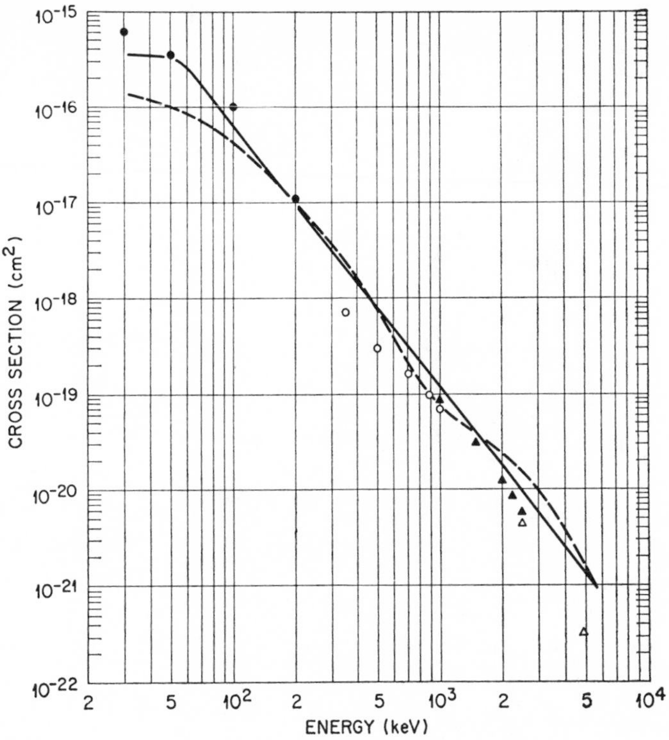

Fig. 35 Electron capture from argon by protons. Classical cross sections $Q(T1)$ and $Q(T2)$ calculated for the processes $H^+ + Ar(3p^6;\ 1S) \rightarrow \sum H(nl) + [Ar^+(mp^5;\ ^2P);\ Ar^+(ns:\ ^2S),\ m = 2, 3, n = 2, 3]$: —— $Q(T1)$ Fermi-Thomas distribution, Bates and Mapleton;[45] - - - $Q(T2)$ Hartree-Fock-Slater distribution, Bates and Mapleton.[45] Experimental results for $H^+ + Ar(3p^6;\ ^1S) \rightarrow \sum H(nl) + \sum Ar^+$: ● Stier and Barnett;[14] ○ Barnett and Reynolds;[31] ▲ Toburen, Nakai, and Langley;[34] △ Welsh, Berkner, Kaplan, and Pyle.[40] Impact energy of H^+ is in keV.

74.[14,30,31] Although the agreement is good, the theoretical predictions underestimate significantly. Later Cheshire used his continuum distorted wave approximation to calculate cross sections for the previous process, but he did not adjust these results to allow for capture into excited states[44] (see Section 3.7). These calculated values are compared with the measured cross sections per gas atom of Eq. 74.[14,31] in Fig. 34. The predictions would be better than those of the impulse approximation if the same adjustment were made to allow for capture into excited states.

Other comparisons which have been used with some success during the past few years involve several classical models. First, Thomas' modifications one and two are compared (see Sections 5.2, 5.3, and 5.4). In Fig. 35 the

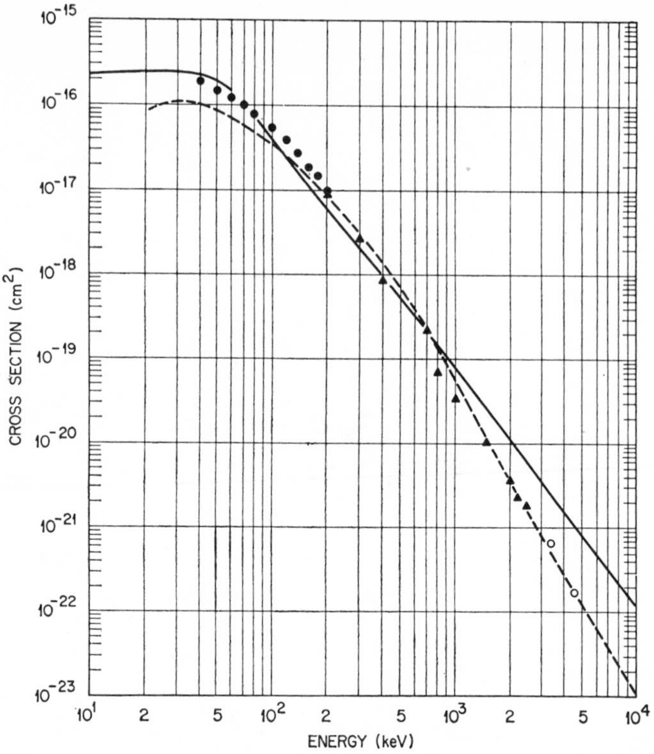

Fig. 36 Electron capture from atomic oxygen by protons. Classical cross sections $Q(T1)$ and $Q(T2)$ (consult this chapter) calculated for the processes $H^+ + O(2p^4: {}^3P) \rightarrow \sum H(nl) + [O^+(2p^3: {}^4S; {}^2D; {}^2P) + O^+ (2s2p^4: {}^2P, {}^4P)]$: —— $Q(T1)$, Mapleton;[46] - - - $Q(T2)$, Mapleton.[46] Experimental results are given for one-half the cross section for $H^+ + O_2(\text{ground}) \rightarrow \sum H(nl) + \sum O_2{}^+$: ● Stier and Barnett;[14] ▲ Toubren, Nakai, and Langley;[34] ○ Schyber.[41] Impact energy of H^+ is in keV.

cross sections $Q(T1)$ and $Q(T2)$ calculated by Bates and Mapleton for the processes

$$H^+ + Ar({}^1S) \rightarrow \sum H(nl) + \begin{cases} Ar^+(mp^5: {}^2P), \\ Ar^+(ns: {}^2S), \\ m = 2, 3;\ n = 2, 3, \end{cases} \tag{81}$$

are compared with experimentally determined total cross sections.[14,31,34,40] The prediction of $Q(T2)$ is good for E between 100 keV and 1 MeV, but $Q(T1)$ is in better accord with the measurements for $E < 100$ keV. It is not understood why $Q(T1)$ predicts so accurately at these low values of E since the assumption of the model are violated at these energies. A brief explanation of these classical cross sections is given next.

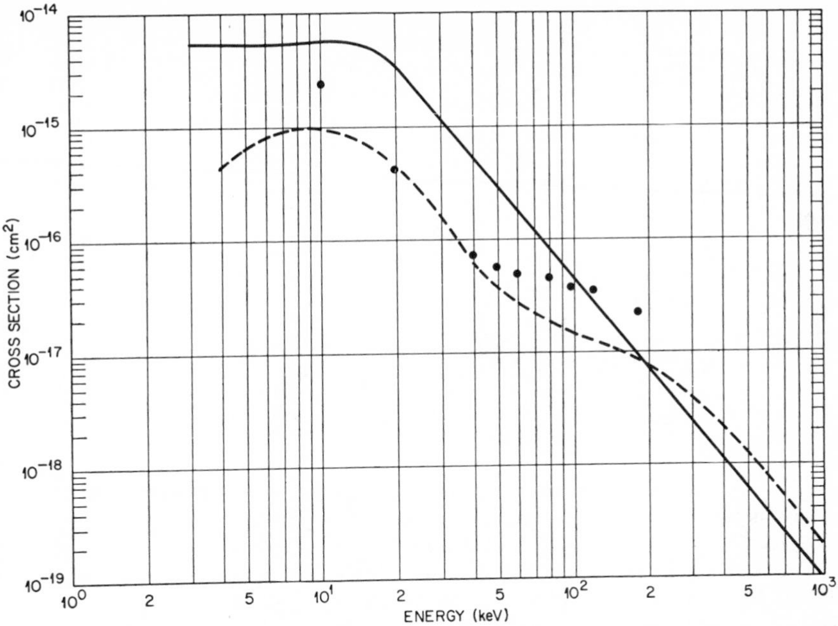

Fig. 37 Electron capture from vaporized atomic sodium by protons. Classical cross sections $Q(T1)$ and $Q(T2)$ are displayed for the processes $H^+ + Na(2p^63s:{}^2S) \rightarrow \sum H(nl) + [Na^+(2p^6:{}^1S);\ Na^+(2p^53s:{}^3P, {}^1P);\ Na^+(2p^62s^3:{}^3S, {}^1S)]$: —— $Q(T1)$, Mapleton;[47] - - - $Q(T2)$, Mapleton.[47] Experimental results represent the total cross section for $H^+ + Na(2p^63s:{}^2S) \rightarrow \sum H(nl) + \sum Na^+$: ● Il'in, Oparin, Solov'ev, and Fedorenko.[48] Impact energy of H^+ is in keV.

Both models allow for capture into all classically allowed bound states of H; however, only $T2$ allows for the shell structure of the atom. Only one ionic state is allowed for each subshell unless they are weighted as explained in a subsequent comparison for capture from atomic oxygen. In the case of $T1$, as explained in Chapter 5, the Thomas-Fermi electronic distribution is used, and only the ionization potential of the outermost subshell ($3p^6$ here) pertains to the subshells. The two most marked differences in the results occur at low impact energies and in the region where a change of curvature reflects the onset of capture from inner subshells, starting near 500 keV in this example. The similarity of the two predictions is remarkable, and the success achieved for this atom is unexpectedly good. In Fig. 36 the measured total cross sections per gas atom for capture from O_2 are compared with the calculated cross sections of Mapleton for the following processes:[14,34,41,46]

$$H^+ + O(2p^4:{}^3P) \rightarrow \sum H(nl) + \begin{cases} O^+(2p^3:{}^4S, {}^2D, {}^2P), \\ O^+(2s2p^4:{}^2P, {}^4P). \end{cases} \qquad (82C)$$

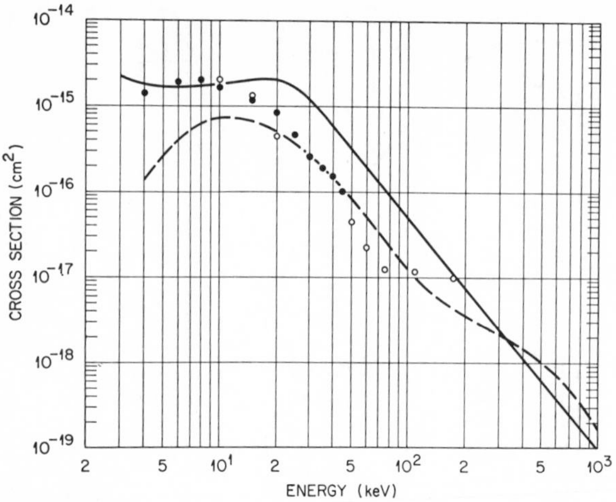

Fig. 38 Electron capture cross sections for protons passing through atomic magnesium vapor. Theoretical curve for reactions $H^+ + Mg(^1S) \rightarrow [Mg^+(3s:{}^2S); Mg^+(2p^53s^2:{}^2P); Mg^+(2s2p^63s^2:{}^2S)]$: —— $Q(T1)$, Mapleton; [47] - - - $Q(T2)$, Mapleton.[47] Experimental results for $H^+ + Mg(3s^2:{}^1S) \rightarrow \sum H(nl) + \sum Mg^+$: ● Futch and Moses;[49] ○ Il'in, Oparin, Solov'ev, and Fedorenko.[48] Impact energy of H^+ is in keV.

Again, the agreement is unexpectedly good, with $T1$ once more being more successful at lower impact energies.

Two alterations are used to obtain $Q(T2)$ for atomic oxygen. First, the three ionic states of the $2p^3$-configuration were accommodated by using the following formula for the effective ionization potential:

$$I_{2p} = \sum (\text{fpc})^2 I_k. \tag{83}$$

The index k refers to the three ionic states. Second, the formula

$$Q(2s) = \tfrac{1}{6}[4Q(^4P) + 2Q(^2P)]$$

is used to allow for capture into two terms of the *same* configuration, the weighting factors being equal to the fraction of the spin states in each ionic multiplet.[46] This is an unnecessary refinement since there is little difference between $Q(2s)$ and either $Q(^2P)$ or $Q(^4P)$; in the subsequent cases only the ground-state term of each ionic configuration is used.

Next, the calculated cross sections $Q(T1)$ and $Q(T2)$, for capture from Na and Mg by protons, are compared with the corresponding measured total cross sections in Figs. 37 and 38, respectively.[47–49] The calculated processes are, respectively,

$$H^+ + Na(^2S) \rightarrow \sum H(nl) + \begin{cases} Na^+(2p^6:{}^1S), \\ Na^+(2p^53s:{}^3P, {}^1P) \\ Na^+(2s2p^63s:{}^3S, {}^1S), \end{cases} \tag{83C}$$

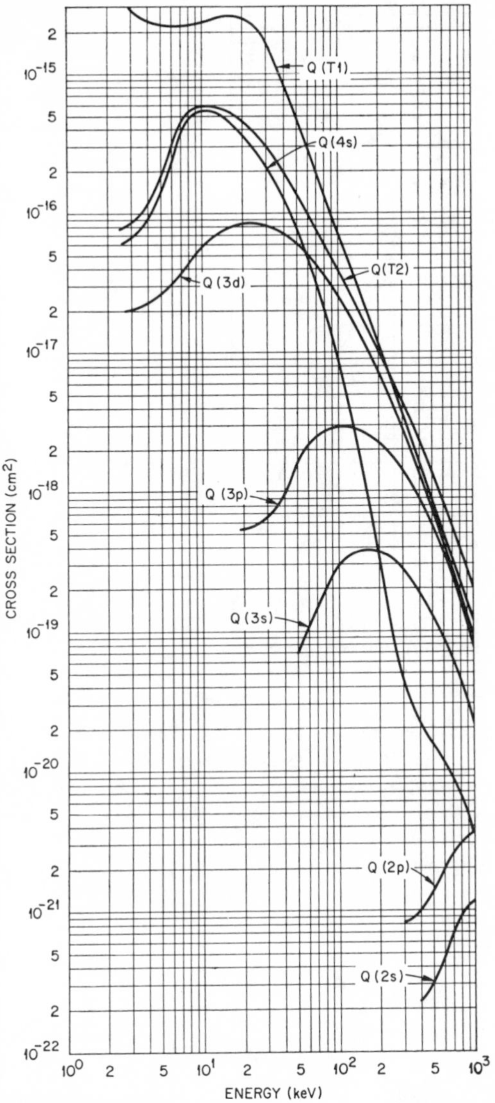

Fig. 39 Electron capture from atomic iron vapor by protons. Theoretical results are for reaction given by Eq. 85 in text. Theoretical values from Mapleton[50] with $Q(T1)$ and $Q(T2)$ defined in Fig. 35. $Q(n)$ denotes capture from atomic subshell n. Impact energy of H^+ is in keV.

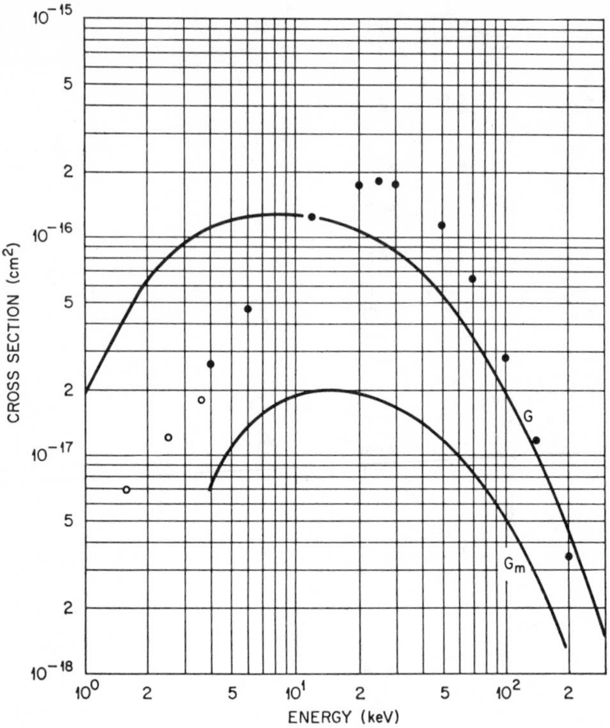

Fig. 40 Electron capture cross sections from helium by protons. Theoretical results for $H^+ + He(1s^2) \rightarrow H(1s) + He^+(1s)$; theoretical values taken from Garcia, Gerjuoy, and Welker.[51] G denotes the classical cross sections obtained using ΔE_l of Eq. 62*a*, and G_m denotes the corresponding cross sections using ΔE_l of Eq. 86. Experimental values represent total cross sections for the processes $H^+ + He(1s^2) \rightarrow \sum H(nl) + \sum He^+(n'l')$: ○ Stedeford and Hasted; [52] ● Stier and Barnett.[14] Impact energy of H^+ is in keV.

and

$$H^+ + Mg(^1S) \rightarrow \sum H(nl) + \begin{cases} Mg^+(3s{:}\,^2S), \\ Mg^+(2p^5 3s^2{:}\,^2P), \\ Mg^+(2s 2p^6 3s^2{:}\,^2S). \end{cases} \qquad (84C)$$

The prediction of $Q(T2)$ for Mg is the more successful of the two cases, although neither agreement is good. Nevertheless, the onset of capture from the inner subshells is in accord with the experimental results; also, again, the following changes in curvature indicate the onset of capture from the corresponding inner subshells: $\approx E = 30$ keV for Na, and 70 keV for Mg. $Q(T1)$ again predicts more accurately at the low-impact-energy end of the curve.

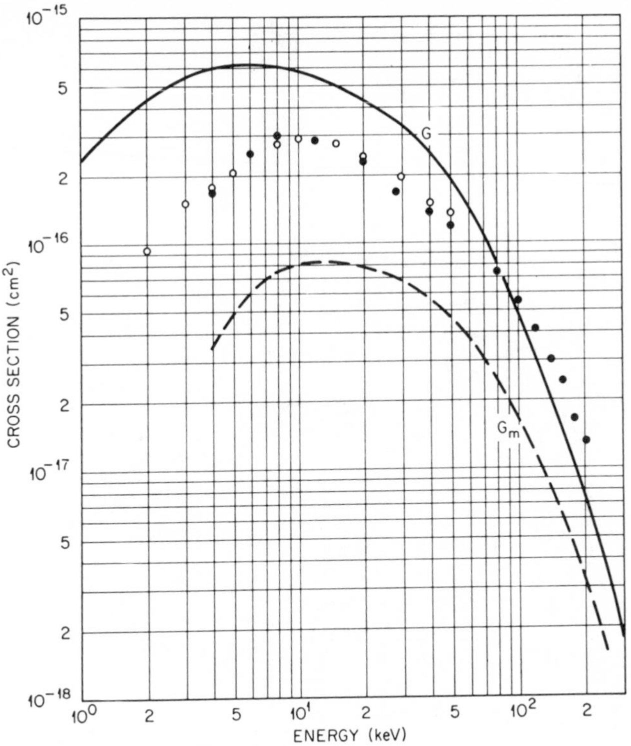

Fig. 41 Electron capture cross sections for protons passing through neon. Theoretical values are for the process $H^+ + Ne(2p^6:{}^1S) \to H(1s) + Ne^+(2p^5:{}^2P)$. Theoretical values taken from Garcia, Gerjuoy, and Welker[51] with G and G_m defined in Fig. 40. Experimental points for the reaction $H^+ + Ne(^1S) \to \sum H(nl) + \sum Ne^+$: ● Stier and Barnett;[14] ○ Williams and Dunbar.[53] Impact energy of H^+ is in keV.

Notwithstanding the absence of measured cross sections for the processes

$$H^+ + Fe(^5D:3d^64s^2) \to \sum H(nl) + \begin{cases} Fe^+(d^6ms:{}^6D, {}^4D), m = 2, 3, 4; \\ Fe^+(d^5:{}^4G, {}^4F, {}^4D, {}^4P, {}^6S); \\ Fe^+(d^6mp^5:{}^6G, {}^4G, {}^6F, {}^4F, {}^6D, {}^4D, \\ {}^6P, {}^4P, {}^6S, {}^4S), m = 2, 3, \end{cases} \tag{85}$$

the cross sections, $Q(T2)$ and $Q(T1)$, and the constituent cross sections for capture from the separate subshells, are given in Fig. 39.[50] $Q(4s)$ is dominant at low values of E, whereas the cross sections $Q(3l)$ ($l = 0, 1, 2$) dominate at high values of E. Capture from $Q(2l)$ ($l = 0, 1$) are unimportant for $E < 1$ MeV. It is not speculated how successful these predictions are.

The last theoretical predictions tested are the modified classical models of Gryzinski (see Section 5.5). In order to avoid divergences in these cross

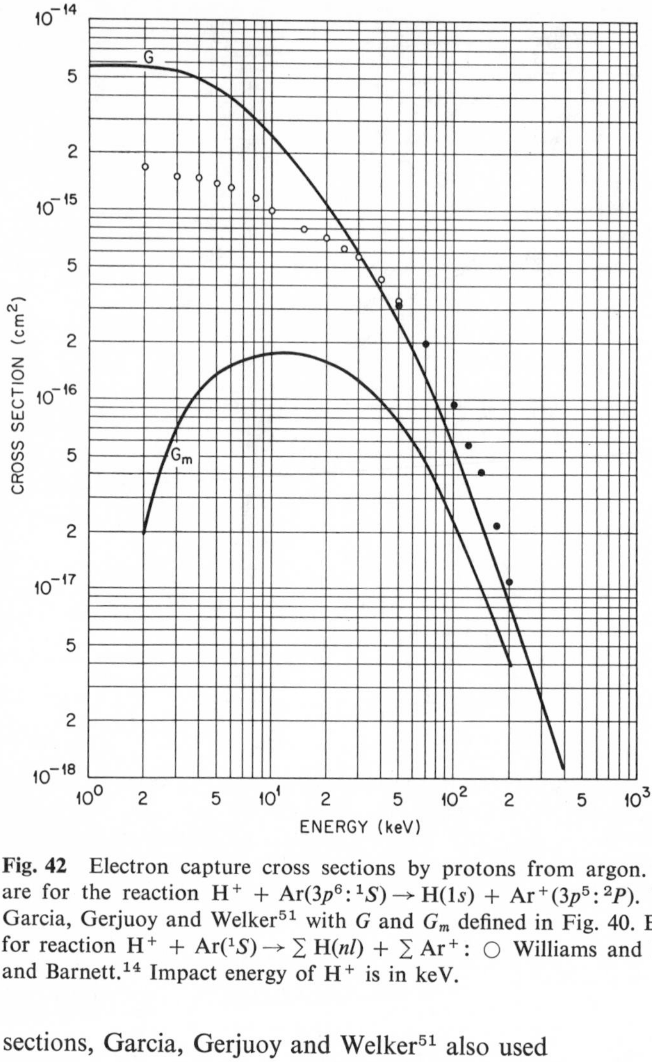

Fig. 42 Electron capture cross sections by protons from argon. Theoretical curves are for the reaction $H^+ + Ar(3p^6{:}\,^1S) \rightarrow H(1s) + Ar^+(3p^5{:}\,^2P)$. Values taken from Garcia, Gerjuoy and Welker[51] with G and G_m defined in Fig. 40. Experimental points for reaction $H^+ + Ar(^1S) \rightarrow \sum H(nl) + \sum Ar^+$: ○ Williams and Dunbar,[53] ● Stier and Barnett.[14] Impact energy of H^+ is in keV.

sections, Garcia, Gerjuoy and Welker[51] also used

$$\Delta E_l = \tfrac{1}{2}mv_1{}^2 + U_A \tag{86}$$

instead of the value given in Eq. 62*a* (Section 5.5). This alternation prevents ΔE_l from becoming too small; the cross section obtained using this revised lower limit in the associated integral is labeled G_m, whereas the cross section corresponding to the unaltered lower limit is denoted by G. Equation 63*b* (Section 5.5) is the formula used to calculate these cross sections, and since ΔE appears in the denominator of the associated terms of Eq. 63*b*, it is clear that G diverges if $U_B > U_A$. In all three cases presented here, cross sections are calculated for capture from the outermost subshell of the process typified by

$$H^+ + A(l^n) \rightarrow H(1s) + A^+(l^{n-1}), \qquad (87)$$

and conservation of spin is not violated.

A comparison of measured total cross sections for capture from He by protons with G and G_m[14,52] is shown in Fig. 40. Although G predicts better at the higher values of E, the experimental values lie roughly midway between G and G_m at the lower values of E. Corresponding results for capture from Ne are presented[14,53] in Fig. 41. The agreement is good above $E = 20$ keV; at lower values of E the experimental points again lie about midway between G and G_m. In Fig. 42, representing capture from Ar, the predictions of G are again the more successful, the more pronounced overestimation occurring for $E < 10$ keV.[14,53]

Although the values of all these calculated cross sections would be increased by contributions from the omitted excited states of H, contributions from inner subshells and excitation of the residual ions are probably negligible in this range of impact energies.[45,51] In view of this likely situation, it is remarkable how well this simple classical model predicts the experimentally determined values.

REFERENCES

1. F. J. Smith, *Proc. Phys. Soc. (London)*, **92**, 866 (1966).
2. A. Dalgarno and H. N. Yadav, *Proc. Phys. Soc. (London)A*, **66**, 173 (1953).
3. W. L. Fite, A. C. H. Smith, and R. F. Stebbings, *Proc. Phys. Soc. (London)A*, **268**, 527 (1962).
4. V. A. Belyaev, B. G. Brezhnev, and E. M. Erastov, *Zh. Eksperim. i Theor. Fiz.*, **52**, 1170 (1967); [*Soviet Phys.-JETP*, **25**, 777 (1967)].
5. D. R. Bates and R. A. Mapleton, *Proc. Phys. Soc. (London)*, **87**, 657 (1966).
6. L. L. Marino, A. C. H. Smith, and E. Caplinger, *Phys. Rev.*, **128**, 2243 (1962).
7. D. V. Chkuaceli, U. D. Nikoleychvili, and A. I. Gouldamachvili, *Izv. Akad. Nauk SSSR, Ser. Fiz.*, **24**, 970 (1960); [*Bull. Acad. Sci. USSR, Phys. Ser.*, **24**, 972 (1960)].
8. J. Perel, R. H. Vernon, and H. L. Daley, *Phys. Rev.*, **138**, A937 (1965).
9. J. B. Hasted and R. A. Smith, *Proc. Roy. Soc. (London)A*, **235**, 354 (1956).
10. T. J. M. Boyd and B. L. Moiseiwitsch, *Proc. Phys. Soc. (London)A*, **70**, 809 (1957).
11. R. McCarroll, *Proc. Roy. Soc. (London)A*, **264**, 547 (1961).
12. G. W. McClure, *Phys. Rev.*, **148**, 47 (1966).
13. T. A. Green, H. E. Stanley, and You-Chien Chiang, *Helv. Phys. Acta*, **38**, 109 (1965).
14. P. M. Stier and C. F. Barnett, *Phys. Rev.*, **103**, 896 (1956).
15. J. B. Hasted, *Proc. Roy. Soc. (London)A*, **227**, 466 (1954).
16. S. E. Lovell and M. B. McElroy, *Proc. Roy. Soc. (London)A*, **283**, 100 (1965).
17. J. E. Bayfield, *Phys. Rev.*, **185**, 105 (1969).

18. R. A. Mapleton, *Phys. Rev.*, **126**, 1477 (1962).
19. D. F. Gallaher and L. Wilets, *Phys. Rev.*, **169**, 139 (1968).
20. M. Rotenberg, *Ann. Phys.* (*N. Y.*), **19**, 262 (1962).
21. L. T. Sin Fai Lam, *Proc. Phys. Soc.* (*London*), **92**, 67 (1967).
22. E. P. Andreev, V. A. Ankudinov, and S. V. Bobashev, *Zh. Eksperim. i Theor. Fiz.*, **50**, 565 (1966); [*Soviet Phys.-JETP*, **23**, 375 (1966)].
23. D. Jaecks, B. Van Zyl, and R. Geballe, *Phys. Rev.*, **137**, A340 (1965).
24. D. Pretzer, B. Van Zyl, and R. Geballe, *Phys. Rev. Letters*, **10**, 340 (1963).
25. F. J. de Heer, J. van Eck, and J. Kistemaker, *Proceedings of the Sixth International Conference on Ionization Phenomenon in Gases* (P. Huber and E. Cremiue-Alcan, eds.), Vol. 1, SERMA, Paris (1963), p. 73.
26. J. Grant and J. Shapiro, *Proc. Phys. Soc.* (*London*), **86**, 1007 (1965).
27. R. H. Bassel and E. Gerjuoy, *Phys. Rev.*, **117**, 747 (1960).
28. J. D. Jackson and H. Schiff, *Phys. Rev.*, **89**, 359 (1953).
29. W. L. Fite, R. F. Stebbings, D. G. Hummer, and R. T. Brackmann, *Phys. Rev.*, **119**, 663 (1960).
30. A. B. Wittkower, G. Ryding, and H. B. Gilbody, *Proc. Phys. Soc.* (*London*), **89**, 541 (1966).
31. C. F. Barnett and H. K. Reynolds, *Phys. Rev.*, **109**, 355 (1958).
32. R. A. Mapleton, *Phys. Rev.*, **122**, 528 (1961).
33. R. A. Mapleton, *Proc. Phys. Soc.* (*London*), **85**, 1109 (1965).
34. L. H. Toburen, M. Y. Nakai, and R. A. Langley, *Phys. Rev.*, **171**, 114 (1968).
35. R. A. Mapleton, *Phys. Rev.*, **130**, 1829 (1963).
36. R. A. Mapleton, *Phys. Rev.*, **117**, 479 (1960).
37. R. A. Mapleton, *Proc. Phys. Soc.* (*London*), **85**, 841 (1965).
38. G. W. McClure, *Phys. Rev.*, **166**, 22 (1968).
39. R. A. Mapleton, *J. Phys. B*, **1**, 529 (1968).
40. L. M. Welsh, K. A. Berkner, S. N. Kaplan, and R. V. Pyle, *Phys. Rev.*, **158**, 85 (1967).
41. U. Schryber, *Helv. Phys. Acta*, **A40**, 1023 (1967).
42. E. Acerbi, M. Castiglioni, G. Dutto, F. Resmini, C. Succi, and G. Tagliaferri, *Nuovo Cimento*, **50B**, 176 (1967).
43. I. M. Cheshire, *Proc. Phys. Soc.* (*London*), **82**, 113 (1963).
44. I. M. Cheshire, *Proc. Phys. Soc.* (*London*), **84**, 89 (1964).
45. D. R. Bates and R. A. Mapleton, *Proc. Phys. Soc.* (*London*), **90**, 909 (1967).
46. R. A. Mapleton, *Phys. Rev.*, **164**, 51 (1967).
47. R. A. Mapleton and N. Grossbard, *Phys. Rev.*, **188**, 228 (1969).
48. R. N. Il'in, V. A. Oparin, E. S. Solov'ev, and N. V. Fedorenko, *Zh. Eksperim. i Teor. Fiz., Pis'ma v Redaktsiyu*, **2**, 310 (1965); [*JEPT Letters*, **2**, 197 (1965)].
49. A. H. Futch and K. G. Moses, *Fifth International Conference on the Physics of Electronic and Atomic Collisions: Abstracts of Papers*, Publishing House Nauka, Leningrad (1967), pp. 12–15.
50. R. A. Mapleton, unpublished calculations.
51. J. D. Garcia, E. Gerjuoy, and J. E. Welker, *Phys. Rev.*, **165**, 72 (1968).
52. J. B. H. Stedeford and J. B. Hasted, *Proc. Roy. Soc.* (*London*)*A*, **227**, 466 (1954).
53. J. F. Williams and D. N. F. Dunbar, *Phys. Rev.*, **149**, 62 (1966).
54. I. M. Cheshire, D. F. Gallaher, and A. J. Taylor, *J. Phys. B*, **3**, 813 (1970).
55. K. H. Berkner, S. N. Kaplan, G. A. Paulikas, and R. V. Pyle, *Phys. Rev.*, **140**, A729 (1965).

APPENDIX 1

EXPANSIONS OF THE MOMENTUM CHANGE VECTORS

In this appendix the momentum change vectors are given in a form suitable for numerical calculations. The meticulous detail occasionally displayed is required in order to avoid excessive loss of significant figures in the calculations of these quantities, particularly at small values of the scattering angle in which domain the major contribution to the cross section often originates. If the numerical values of the ratios of the masses and the values of the atomic binding energies both were known to an infinite number of significant figures, the expansions of this appendix would be unnecessary. This loss of significant figures is avoided by effecting as much cancellation as possible before the parameters are evaluated.

Units that have been used throughout this book are recorded again for convenient reference:

$$\epsilon_1 = \frac{me^4}{2\hbar^2}, \qquad \text{the unit of energy,}$$

$$v_0 = \frac{e^2}{\hbar}, \qquad \text{the unit of velocity,} \tag{A1.1a}$$

$$a_0^{-1} = \frac{me^4}{\hbar^2}, \qquad \text{the unit of inverse length.}$$

The relations among several units used for the impact energy E are repeated:

$$\frac{E(\text{eV})}{\epsilon_1} \approx \frac{M}{m} E(25 \text{ keV}) = \frac{4M}{m} E(100 \text{ keV}). \tag{A1.1b}$$

Although these last two relations are approximate, they are adequate for most applications. Rather than refer to previous equations, the process for single electronic capture and the notations for the associated masses, reduced masses, and momentum-change vectors are collected in Eq. A1.1c:

$$
\begin{aligned}
&P^+(M_f) + A(M_{i1}) \to P(M_{f1}) + A^+(M_i), \\
&M_i = M_a + (n_p - 1)m, \qquad M_{i1} = M_i + m, \\
&M_f = M_p + (n_a - 1)m, \qquad M_{f1} = M_f + m, \\
&M_t = M_{i1} + M_f = M_i + M_{f1}, \\
&\mu_i = \frac{M_f M_{i1}}{M_t}, \qquad \mu_f = \frac{M_{f1} M_i}{M_t}, \\
&\mathbf{A}_i = \frac{M_i}{M_{i1}} \mathbf{K}_i - \mathbf{K}_f, \qquad \mathbf{A}_f = \frac{M_f}{M_{f1}} \mathbf{K}_f - \mathbf{K}_i.
\end{aligned}
\tag{A1.1c}
$$

Next, conservation of energy, together with Eq. 9a for the relations among the energies and velocities in the center-of-mass and laboratory systems expressed in the units of Eq. A1.1a, is used to relate K_i and K_f to E and the change of internal energy $\Delta\epsilon$:

$$
\begin{aligned}
E &= \text{impact energy}, \\
\mathbf{v}_{i,f} &= \text{respective initial and final velocities}, \\
\mathbf{K}_{i,f} &= \text{respective initial and final wave vectors of relative motion},
\end{aligned}
$$

$$
\begin{aligned}
&\frac{\mu_i}{m} v_i^2 - \epsilon_i = \frac{\mu_f}{m} v_f^2 - \epsilon_f, \qquad E - \epsilon_i = \frac{M_f}{m} v_i^2 - \epsilon_i, \\
&\mathbf{K}_i = \frac{\mu_i}{m} \mathbf{v}_i, \qquad \mathbf{K}_f = \frac{\mu_f}{m} \mathbf{v}_f, \\
&K_i^2 = \frac{\mu_i M_{i1}}{m M_t} E, \qquad K_f^2 = \frac{\mu_f M_{i1}}{m M_t} E - \frac{\mu_f}{m} \Delta\epsilon, \qquad \Delta\epsilon = \epsilon_i - \epsilon_f.
\end{aligned}
\tag{A1.1d}
$$

Since the scattering angle is measured from $\mathbf{K}_i$ to $\mathbf{K}_f$ and never exceeds the value π in the center-of-mass system, the momentum change vectors are

$$
\begin{aligned}
A_i^2 &= \left(\frac{M_i K_i}{M_{i1}}\right)^2 + K_f^2 - 2\frac{M_i}{M_{i1}} K_i K_f \cos\theta, \\
A_f^2 &= \frac{M_{i1} M_f}{M_i M_{f1}} A_i^2 + \frac{M_f}{M_{f1}} \Delta\epsilon.
\end{aligned}
\tag{A1.1e}
$$

The relation between A_f^2 and A_i^2 in Eq. A1.1e is deduced from Eq. A1.1c and from the relations between K_f^2 and K_i^2 given by Eq. A1.1d.

Now the expansion for A_i^2 is written, giving only enough of the details to elucidate the method of expansion. The relations of Eq. A1.1d are used to express $K_{i,f}$ in terms of E, and a convenient parameter x is defined as

$$
x = \frac{M_t \Delta\epsilon}{2M_{i1}}; \tag{A1.1f}
$$

next expand the square root that stems from the definition of K_f and add and subtract some terms to get, finally,

$$A_i^2 = \left[\left(\frac{M_i^2 M_f}{mM_t^2} + \frac{M_i M_{i1} M_{f1}}{mM_t^2}\right)E - \frac{M_i M_{f1}}{mM_t}\,\Delta\epsilon\right](1 - \cos\theta)$$
$$+ \left[\left(\frac{M_i^2 M_f}{mM_t^2} + \frac{M_i M_{i1} M_{f1}}{mM_t^2}\right)E - \frac{M_i M_{f1}}{mM_t}\,\Delta\epsilon\right.$$
$$- 2\,\frac{M_i M_{i1} M_f}{mM_t^2}\left(\frac{M_i M_{f1}}{M_{i1} M_f}\right)^{1/2} E + \frac{M_i M_f}{mM_t}\left(\frac{M_i M_{f1}}{M_{i1} M_f}\right)^{1/2}\Delta\epsilon \qquad \text{(A1.1g)}$$
$$\left. + \frac{M_i M_f (\Delta\epsilon)^2}{4mM_{i1}E}\left(\frac{M_i M_{f1}}{M_{i1} M_f}\right)^{1/2}\left(1 + \frac{x}{E} + \frac{5}{4}\left(\frac{x}{E}\right)^2 + \cdots\right)\right]\cos\theta.$$

The significant cancellation is effected for the factors of E and $\Delta\epsilon$ in the bracket, which is the factor of $\cos\theta$; this is the factor that provides the chief contribution to A_i^2 near the direction of incidence, the sensitive angular region. The expansion used for the factors of E is illustrated; the factors are

$$\frac{M_{i1}^2 M_f}{mM_t^2}\left[\frac{M_i}{M_{i1}} - \left(\frac{M_i M_{f1}}{M_{i1} M_f}\right)^{1/2}\right]^2$$
$$= \frac{M_{i1}^2 M_f}{mM_t^2}\left[1 - \frac{m}{M_{i1}} - \left(1 + \frac{m}{2M_f} - \frac{m^2}{8M_f^2}\cdots\right)\left(1 - \frac{m}{2M_{i1}} - \frac{m^2}{8M_{i1}^2}\cdots\right)\right]^2$$
$$\approx \frac{M_{i1}^2 M_f}{mM_t^2} \times \frac{m^2 M_t^2}{4M_{i1}^2 M_f^2} = \frac{m}{4M_f}. \qquad \text{(A1.1h)}$$

The simplified expansion of A_i^2 is given by

$$A_i^2 = \left[\left(\frac{M_i^2 M_f}{mM_t^2} + \frac{M_i M_{i1} M_{f1}}{mM_t^2}\right)E - \frac{M_i M_{f1}}{mM_t}\,\Delta\epsilon\right](1 - \cos\theta)$$
$$+ \left[\frac{m}{4M_f}E - \frac{M_i}{2M_{i1}}\left(1 + \frac{mM_t}{4M_{i1}M_f}\right)\Delta\epsilon \right. \qquad \text{(A1.1i)}$$
$$\left. + \frac{M_i M_f (\Delta\epsilon)^2}{4mM_{i1}E}\left(\frac{M_i M_{f1}}{M_{i1} M_f}\right)^{1/2}\left(1 + \frac{x}{E} + \frac{5}{4}\left(\frac{x}{E}\right)^2 + \cdots\right)\right]\cos\theta.$$

The same methods of expansion are used to evaluate the remaining derived quantities:

$$R^2 = |\mathbf{A}_i + \mathbf{A}_f|^2 = \frac{mM_i}{M_t^2} E\left\{\frac{M_{i1}}{M_{f1}} + \frac{M_f}{M_i} - \frac{M_t}{M_{f1}}\frac{\Delta\epsilon}{E} + 2\left[\frac{M_{i1}M_f}{M_iM_{f1}}\left(1 - \frac{M_t\Delta\epsilon}{M_{i1}E}\right)\right]^{1/2} \cos\theta\right\}, \tag{A1.1j}$$

$$\begin{aligned}\mathbf{A}_i \cdot \mathbf{A}_f = {} & \left[-\frac{2M_iM_{i1}M_f}{mM_t^2} E + \frac{M_iM_f}{mM_t}\Delta\epsilon\right](1 - \cos\theta) \\ & + \left[\frac{mM_f(M_t + m)^2}{4M_{f1}^2M_t^2} E - \frac{mM_f(M_t + m)^2}{8M_{i1}M_{f1}^2M_t}\Delta\epsilon \right. \\ & - \frac{M_f(\Delta\epsilon)^2}{8mE}\left(1 + \frac{M_iM_f}{M_{i1}M_{f1}}\right)\left(\frac{M_iM_{f1}}{M_{i1}M_f}\right)^{1/2} \\ & \left. \times \left(1 + \frac{x}{E} + \frac{5}{4}\left(\frac{x}{E}\right)^2 + \cdots\right)\right] \cos\theta.\end{aligned} \tag{A1.1k}$$

$$\begin{aligned}|\mathbf{A}_i - \mathbf{A}_f|^2 = {} & \left[\frac{M_iM_{i1}M_{f1}}{mM_t^2}\left(1 + \frac{M_f}{M_{f1}}\right)^2 E + \frac{M_{i1}^2M_f}{mM_t^2}\left(1 + \frac{M_i}{M_{i1}}\right)^2 E \right. \\ & \left. - \frac{M_{i1}M_f}{mM_t}\left(1 + \frac{M_f}{M_{f1}}\right)^2 \Delta\epsilon\right](1 - \cos\theta) \\ & + \left\{\frac{m^3M_fM_{i1}^2}{16M_t^2M_{f1}^4}\left[1 - \left(\frac{M_{f1}}{M_{i1}}\right)^2\right]^2 E \right. \\ & + \frac{mM_iM_{f1}}{4M_tM_{i1}^2}\left[1 + \frac{M_f}{M_{f1}}\right]\left[1 - \left(\frac{M_{i1}}{M_{f1}}\right)^2\right]\Delta\epsilon \\ & + \frac{M_f}{4m}\left(1 + \frac{M_i}{M_{i1}}\right)\left(1 + \frac{M_f}{M_{f1}}\right)\left(\frac{M_iM_{f1}}{M_{i1}M_f}\right)^{1/2}\frac{(\Delta\epsilon)^2}{E} \\ & \left. \times \left(1 + \frac{x}{E} + \frac{5}{4}\left(\frac{x}{E}\right)^2 + \cdots\right)\right\} \cos\theta.\end{aligned} \tag{A1.1l}$$

In the evaluation of the asymptotic variation of the cross section as a function of the impact energy it is frequently convenient to express the energy in units of 100 keV; this result is given in Eq. A1.1*b* and is used in the approximations for the dominant parts of A_i^2 and R^2 at two values of θ:

$$\theta = 0$$

$$A_i^2 \approx \frac{M}{M_f} E$$

$$R^2 \approx \frac{4MM_i}{M_t^2} E\left[\left(\frac{M_{i1}}{M_{f1}}\right)^{1/2} + \left(\frac{M_f}{M_i}\right)^{1/2}\right]^2; \tag{A1.1m}$$

$$\theta = \pi$$

$$A_i^2 \approx \frac{16MM_fM_i^2}{m^2M_t^2} E,$$

$$R^2 \approx \frac{4MM_i}{M_t^2} E\left[\left(\frac{M_{i1}}{M_{f1}}\right)^{1/2} - \left(\frac{M_f}{M_i}\right)^{1/2}\right]^2. \tag{A1.1n}$$

The quantities inside the brackets in the two expressions for R^2 are exact; only the external factors are approximated in the conversion to units of 100 keV. The important features, however, are the rapid increase of A_i^2 with the increasing θ and the vanishing of $R^2(\theta = \pi)$ if the masses M_i and M_f are equal.

APPENDIX 2

FRACTIONAL PARENTAGE COEFFICIENTS FOR EQUIVALENT p ELECTRONS

The material in this appendix pertains to the coefficient of fractional parentage for equivalent p electrons (fpcs) and is taken from the papers of Racah*[1,2] and Chapters 26 and 31 of the book by de Shalit and Talmi.[3] The discussion is restricted to the LSM_lM_s representation, and the formulas of Racah that are used are diagonal in M_l and M_s; moreover, as in all of this appendix the LS-coupling approximation is used.

In Racah's paper the functions $\Psi_R(p^{6-n}vSL)$ and $\Psi_L(p^nvSL)$ are introduced, and these functions are called self-corresponding for the middle of the subshell $n = 3$, in which case they satisfy the equation

$$\Psi_L(p^3vSLM_sM_l) = \pm\Psi_R(p^3vSLM_sM_l). \qquad \text{(RII.76; A2.1)}$$

In this relation the symbol v denotes the seniority number. This seniority number, as used here, represents the minimum number of equivalent p electrons for a term first to occur, and the values of v and the corresponding terms are given in Table A2.1.

These tabulated values exhaust the possibilities for equivalent p electrons. The $\pm$ signs of Eq. A2.1 are next expressed in terms of v:

$$\begin{aligned} \Psi_R(p^3vSL) &= (-1)^{(v-1)/2}\Psi_L(p^3vSL), \\ \Psi_R(p^4vSL) &= (-1)^{v/2}\Psi_L(p^4vSL). \end{aligned} \qquad \text{(RIII.65; A2.2)}$$

According to the conventions of Racah, $\Psi = \Psi_L$ for $n \leq 3$ and $\Psi = \Psi_R$ for $n \geq 4$, and as a consequence the factor

$$(-1)^{(v-1)/2}$$

* In this appendix the equations of the papers by Racah (II) and (III) are designated *R*II— and *R*III—, respectively.

TABLE A2.1 SENIORITY NUMBERS

v	Term	Number of p-Electrons
v	${}^{(2S+1)}{}_{v}T$	v
0	${}^{1}_{0}S = {}^{1}S$	0
1	${}^{2}_{1}P = {}^{2}P$	1
2	${}^{3}_{2}P = {}^{3}P;\ {}^{1}_{2}D = {}^{1}D$	2
3	${}^{4}_{3}S = {}^{4}S;\ {}^{2}_{3}D = {}^{2}D$	3

of Eq. A2.2 must be used with Eq. A2.4 which is the formula used to calculate the fpcs in this appendix:

$$(p^{5-n}(S'L')pSL|\}p^{6-n}SL)$$

$$= \delta(-1)^{(S+S'+L+L'-3/2)}\left[\frac{(n+1)(2S'+1)(2L'+1)}{(6-n)(2S+1)(2L+1)}\right]^{1/2} \times (p^{n}(SL)pS'L'|\}p^{n+1}S'L'), \quad \text{(RIII.19; A2.4)}$$

$$\delta = 1, \quad n \neq 2; \qquad \delta = (-1)^{(v-1)/2}, \quad n = 2.$$

The value of v corresponding to the $(S'L')$ term is used in this formula. There is a relation among the fpcs of a configuration called the orthogonality relation, which is given by Eq. A2.5:

$$\sum_{(S'L')} \overline{(p^{n-1}(S'L')pSL|\}p^{n}SL)}(p^{n-1}(S'L')pSL|\}p^{n}SL) = 1. \quad \text{(RIII.13; A2.5)}$$

The notation used with the summation sign in Eq. A2.5 means that the term-values $(S'L')$ of the configuration are summed. In connection with these relations among the fpcs the reader is reminded that the fpcs are the weighting coefficients used to couple p-orbitals to the orthonormal terms of the p^{n-1} configuration in order to construct the orthonormal terms of the p^n configuration; we must also realize that the ordinary Clebsch-Gordan, or vector-coupling, formulas for the addition of two angular momenta are used in conjunction with the fpcs.[3,4]

Since the fpcs calculated by Racah are used with Eq. A2.4 to calculate other fpcs, we reproduce in Table A2.2 those calculated by Racah.[2] Included at the bottom of this table are the additional two fpcs needed to supplement Eq. A2.4, which is used to calculate the values in Tables A2.3 and A2.4. In the evaluation of Table A2.3 we must use the factor $(-1)^{(v-1)/2}$ since $n = 2$. Included in Table A2.4 is the fpc from $p^5(^2P)$ to $p^6(^1S)$, whose value is obvious.

TABLE A2.2 $(p^2(S'L')pSL|\}p^3SL)$

	p^2		
p^3	^3P	^1D	^1S
^4S	1	0	0
^2D	$\left(\frac{1}{2}\right)^{1/2}$	$-\left(\frac{1}{2}\right)^{1/2}$	0
^2P	$-\left(\frac{1}{2}\right)^{1/2}$	$-\frac{1}{6}(10)^{1/2}$	$\frac{1}{3}(2)^{1/2}$

$(p^0(S'L')pSL|\}pSL) = (p(S'L')pSL|\}p^2SL) = 1$

Racah[2] also gives the formulas for the calculation of fpcs that couple two equivalent p-electrons at a time, and these are related to those of this appendix; however, only those given here are needed for the applications of this appendix.

Next we study the relations of the amplitude to the fpcs connecting the term of the atom with the possible terms of the residual ions formed by capture. Only capture of p-orbital electrons by protons into s-states of atomic hydrogen from configurations of equivalent p-electrons is considered.

The normalized antisymmetrical wave function for the configuration of p^n is expressed in terms of the configuration p^{n-1} by Eq. A2.6:

$$\begin{aligned}\psi(p^nSL) &= \sum_{(S'L')} \psi[p^{n-1}(S'L')pSL](p^{n-1}(S'L')pSL|\}p^nSL) \\ &= \sum_{(S'L')}\sum_{(S''L'')} \psi[p^{n-2}(S''L'')p(S'L')pSL] \qquad \text{(RIII.10; A2.6)} \\ &\quad \times (p^{n-2}(S''L'')pS'L'|\}p^{n-1}S'L')(p^{n-1}S'L'pSL|\}p^nSL).\end{aligned}$$

The sums in Eq. A2.6 are over the terms of the p^{n-1} and p^{n-2} configurations, and the equation is diagonal in M_l and M_s as explained previously; the second form of the wave function expressed as a double sum is used in conjunction

TABLE A2.3 $(n = 2)$: $(p^3(S'L')pSL|\}p^4SL)$

	p^3		
p^4	^4S	^2D	^2P
^3P	$-\left(\frac{1}{3}\right)^{1/2}$	$\left(\frac{5}{12}\right)^{1/2}$	$-\frac{1}{2}$
^1D	0	$-\left(\frac{3}{4}\right)^{1/2}$	$-\frac{1}{2}$
^1S	0	0	1

TABLE A2.4 $(n = 1)$: $(p^4(S'L')pSL|\}p^5SL)$

	p^4		
p^5	^3P	^1D	^1S
^2P	$\left(\frac{3}{5}\right)^{1/2}$	$\left(\frac{1}{3}\right)^{1/2}$	$\left(\frac{1}{15}\right)^{1/2}$

$$(p^5(S'L')pSL|\}p^6SL) = 1$$

with the passive interactions. (Although Eq. A2.6 was discussed in connection with Eq. 44f in Section 3.4, many of the omitted details of that account are included in this appendix. The sums over the M_l- and M_s-values are not given in Eq. A2.6 since this feature is explained in the applications of the formulas.) We have adopted the convention of labeling the electrons in p^n as n, $n - 1$, $\ldots$, 1; p^{n-1} as n, $n - 1, \ldots, 2$; p^{n-2} as n, $n - 1, \ldots, 3$; etc.; the p-orbital, which is vector-coupled to p^{n-1} in the first sum of Eq. A2.6 is labeled 1, and in the second double sum the corresponding two p-orbitals are labeled 2, 1 from the left to right. This procedure is permissible since each wave function is antisymmetrical in its group of equivalent p-electrons, and Eq. A2.6 is antisymmetrical in all the equivalent p-electrons by construction.

This is an appropriate place to discuss the dependence of the phase, ± 1, of a term upon the coupling procedure. For example, in the formation of the wave function representing a p^n-group of equivalent electrons, $\phi(p, 1)$ can be coupled to $\psi(p^{n-1}; 2, 3, \text{---} n)$ to form the antisymmetrical function $\Psi(p^n; 1, 2, \text{---} n)$ or $\pm\Psi(p^n; 1, 2, \text{---} n)$ can be formed by coupling $\phi(p; n)$ to $\Psi(p^{n-1}; 1, 2, \text{---} n - 1)$. We digress from the difficulties of equivalent p-electrons first to consider two simple cases of the formation of an n-electron wave function as just described. The only equivalent electrons admitted in this example are the configurations $(ms)^2$, and it is assumed that the n-electron wave function can be represented by a single determinant. An elementary calculation reveals that the two antisymmetrical wave functions resulting from the two coupling procedures differ by the factor $(-1)^{n-1}$. Only if $(n - 1)$ is odd do the signs differ.

Now return to the more difficult case of equivalent p-electrons. First, there is the relation for the coupling of two angular momenta,[5]

$$\psi(j_2 j_1 jm) = (-1)^{(j_1 + j_2 - j)}\psi(j_1 j_2 jm), \qquad \text{(A2.7)}$$

and second there is the relation between two fpcs,

$$(p, p^{n-1}(S'L'), SL|\}p^nSL) = (-1)^{(S'+L'+S+L-3/2)}(p^{n-1}(S'L')pSL|\}p^nSL). \qquad \text{(RIII.29; A2.8)}$$

If Eqs. A2.7 and A2.8 are applied to the construction of the wave functions for the terms $p^3(^4S)$, $p^3(^2D)$, and $p^3(^2P)$; it is found that no difference in signs occurs; this is in accord with $(n-1)=2$ being even. However, the corresponding calculation for the terms $p^4(^3P)$, $p^4(^1D)$, and $p^4(^1S)$ reveals a difference in sign for the two methods of coupling, and indeed $(n-1)$ is odd in this case. It is clear that the same coupling procedure should be used in a given calculation for purposes of consistency.

In the application of Eq. A2.6 to the calculation of the OBK amplitude, only the single sum is needed, and the calculation for $n = 4$ will be illustrated. In this clarifying calculation it is necessary to insert the Clebsch-Gordan coefficients (CGC or vector-coupling coefficients, VCC) into the sum of Eq. A2.6, and it is important to recognize that the coordinate of the OBK interaction V_a is not contained among the coordinates of the atom and ion:

$$\begin{aligned}\psi[p^4(LS); 4321] = \sum_{(S'L')} \sum_{m_l'' M_l''} \sum_{m_s'' M_s''} \psi\{&p^3(S'L'M_s''M_l''; 432) \\ &\times p(1, m_l''; 1)s(\tfrac{1}{2}, m_s''; 1)(L'M_l''1m_l''|L'1LM_l) \\ &\times (S'M_s''\tfrac{1}{2}m_s''|S'\tfrac{1}{2}SM_s)(p^3(S'L')pSL|\}p^4SL)\}.\end{aligned} \tag{A2.9}$$

In Eq. A2.9, $p(1)$, $s(1)$ denote the respective spatial and spin-orbitals of electron 1: lower-case subscripts are used throughout since the meaning of the matrix elements and fpcs is sufficiently standard so that the significance of these labels should be evident. The lower case letters l, s, m_l, and m_s are used to denote the orbital and spin angular momenta, and their respective components, of one-electron terms (only the active electron here). Equation A2.9 is next multiplied by the complex conjugate of the ionic term of interest. The result is integrated and using the orthogonality properties of these three-electron terms, the relations of Eq. A2.10 are obtained.[6] Although a bar is used to denote the complex conjugate of the wave function, the bar also means transposed complex conjugate when it applies to spin functions and CGC.

$$\int \overline{\psi[p^3(L_3S_3); 432]}\psi[p^3(L'S'); 432]\, d\mathbf{x}_4\, d\mathbf{x}_3\, d\mathbf{x}_2 = \delta(S'L', S_3L_3)\delta(M_{l_3}, M_l')\delta(M_{s_3}, M_s'),$$

$$\begin{aligned}\int \overline{\psi[p^3(L_3S_3); 432]}&\psi[p^4(LS); 4321]\, d\mathbf{x}_4\, d\mathbf{x}_3\, d\mathbf{x}_2 \\ &= \sum_{m_l''} \sum_{m_s''} p(1, m_l''; 1)s(\tfrac{1}{2}, m_s''; 1)(L_3M_{l_3}1m_l''|L_31LM_l) \\ &\quad \times (S_3M_{s_3}\tfrac{1}{2}m_s''|S_3\tfrac{1}{2}SM_s)(p^3(S_3L_3)pSL|\}p^4SL),\end{aligned} \tag{A2.10}$$

$$s'' = \tfrac{1}{2}, \qquad m_s'' = \pm\tfrac{1}{2}, \qquad l'' = 1, \qquad m_l'' = 0, \pm 1.$$

The delta functions of this equation are the Kronecker deltas; in these integrals and other similar integrals of this appendix factors of order unity are neglected; that is, factors which arise from the small differences between the radial functions of corresponding terms of the atom and ion. The salient feature of Eq. A2.10, however, is the appearance of the fpc as a factor of the sum of the product of the orbital and spin wave functions of electron 1 and the two VCCs. Since the coordinate of V_a is not contained among those of Eq. A2.10, it is evident that the OBK amplitude is proportional to the fpc, a fact noted in Eq. 44*i* (Section 3.4).

In these cases, in which the *s*-orbitals of the closed subshells are omitted as in the foregoing example, the integrations over these other orbitals, space and spin, are assumed since these calculations can be effected separately as explained in Section 3.4. Next, the corresponding matrix elements of the prior and post interactions are analyzed for the case $n = 4$.

The reader is reminded that the respective coordinates of the hydrogen atom, active electron, ion, and atom are (x_{n+1}); (x_1); $(x_2 \cdots x_n)$; $(x_1 \cdots x_n)$, respectively; moreover, as in Section 3.4, the unsymmetrized amplitude is investigated first. Among the interactions given by Eq. 45*a* for the case $n = 8$ are the proton-nucleus term $|\mathbf{x}_1 - \mathbf{x}_9|^{-1}$ and the post interaction term $|\mathbf{x}_1|^{-1}$. Since x_1 only occurs in the $p(1)$-orbital that is coupled to the $p^{n-1}(432)$-terms of Eq. A2.9, the integral of Eq. A2.10 is unchanged, and the matrix element of each of these two interactions is again proportional to the fpc as we found for the OBK matrix element. It is noted that the details of the coupling of the spin part of atomic hydrogen to the spin part of the residual ion are not needed here. The other two types of interactions in $V_{i,f}$ are $|\mathbf{x}_1 - \mathbf{x}_2 - \mathbf{x}_{n+1}|^{-1}$ and $|\mathbf{x}_1 - \mathbf{x}_2|^{-1}$, and there are three types of matrix elements common to both of these interactions, elements which are denoted by

$$PP\bar{P} = P(1)P(2)\overline{P(2)}, \qquad PST = P(1)S(2)\overline{T(2)}, \qquad SPT = S(1)P(2)\overline{T(2)},$$

in agreement with the notation of the original version of this research.[7] The letters P and S denote the respective p and s orbitals of the atom, and $\bar{P}$, $\bar{T}$, the complex conjugates of the corresponding ionic orbitals. The first two orbitals, reading from left to right, always belong to the atom.

The purpose of this investigation of the passive electronic matrix elements (which occur in the wave version of the first Born approximation using the prior, post, or distorted wave interactions) is to exemplify the conditions for the proportionality of the amplitude to the fpc associated with the residual ion formed by the capturing process. Since the SPT and PST matrix elements are proportional to the relevant fpc, only the $PP\bar{P}$ case is analyzed. Although p-orbital capture is studied here, similar variations occur in processes of s-orbital capture.

This is the most difficult case to analyze, and here we first specialize to the term ^4S of the three possible O^+ ions. As the first step the p^4 and p^3 configurations are written in terms of the p^2 configuration:

$$\begin{aligned}\psi[p^4(SL); 4321] = & \sum_{(S'L'432)} \sum_{(S''L''43)} \sum_{M'_{l43}} \sum_{M'_{s43}} \sum_{m_{s2}} \sum_{m_{l2}} \sum_{M''_{l432}} \sum_{M''_{s432}} \sum_{m_{l1}} \sum_{m_{s1}} \\ & \times \psi[p^2(L''_{43}M'_{l43}S''_{43}M'_{s43})]p(1, m_{l2})s_i(\tfrac{1}{2}, m_{s2}) \\ & \times (L''_{43}M'_{l43}1m_{l2}|L''_{43}1L'_{432}M''_{l432}) \\ & \times (S''_{43}M'_{s43}\tfrac{1}{2}m_{s2}|S''_{43}\tfrac{1}{2}S'_{432}M''_{s432})p(1, m_{l1})s_i(\tfrac{1}{2}, m_{s1}) \\ & \times (L'_{432}M''_{l432}1m_{l1}|L'_{432}1L_{4321}M_{l4321}) \\ & \times (S'_{432}M''_{s432}\tfrac{1}{2}m_{s1}|S'_{432}\tfrac{1}{2}S_{4321}M_{s4321}) \\ & \times (p^3_{432}(S'L')p_1SL|\}p^4_{4321}(SL)) \\ & \times (p^2(S''_{43}L''_{43})p_2S'_{432}L'_{432}|\}p^3_{432}(S'_{432}L'_{432})), \\ S_{4321} = S = 1, & \qquad L_{4321} = L = 1.\end{aligned} \tag{A2.11}$$

$$\begin{aligned}\psi[p^3(S_3L_3); 432] = & \sum_{(SL43)} \sum_{M''_{l43}} \sum_{M''_{s43}} \sum_{m''_{l2}} \sum_{m''_{s2}} \\ & \times \overline{\psi[p^2(L_{43}M''_{l43}S_{43}M''_{s43})]}\,\overline{p(1m''_{l2})s_f(\tfrac{1}{2}, m''_{s2})} \\ & \times \overline{(L_{43}M''_{l43}1m''_{l2}|L_{43}1L_{432}M_{l432})} \\ & \times \overline{(S_{43}M''_{s43}\tfrac{1}{2}m''_{s2}|S_{43}\tfrac{1}{2}S_{432}M_{s432})} \\ & \times \overline{(p^2_{43}(S_{43}L_{43})pS_{432}L_{432}|\}p^3_{432}S_3L_3)}, \\ S_{432} = S_3 = \tfrac{3}{2}, & \qquad L_{432} = L_3 = 0.\end{aligned} \tag{A2.12}$$

Next Eq. A2.11 is multiplied by Eq. A2.12 and the orthonormal properties of the spin functions of particle 2, $s_i(\frac{1}{2}, m_{s2})$ and $s_f(\frac{1}{2}, m''_{s2})$, together with the properties of Eq. A2.10 resulting from the integration over $d\mathbf{x}_3$ and $d\mathbf{x}_4$, are used to give the intermediate results:

$$\begin{aligned}\int d\mathbf{x}_3\, d\mathbf{x}_4 & \overline{\psi[p^2(L_{43}M''_{l43}S_{43}M''_{s43})]}\psi[p^2(L''_{43}M'_{l43}S''_{43}M'_{s43})] \\ & = \delta(S''_{43}L''_{43}, S_{43}L_{43})\delta(M'_{l43}, M''_{l43})\delta(M'_{s43}, M''_{s43}), \\ & \overline{s_f(\tfrac{1}{2}, m''_{s2})}s_i(\tfrac{1}{2}, m_{s2}) = \delta(m''_{s2}, m_{s2}).\end{aligned} \tag{A.2.13}$$

With the aid of Eq. A2.13 and the integration over the spin coordinates of particle 2, the two CGCs that couple the spin of particle 2 become

$$(S_{43}M''_{s43}\tfrac{1}{2}m''_{s2}|S_{43}\tfrac{1}{2}S'_{432}M''_{s432})$$

and

$$\overline{(S_{43}M''_{s43}\tfrac{1}{2}m''_{s2}|S_{43}\tfrac{1}{2}S_{432}M_{s432})},$$

and consequently the sum over the product of these two CGCs can be calculated:[6]

$$\begin{aligned}\sum_{m''_{s2}} \sum_{M''_{s43}} & \overline{(S_{43}M''_{s43}\tfrac{1}{2}m''_{s2}|S_{43}\tfrac{1}{2}S_{432}M_{s432})}(S_{43}M''_{s43}\tfrac{1}{2}m''_{s2}|S_{43}\tfrac{1}{2}S'_{432}M''_{s432}) \\ & = \delta(S_{432}, S'_{432})\delta(M_{s432}, M''_{s432})\end{aligned} \tag{A2.14}$$

In order to effect this last sum we used the fact that two fpcs are diagonal in M''_{s43}. Since $S_{432} = S_3 = \frac{3}{2}$ and since only one term of $(S'_{432}L'_{432})$ has this value in the sum over the three-particle states in Eq. A2.11, this means that only the quartet term remains after the preceding integrations and sums are accomplished. This is a crucial point since in other cases such isolation is not possible; for example, if our ionic state (S_3L_3) is a ${}^2P$ or ${}^2D$ term, all that can be determined is that $(S'_{432}L'_{432})$, originating from Eq. A2.11 is a doublet, but the value of L'_{432} cannot be specified.

In the present instance the result can be simplified further, as we now show. For this purpose we use the relations of Eq. A2.13 and the conditions $L'_{432} = L_{432} = 0$ in Eqs. A2.11 and A2.12. The two orbital CGC that couple particle 2 vanish unless $M'_{l43} + m_{l2} = M''_{l432}$, and $M''_{l43} + m''_{l2} = M_{l432}$. It is clear, hence, that $m''_{l2} = m_{l2}$,[6] and

$$(L''_{43}M'_{l43}1m_{l2}|L''_{43}1L'_{432}M''_{l432}) = (L_{43}M''_{l43}1m''_{l2}|L_{43}1L_{432}M_{l432}). \qquad \text{(A2.15)}$$

In our integral over $d\mathbf{x}_3 d\mathbf{x}_4$, however, we still cannot sum over M''_{l43} since m_{l2} cannot be varied owing to the dependence of $p(1, m_{l2})$ upon this variable. The integral of interest can be reduced further, nevertheless, since $L_3 = 0$, $M_{l3} = 0$, $S_3 = \frac{3}{2}$, $S = 1$, $L = 1$, and in this case the square of the CGC,[8]

$$|(L_{43}M''_{l43}1m_{l2}|L_{43}100)|^2, \qquad \text{(A2.16)}$$

vanishes unless $L_{43} = 1$,

$$\begin{aligned}
&\int d\mathbf{x}_3\, d\mathbf{x}_4 \overline{\psi[p^3(S_3L_3); 432]}\psi[p^4(SL); 4321] \\
&\quad = \sum_{(S_{43}L_{43})} \sum_{m_{l1}} \sum_{m_{s1}} \sum_{m_{l2}} \sum_{M''_{l43}} p(1m_{l2}; 2)\overline{p(1m_{l2}; 2)}p(1m_{l1}; 1) \\
&\qquad \times s_i(\tfrac{1}{2}m_{s1})(L_3M_{l3}1m_{l1}|L_31LM_l)(S_3M_{s3}\tfrac{1}{2}m_{s1}|S_3\tfrac{1}{2}SM_s) \\
&\qquad \times |(L_{43}M''_{l43}1m_{l2}|L_{43}1L_{432}M_{l432})|^2(p^3(S_3L_3)p_1SL|\}p^4SL) \\
&\qquad \times \overline{(p^2(S_{43}L_{43})p_2S_3L_3|\}p^3S_3L_3)}(p^2(S_{43}L_{43})p_2S_3L_3|\}p^3S_3L_3),
\end{aligned} \qquad \text{(A2.17)}$$

$$L_{43} = 1, \quad S_{43} = 1, \quad L_3 = L_{432} = L'_{432} = 0, \quad M_{l432} = M''_{l432} = 0,$$
$$S_{432} = S_3 = \tfrac{3}{2}, \quad S = 1, \quad L = 1.$$

In the display of Eq. A2.17 the values of the parameters found up to this point of the calculation have been inserted. Since the p^3-term is ${}^4S$, it is clear from Table A2.2 that each of the last two fpcs connecting $p^2({}^3P)$ and $p^3({}^4S)$ each equals unity, and the remaining fpc is a factor of the sum, a salient feature of this result. If we remove the interaction, integrate Eq. A2.17 over $d\mathbf{x}_2$, and insert the values of the $(\text{VCC})^2$ that are all equal to $\frac{1}{3}$ in this case, the sums over $(S_{43}L_{43})$, m_{l2}, and M''_{43} are easily effected, and the right member of Eq. A2.10 is obtained as it should be.

In this example the prior, post, and distorted wave Born amplitudes are each proportional to the fpc connecting the residual ion and the atom from

TABLE A2.5 TERM VALUES OF THE ATOM OR ION FOR WHICH THE REACTION IS PROPORTIONAL TO THE SQUARE OF THE FPC.

Atom + H^+	Ion + H(*ns*)
$p^6(^1S)$	$p^5(^2P)$
$p^4(^1S)$	$p^3(^2P)$
$p^4(^3P)$	$p^3(^4S)$
$p^3(^4S)$	$p^2(^3P)$
$p^2(^3P_1\, ^1D,\, ^1S)$	$p(^2P)$

which this ion was formed by electronic capture. For those other cases characterized by only one term of the ion with a given value of the multiplicity, we can express the result similarly to the one given by Eq. A2.17; in other cases the corresponding expressions are more complicated and are not written.

It is emphasized that only the $PP\bar{P}$ matrix element prevents the direct proportionality of the total matrix element to the fpc in these other cases of *p*-orbital capture. Although the preceding analysis was applied to capture from a $(2p)^n$ outermost subshell, it is applicable to *p*-orbital capture from any $(mp)^n$ subshell. However, the reader is reminded that the validity of these results depends upon (1) the passive momentum-change vectors being neglected, an excellent approximation, and (2) the assumed orthogonality of the one-electron orbitals of different subshells. The information on fpcs is now used to construct a table of reactions for *p*-orbital capture; each of the cross sections is proportional to the square of the fpc.

The complication of the differential cross sections depending upon the orientation of the atomic target can be traced to the $PP\bar{P}$ matrix element in the case of *p*-orbital capture. Examples of such capturing processes are given in Section 3.5, and the averaging process of Eq. 45*b* is used to eliminate the dependence upon the orientation. It should be emphasized, nevertheless, that this mixing effect is the product of a particular approximating procedure, and it is not known whether this dependence of the cross section upon orientation would occur if a more refined approximation were used.

APPENDIX 3

FIRST AND SECOND BORN APPROXIMATIONS: SUPPLEMENTARY NOTES

A short supplement to Sections 3.5 and 4.4 on the first Born approximation is presented first. The problem of interest is the variation of the prior, or post, cross section with high impact energies for the case of exact resonance. According to Eq. 55*g* of Section 4.4, the cross section for charge transfer between protons and atomic hydrogen in the 1*s*-state for sufficiently large E_i is given by

$$Q = \frac{1}{12}\left(\frac{m}{M}\right)^2 \frac{\pi a_0^2}{E_i^3} = \frac{16}{3}\left(\frac{m}{M}\right)^2\left(\frac{v_0}{v_i}\right)^6 \pi a_0^2, \qquad E_i = E_i(100\text{ keV}), \quad \text{(A3.1}a\text{)}$$

if the protons are treated as distinguishable; this cross section originates from the angular region near $\theta = \pi$. Since the associated classical cross section of Eq. 65*p* for capture into all final states of the classical hydrogen atom equals 1.25 Q of Eq. A3.1*a*, it is pertinent to inquire whether or not there are contributions from capture into excited states provided by the corresponding quantum mechanical approximation. The result of the subsequent demonstration suggests that the total E_i^{-3} dependence to the cross section is contained in Eq. A3.1*a*. Consistent with the other chapters, protons are considered as distinguishable particles throughout this appendix.

It has been established from earlier research that the E_i^{-3}-dependence of the cross section must originate from the proton-proton interaction,[9] and, accordingly, the point of departure for the calculation is the corresponding matrix element,

$$I(\beta, \theta) = \frac{\mu a_0}{2\pi m}\int \frac{d\mathbf{x}_i\, d\mathbf{x}_f}{|\mathbf{x}_i - \mathbf{x}_f|} \frac{\exp\left[-\beta x_f - x_i + i(\mathbf{A}_f \cdot \mathbf{x}_f + \mathbf{A}_i \cdot \mathbf{x})\right]}{\pi}, \qquad \mu = \frac{M}{2}. \quad \text{(A3.1}b\text{)}$$

The parameter β will be used to generate final *ns*-states of hydrogen, the first processes investigated; moreover, the normalizing factors associated with these *ns*-states are discarded since they are unnecessary for the present application. It is calculationally convenient to express Eq. A3.1*b* by the Feynman integral,[10]

$$I(\beta, \theta) = \frac{8Ma_0}{m}\beta\int_0^1 dxx(1-x)\left[\frac{2}{U^3V^{1/2}} + \frac{1}{U^2V^{3/2}} + \frac{3}{4UV^{5/2}}\right],$$

$$U = (\beta^2 + A_f^2)(1-x) + x(1 + A_i^2),$$
$$V = \beta^2(1-x) + x + R^2x(1-x),$$

$$R^2 = |\mathbf{A}_i + \mathbf{A}_f|^2 = 2E_i\left(1 - \frac{m\Delta\epsilon}{4ME_i}\right)(1 + \cos\theta) - \left(\frac{m\Delta\epsilon}{4M}\right)^2\frac{\cos\theta}{E_i},$$

$$A_f^2 \approx A_i^2, \qquad \Delta\epsilon = \left(1 - \frac{1}{n^2}\right), \tag{A3.1c}$$

$$R^2(\theta = \pi) = \left(\frac{m\Delta\epsilon}{4M}\right)^2\frac{1}{E_i}, \qquad A_i^2(\pi) = \left(\frac{2M}{m}\right)^2 E_i.$$

Here the impact energy is again expressed in units of 100 keV, and values of A_i^2, A_f^2, and R^2 are given for the center-of-mass scattering angle equal to π; note that $R^2(\pi)$ does not vanish except for $n = 1$, although it is very small for moderately large values of E_i, and of course $A_i^2(\pi)$ is contrastingly large for the same values of E_i. Although it is known[11] that the term $(UV^{5/2})^{-1}$ of this integrand is dominant for large E_i and small θ for *ns*-capture, it dominates only for $n = 1$ in the neighborhood of $\theta = \pi$. For *ns*-capture with $n \geq 2$, however, the second term of the integrand of Eq. A3.1*c* contributes nearly equally to the last term just mentioned. This is not obvious, and this point is elucidated after the contribution from the last term is evaluated for $n = 2$.

An examination of the structure of U in Eq. A3.1*c* reveals that the terms containing x as a factor can be discarded in this limit of large E_i, a conjecture supported by an exact integration. This approximation simplifies the integration considerably. (As noted in the discussion of Eq. 54*f* in Section 4.4, $\beta^2 + A_f^2 = 1 + A_i^2$ with negligible error.)

$$I(\beta, R^2) = \frac{L}{\beta^2 + A_f^2}\frac{(1+\beta)}{q^2},$$

$$L = \frac{8Ma_0}{m}, \qquad q = R^2 + (1+\beta)^2, \tag{A3.1d}$$

$$\left[\left(2 + \frac{\partial}{\partial\beta}\right)I\right]_{\beta=1/2} = I_3(2s) = \left[\left(\frac{L}{A_f^2 + \beta^2}\right)\frac{4R^2}{q^3}\right]_{\beta=1/2}.$$

In the last line of this equation the operator that generates the 2*s*-state from $e^{-\beta r}$ is used (normalizing factor omitted), and the term containing $(\beta^2 + A_f^2)^{-2}$

is discarded, as will be done henceforth, since it is dominated by the retained term. (This interchanging of the order of integration and parametric differentiation is valid as it may be corroborated by doing the calculation exactly.) Next, a good approximation to I_3 is deduced.

It is evident from a study of the structure of R^2 that sufficiently near $\theta = \pi$ the numerator of $I_3(2s)$ varies as E_i^{-1}, whereas q^3 is constant; consequently, $I_3(2s)$ varies as E_i^{-2} in this smaller angular region. As θ decreases toward zero, however, the numerator and q^3 vary as E_i and E_i^3, respectively, so that $I_3(2s)$ varies as E_i^{-3}. Therefore contributions to the cross section from the angular region not near π can be discarded as relatively negligible, and the numerator of $I_3(2s)$ and A_f^2 may be approximated by their values at $\theta = \pi$. We return to the contribution from $(U^2V^{3/2})^{-1}$ and denote the corresponding integral by $I_2(2s)$. An approximate calculation of this term yields the result

$$I_2(2s) = \frac{LN}{qA_f^4(\pi)},$$

in which N is a constant; thus the resultant amplitude is

$$I(2s) = \frac{L}{A_f^2(\pi)q}\left[\frac{3R^2(\pi)}{q^2} + \frac{N}{A_f^2(\pi)}\right]. \tag{A3.1e}$$

From Eq. A3.1*c* it is clear that the last two terms in the square brackets of the integrand both vary as E_i^{-1} near $\theta = \pi$, and the change of variable

$$x = R^2, \qquad d\theta \sin\theta \approx -\frac{dx}{2E} \tag{A3.1f}$$

is used to facilitate the integration of $|I(2s)|^2$, the resulting cross section varying as E_i^{-5}. Thus it has been demonstrated that $Q(2s)$ is negligible in comparison to $Q(1s)$. Capture into H(ns), $n \geq 2$, is treated next.

Only $I_3(ns)$ is evaluated since the evaluation of $I_2(ns)$ is similarly accomplished; moreover, $I_2(ns)$ is related to $I_3(ns)$ roughly the same as $I_2(2s)$ is related to $I_3(2s)$. First, it is observed that $I_3(\beta, R^2) \equiv I_3$ can be displayed as a rational function, a quotient of polynomials in either β or R^2, in the domains $0 \leq R^2$, $0 \leq \beta < 1$, and the denominator of I_3 does not vanish in these ranges of R^2 and β. Consequently, I_3 is continuous in both R^2 and β, and I_3 is differentiable with respect to β to any order; therefore it is meaningful to operate on I_3 with

$$L^{(1)}_{(n-1)}\left(-\frac{2}{n}\frac{\partial}{\partial\beta}\right) = \sum_{s=1}^{n} \frac{n!}{(n-s)!\,s!}\,\frac{[(2/n)(\partial/\partial\beta)]^{s-1}}{(s-1)!}, \tag{A3.1g}$$

which operator generates the ns-radial wave functions from the function, $e^{-\beta r}$, normalizing factors again being omitted.[12] By reason of continuity with respect to R^2, however, we can operate on $I_3(\beta, 0) = (1 + \beta)^{-3}$ and predict the important dependence upon R^2 since we know that $L^{(1)}_{(n-1)}I_3(\beta, R^2)$ is still a quotient of polynomials in either R^2 or β. We operate on $I_3(\beta, 0)$, again omitting this operation on the factor $(\beta^2 + A_f^2)^{-1}$, to get

$$L^{(1)}_{(n-1)}\left(-\frac{2}{n}\frac{\partial}{\partial\beta}\right)(1+\beta)^{-3}]_{\beta=n^{-1}} = \frac{n^4}{(1+n)^3}\sum_{s=1}^{n}\frac{(1+s)(n-1)!}{2(n-s)!(s-1)!}\left(-\frac{2}{1+n}\right)^{s-1}, \tag{A3.1h}$$

and the value of several of the partial sums are recorded. Examples in which Σ_k denotes the sum of the first k terms are

$$\Sigma_2 = \frac{-2(n-2)}{(n+1)}, \qquad \Sigma_3 = \frac{2^2(n-2)(n-3)}{2!\,(n+1)^2},$$
$$\Sigma_4 = \frac{-2^3(n-2)(n-3)(n-4)}{3!\,(n+1)^3}, \tag{A3.1i}$$
$$\Sigma_k = \frac{(-2)^{k-1}(n-2)(n-3)\cdots(n-k)}{(k-1)!\,(n+1)^{k-1}}.$$

If the $(k+1)$st term is added to Σ_k, the result obtained is identical to that represented by $\Sigma_{(k+1)}$ in Eq. A3.1i; therefore induction can be applied to get the sum

$$\Sigma_n = (-2)^{n-1}\frac{(n-2)(n-3)\cdots(n-n+1)(n-n)}{(n-1)!\,(n+1)^{n-1}} = 0. \tag{A3.1j}$$

Continuity with respect to R^2 and the property of factorization of polynomials are now invoked to prove that $I_3(ns)$ is equal to

$$\frac{L}{n^{-2}+A_f^2}L^{(1)}_{(n-1)}\left(-\frac{2}{n}\frac{\partial}{\partial\beta}\right)I_3]_{\beta=n^{-1}}$$
$$= L(n^{-2}+A_f^2)^{-1}R^{2m}P(R^2; n^{-1})q(R^2; n^{-1})^{-(n+1)}. \tag{A3.1k}$$

In this formula P is a polynomial, and it may contain a factor that cancels some power of q with $q(\beta = n^{-1})$ being defined in Eq. A3.1d. However, the value of $m = 1, 2, \ldots,$ in the exponent of R is not known, but when m exceeds unity a multiple zero occurs at $R^2 = 0$. A test of a multiple zero is the vanishing of

$$L^{(1)}_{(n-1)}\left(-\frac{2}{n}\frac{\partial}{\partial\beta}\right)\frac{\partial}{\partial R^2}I(\beta, R^2)]_{\beta=1/n};$$

accordingly, these functions were calculated and found not to vanish. Thus the value of m is unity in Eq. A3.1k. Therefore capture into H(ns) is dominated by $n = 1$ in Eq. A3.1a.

Next the cases $l \geq 1$ are studied, and for this purpose the operators

$$\mathrm{O}_m = \frac{\partial}{\partial A_{fz}}, \qquad \frac{1}{\sqrt{2}}\left(\frac{\partial}{\partial A_{fx}} \pm i\frac{\partial}{\partial A_{fy}}\right) \tag{A3.1l}$$

are applied to $I(\beta, R^2)$ of Eq. A3.1*d*. Only the term resulting from the variation of q^{-2} is retained, and the corresponding differential cross section is proportional to

$$\sum_m |\mathrm{O}_m I(\beta, R^2)|^2 = \frac{16L^2}{(\beta^2 + A_f^{\,2})^2} \frac{(A_{fx} + A_{ix})^2 + (A_{fy} + A_{iy})^2 + (A_{fz} + A_{iz})^2}{q^6}. \tag{A3.1m}$$

The purpose of displaying R^2 in terms of its rectangular components is to emphasize the origin of this factor. The rest of $I(\beta, \theta)$ of Eq. A3.1*c* does not contribute near $\theta = \pi$. Clearly Eq. A3.1*m* produces a cross section that varies as E_i^{-4} and is thus negligible. Cross sections for capture into other states for $l \geq 2$ are similarly dominated by capture into H(1*s*), and this completes the demonstration.

In summary, the quantum mechanical cross section with the E_i^{-3}-dependence originates entirely from capture into H(1*s*) and thus represents capture into all H-states similar to the corresponding classical result of Eq. 65*p*, which also depends upon capture into all final states of the classical hydrogen atom. In analogy to this result we recall that the Rutherford differential cross section is equal to the corresponding cross section derived from the first Born approximation. The next topic of this appendix relates to the approximations used to evaluate the constituent parts of the second Born amplitude, which originate from G_{fa} of Eq. 50*g* in Section 4.2.

The purpose of this supplementary analysis is to establish the credibility of the asserted asymptotic variations that are described between Eqs. 51*m* and 51*n*. For this purpose the integral of Eq. 51*g* is written in a form which is more suitable to the present analysis:

$$\begin{aligned}
I(\Gamma) &= \int \frac{d\mathbf{U}}{K^2 + i\delta - U^2} |\mathbf{U} - \mathbf{K}_f|^{-2}[\Gamma^2 + |\mathbf{U} - \mathbf{V}|^2]^{-2} \\
&= \pi^2 \frac{\partial}{\partial R} \int_0^1 \frac{dx}{D}\left[2iK + \frac{Rx + S - K^2}{(Rx + S - Q^2)^{1/2}}\right] = I_1(\Gamma) + I_2(\Gamma), \\
D &= (Rx + K^2 + S)^2 - 4K^2Q^2, \qquad R = \Gamma^2, \\
K_f^{\,2} - K^2 &= \frac{\mu a}{m}\left(1 - \frac{1}{n^2}\right), \\
K_f - K &= \frac{c(1 - 1/n^2)}{\sqrt{E_i}}, \qquad S = K_f^{\,2}(1 - x) + xV^2, \\
\mathbf{Q} &= \mathbf{K}_f(1 - x) + x\mathbf{V}.
\end{aligned} \tag{A3.2a}$$

In this equation K is defined differently from the K of Eq. 51g ($n = 1$); it depends upon E_i through the relation of K_f to E_i in Eq. 50f of Section 4.2, and although $(K_f^2 - K^2)$ is independent of impact energy E_i, $(K_f - K_i)$ depends upon E_i; this latter relation is obtained by neglecting $\Delta\varepsilon$ relative to E_i. The value of c is not needed for this study. The variation of the contribution from ns-intermediate states is analyzed first, but the case $a = b$ (exact resonance) is treated separately.

The functions corresponding to Eqs. 51b and 51c are designated

$$G(ns) = \frac{256\mu}{\pi m} N^2(ns) \int \frac{d\mathbf{U}L^{(1)}_{(n-1)}[-(2/n)(\partial/\partial\beta)](\lambda + \beta)}{[K^2 + i\delta - U^2]|\mathbf{U} - \mathbf{K}_f|^2}$$
$$\times \left[\frac{1}{(1-a)^4(\Gamma_1{}^2 + |\mathbf{U} - \mathbf{K}_f|^2)^2} - \frac{1}{a^4(\Gamma_2{}^2 + |\mathbf{U} - \mathbf{K}_f|^2)^2}\right] \quad \text{(A3.2}b\text{)}$$
$$\times L^{(1)}_{(n-1)}\left(\frac{2}{n}\frac{\partial}{\partial\beta'}\right) \frac{\lambda'\beta'}{[(\beta')^2 + |a\mathbf{U} - \mathbf{K}_i|^2][(\lambda')^2 + |\mathbf{U} - bK_i|^2]^2},$$
$$\Gamma_1 = (\lambda + \beta)(1 - a)^{-1}, \qquad \Gamma_2 = (\lambda + \beta)a^{-1};$$

$$F(ns) = \frac{256\mu N^2(ns)}{\pi m} \int \frac{d\mathbf{U}L^{(1)}_{(n-1)}[-(2/n)(\partial/\partial\beta)](\lambda + \beta)}{(K^2 + i\delta - U^2)|\mathbf{U} - \mathbf{K}_f|^2}$$
$$\times \left[\frac{1}{(1-a)^4(\Gamma_1{}^2 + |\mathbf{U} - \mathbf{K}_f|^2)^2} - \frac{1}{a^4(\Gamma_2{}^2 + |\mathbf{U} - \mathbf{K}_f|^2)^2}\right]$$
$$\times L^{(1)}_{(n-1)}\left(-\frac{2}{n}\frac{\partial}{\partial\beta'}\right) \frac{\lambda'\beta'}{\pi^2} \quad \text{(A3.2}c\text{)}$$
$$\times \int \frac{d\mathbf{V}}{(V^2 + \varepsilon^2)[(\lambda')^2 + |\mathbf{V} + \mathbf{U} - b\mathbf{K}_i|^2]^2[(\beta')^2 + |\mathbf{V} + a\mathbf{U} - \mathbf{K}_i|^2]^2}$$

The operators $L^{(1)}_{(n-1)}$ are defined by Eq. A3.1g and are evaluated at $\beta = n^{-1}$, $\beta' = n^{-1}$. (The variables of integration, $d\mathbf{V}$, are *not* related to V^2 of Eq. A3.2a!) First the case for $n = 1$ is reviewed; in this instance $L^{(1)}_{(n-1)}$ is the identity operator, and the normalizing factor, $N(1s)$, is $\pi^{-1/2}$. According to Section 4.2 the dominant part of $G(1s)$ is obtained by equating $\mathbf{U}$ to $\mathbf{K}_f$ in the last two factors of Eq. A3.2b and evaluating the difference of the remaining two integrals. The dominant part of this result, displayed in Eq. 51m, varies as $E_i^{-1/2}$, and the associated partial amplitude varies as $E_i^{-3.5}$. However, if the peak at $\mathbf{U} = b\mathbf{K}_i$ is used ($b \neq a$ or $b = a$), Eq. 51d is applicable to the entire integrand, and an $E_i{}^{-5}$-variation ensues. In order to evaluate $F(1s)$, $\mathbf{U}$ is equated to $\mathbf{K}_f$ in the $d\mathbf{V}$-integrand, and Eq. 51d is used for the

$d\mathbf{V}$-integration. This factor of $F(1s)$ yields an E_i^{-3}-dependence, and the preceding result is used for the remaining two integrals to obtain the $E^{-3.5}$-dependence again. Alternatively, the peak at $\mathbf{U} = -(1 - a)^{-1}(1 - b)\mathbf{K}_i$ is used in the other two peaking factors, and the remaining integral is evaluated (see Eq. 51k). This part of the result displayed in Eq. 51m varies as E_i^{-5} for $b \neq a$, and as $E_i^{-4.5}$ for $b = a$. We see that the dominant part involves the integral $I(\Gamma)$ in each case, and the purpose of this short review has been to emphasize this fact. Thus in order to calculate contributions from the ns-states ($n \geq 2$), the effect of the operator $L^{(1)}_{(n-1)}$ on $I(\Gamma)$ must first be determined, and this feature is analyzed next.

Again $\mathbf{U}$ is equated to $\mathbf{K}_f$ in the last two factors of $G(ns)$ and in the $d\mathbf{V}$-integrand of $F(ns)$; in the latter integral the integration over $d\mathbf{V}$ is effected using the peaking approximation. According to the definition of

$$L^{(1)}_{(n-1)}\left(-\frac{2}{n}\frac{\partial}{\partial \beta'}\right)$$

in Eq. A3.1g the sum of the terms $s = 1$ and $s = 2$ is

$$n + (n - 1)\frac{\partial}{\partial \beta'};$$

consequently, this term does not alter the dependence on energy of the integrated factors if $\partial/\partial\beta'$ is applied to the numerators only! [The terms $s > 2$ are dominated by these terms since $(\partial/\partial\beta')$ acts on the denominators in these cases.] These terms that belong to $G(ns)$ and $F(ns)$ are, respectively,

$$\frac{\lambda' n}{(n^{-2} + A_f^2)[(\lambda')^2 + A_i^2]^2} \tag{A3.2d}$$

and

$$\frac{n}{A_i^2[n^{-2} + |(1 - a)\mathbf{K}_f + (1 - b)\mathbf{K}_i|^2]^2} + \frac{\lambda' n}{A_f^2[(\lambda')^2 + |(1 - a)\mathbf{K}_f + (1 - b)\mathbf{K}_i|^2]^2}. \tag{A3.2e}$$

Both these factors possess the asymptotic E_i^{-3}-variation. The parameters of the integral of Eq. A3.2a (for ns-states) are

$$R = \Gamma_1^2, \Gamma_2^2, \qquad \mathbf{V} = \mathbf{K}_f, \qquad \mathbf{Q} = \mathbf{K}_f, \qquad S = K_f^2,$$

$$D = (R{+}{+})(R{+}{-}), \qquad R{+}{+} = Rx + (K_f + K)^2, \tag{A3.2f}$$

$$R{+}{-} = Rx + (K_f - K)^2.$$

Since $(K_f + K)^2$ is proportional to E_i, the dominant part of the integral [still designated $I(\Gamma)$] is

$$I_1(\Gamma) = \frac{2i\pi^2 K}{(K_f + K)^2}\frac{\partial}{\partial R}\int_0^1 \frac{dx}{(R+-)}, \qquad \Gamma = \frac{\lambda+\beta}{1-a}, \quad \Gamma_2 = \frac{\lambda+\beta}{a};$$

$$I_2(\Gamma) = \frac{\pi^2}{(K_f + K)^2}\frac{\partial}{\partial R}\frac{1}{\sqrt{R}}\int_0^1 \frac{dx(Rx + K_f^2 - K^2)}{x^{1/2}(R+-)}. \tag{A3.2g}$$

These integrals are easily evaluated to give

$$I_1(\Gamma) = \frac{2\pi^2 iK}{(K_f + K)^2}\left\{-\frac{1}{R^2}\log\frac{R + (K_f - K)^2}{(K_f - K)^2} + \frac{1}{R[R + (K_f - K)^2]}\right\},$$

$$I_2(\Gamma) = \frac{2\pi^2}{(K_f + K)^2}\left\{\frac{-1}{2R^{3/2}} + \frac{K(K_f - K)}{R^{3/2}[R + (K_f - K)^2]} - \frac{2K}{R^2}\tan^{-1}\left(\frac{\sqrt{R}}{K_f - K}\right)\right\}. \tag{A3.2h}$$

An expansion in a Taylor's series about $(K_f - K) = 0$ is used next, and the uncanceled remaining terms of

$$(\lambda + \beta)[(1 - a)^{-4}I(\Gamma_1) - a^{-4}I(\Gamma_2)]$$

are

$$= \frac{2\pi^2 iK}{(K_f + K)^2}\frac{1}{(\lambda+\beta)^3}\left[\log\left(\frac{1-a}{a}\right)^2 - \frac{2(K_f - K)^2}{\Gamma_1^2} + \frac{2(K_f - K)^2}{\Gamma_2^2}\right]$$
$$+ \frac{2\pi^2}{(K_f + K)^2}\left\{\frac{1}{2(\lambda+\beta)^2}\left[\frac{-1}{(1-a)} + \frac{1}{a}\right]\right.$$
$$\left. + \frac{K(K_f - K)}{(\lambda+\beta)^4}[3(1-a) - 3a]\right\} + \cdots, \qquad \lambda = 1. \tag{A3.2i}$$

Reference is made to Eq. A3.1i to show that the operator

$$L^{(1)}_{(n-1)}\left(-\frac{2}{n}\frac{\partial}{\partial \beta}\right)\Bigg]_{\beta = n-1}$$

annihilates the term $\log[(1 - a)/a]^2$ of Eq. A3.2i; consequently, the dominant part of this expression varies as E_i^{-1}, and with the use of Eqs. A3.2d and A3.2e it is clear that the asymptotic contributions from all ns-states, $n \geq 2$, vary as E_i^{-4}.

We next investigate the case of exact resonance, $a = b$, for scattering angle $\theta = \pi$. From the results just obtained the variation of $G(ns)$ is unaltered at $\theta = \pi$; however, the dependence of Eq. A3.2e upon E_i changes from E_i^{-3} to E_i^{-1} at $\theta = \pi$ ($b = a$) with the result that this part of $F(ns)$ varies as E_i^{-2}. This alteration is due to the vanishing of

$$R_1 = (1 - a)(\mathbf{K}_i + \mathbf{K}_f) = \mathbf{A}_i + \mathbf{A}_f$$

in the backward direction. However, the dominant dependence of $F(ns)$ is obtained by first peaking in **V**-space and using the resulting value of **U** at the peak in **U**-space in all factors of the integrand except in the peaking function and $(K^2 + i\delta - U^2)^{-1}$. The remaining integrals are evaluated to get

$$F(ns) = -\frac{256\mu\pi}{m} N^2(ns) L^{(1)}_{(n-1)}\left(-\frac{2}{n}\frac{\partial}{\partial\beta'}\right)(1 - a^2)^{-2} K_i^{-2}$$

$$\times \left\{\frac{1}{\beta'[\beta'/(1 - a) + 2iK + (1 - a)(K_i^2 - K^2)/\beta']}\right.$$

$$\left. + \frac{1}{\lambda'[\lambda'/(1 - a) + 2iK + (1 - a)/\lambda'(K_i^2 - K^2)]}\right\} L^{(1)}_{(n-1)}\left(-\frac{2}{n}\frac{\partial}{\partial\beta}\right)$$

$$\times \frac{2(\lambda + \beta)\{(\lambda + \beta)^2[a^2 - (1 - a)^2] + R_1^2[a^4(1 - a)^{-2} - (1 - a)^2]\}}{[(\lambda + \beta)^2 + R_1^2]^2[(\lambda + \beta)^2 + a^2(1 - a)^{-2}R_1^2]^2},$$

$$K_i^2 - K^2 = K_f^2 - K^2 = \frac{\mu}{a} m(1 - n^{-2}). \qquad \text{(A3.2}j\text{)}$$

From our previous study of the operators $L^{(1)}_{(n-1)}$ we see that $F(ns)$ varies as $E_i^{-1.5}$ at $\theta = \pi$ since $R_1^2(\pi) = 0$. Nevertheless, if $|\mathbf{U} - \mathbf{K}_f|^{-2}$ is included in the integral, two integrals (one containing λ' and the other, β') which are defined by Eq. A3.2a occur and the integral corresponding to $I_1(\Gamma)$ of Eq. A3.2g varies asymptotically as $E^{-1/2} \log E$; therefore this part of $F(ns)$ varies as $E^{-1.5} \log E$, but its value is zero! The last result vanishes since the terms containing Γ_1 and Γ_2 cancel identically at $\theta = \pi$. These results are summarized: if the procedure adopted for nonresonance is used, the variation is E_i^{-2}, but if $|\mathbf{U} - \mathbf{K}_f|^{-2}$ is canceled and the associated factors are peaked at $\mathbf{U} = -\mathbf{K}_i$, the variation is $E_i^{-1.5}$; if $|\mathbf{U} - \mathbf{K}_f|^{-2}$ is included in the integrals, the result vanishes for $n \geq 2$, but an indeterminancy occurs for $n = 1$ as explained following Eq. 51m in Section 4.2. Although the other two results are well defined for all n, these mathematical vagaries should be noted. This concludes the analysis of the contributions from ns-intermediate states, and the corresponding calculations for nl-states, $l \geq 1$ are effected next.

The nl-states are derived from approximately parameterized $1s$-states, and for this purpose the exponentials $\exp[-i\mathbf{K}_i \cdot \mathbf{x}_f']$ and $\exp[ia\mathbf{K}_f \cdot \mathbf{x}_f]$ of Eq. 51a in Section 4.2 are replaced by $\exp[-i\mathbf{B} \cdot \mathbf{x}_f']$ and $\exp[i\mathbf{A} \cdot \mathbf{x}_f]$, respectively. With the use of the appropriate linear combinations of products of the operators,

$$\frac{\partial}{\partial A_z}, \qquad \frac{\partial}{\partial A_x} \pm i\frac{\partial}{\partial A_y} = \frac{\partial}{\partial A_\pm}, \qquad \frac{\partial}{\partial B_z}, \qquad \frac{\partial}{\partial B_\pm},$$

all nl-states can be constructed. The matrix elements corresponding to $F(1s)$ and $G(1s)$ are calculated, and the appropriate parametric differentiations are effected, after which operation $\mathbf{A}$ and $\mathbf{B}$ are equated to $a\mathbf{K}_f$ and $\mathbf{K}_i$, respectively, these relations being assumed for each component of the associated vectors. These matrix elements, designated F and G, are

$$G = \frac{256\mu}{\pi m} N^2(nl) \int \frac{d\mathbf{U}}{(K^2 + i\delta - U^2)|\mathbf{U} - \mathbf{K}_f|^2} F(S_1, S_2)$$
$$\times \frac{\lambda'(\lambda + \beta)}{[(\beta')^2 + |\mathbf{B} - a\mathbf{U}|^2][(\lambda')^2 + |\mathbf{U} - b\mathbf{K}_i|^2]^2}, \qquad \lambda' = 1; \tag{A3.3a}$$

$$F(S_1, S_2) = (1 - a)^{-4}[\Gamma_1{}^2 + S_1{}^2]^{-2} - a^{-4}[\Gamma_2{}^2 + S_2{}^2]^{-2},$$

$$\mathbf{S}_1 = \mathbf{U} - \frac{\mathbf{K}_f - \mathbf{A}}{1 - a}, \qquad \mathbf{S}_2 = \mathbf{U} - \frac{\mathbf{A}}{a}.$$

$$F = \frac{256\mu N^2(nl)(\lambda + \beta)}{\pi m} \int \frac{d\mathbf{U}}{(K^2 + i\delta - U^2)|\mathbf{U} - \mathbf{K}_f|^2} F(S_1, S_2) I_V, \tag{A3.3b}$$

$$I_V = \frac{1}{\pi^2} \int \frac{dV}{V^2 + \varepsilon^2} \frac{\lambda'\beta'}{[(\beta')^2 + |\mathbf{V} + a\mathbf{U} - \mathbf{B}|^2]^2[(\lambda')^2 + |\mathbf{V} + \mathbf{U} - b\mathbf{K}_i|^2]^2}.$$

In the preceding two equations $N(nl)$ represents the normalizing factor appropriate to the intermediate state whose quantum numbers are nlm. There are $(2l + 1)$ of these states for each l-value $(|m_l| \leq l)$, and the simplest structure to reproduce are those belonging to $|m_l| = l$, the angular part being

$$(\sin\theta\, e^{\pm i\phi})^l.$$

However, only for the case of np-states has it been determined that the asymptotic variation with E_i is the same for each m_l-value; furthermore, on the basis of several exemplifying calculations, it is argued that the constituent amplitudes from states with $l \geq 2$ decrease with E_i more strongly than those of *np*-states. In case of the states with quantum number $m_l = 0$, the angular dependence of the ϕ_{ml0} can be represented by a sum of cosine functions $(\cos\theta)^l$, $(\cos\theta)^{l-2}, \ldots$, and the associated operators required to produce these functions are

$$\left(\frac{\partial}{\partial A_z}\right)^l, \quad \left(\frac{\partial}{\partial A_z}\right)^l \frac{\partial^2}{\partial \beta^2}, \ldots,$$

the operator in β being needed to produce the correct power of the radial variable. Thus the program of this analysis is to study the effect of the operators

$$\left(\frac{\partial}{\partial A_z}\frac{\partial}{\partial B_z}\right)^l,$$

which contribute to the state $m_l = 0$, and the operators

$$\left(\frac{\partial}{\partial A_\pm}\frac{\partial}{\partial B_\mp}\right)^l,$$

which generate the states $|m_l| = l$. Consequently, the result for $|m_l| = l$ is precise, but the answers for $m_l = 0$ are only suggestive. The constituent parts originating from G of Eq. A3.3*a* are calculated first.

We use the peak at $\mathbf{U} = b\mathbf{K}_i$ and effect the parametric differentiations to get some of the dominant terms:

$$\left(\frac{\partial}{\partial B_-}\frac{\partial}{\partial A_+}\right)^l G(\mathbf{A} = a\mathbf{K}_f, \mathbf{B} = \mathbf{K}_i) = \frac{256\pi\mu}{m} N^2(nl)(\lambda+\beta)$$
$$\times \frac{C(l)(\lambda+\beta)(1-ab)^l K_i^{\,l}}{(K^2 - b^2K_i^{\,2})A_i^{\,2}[\beta'^2 + (1-ab)^2K_i^{\,2}]^{l+1}}(A_i^{\,+})^l \quad \text{(A3.3}c\text{)}$$
$$\times \left(\frac{(1-a)^{-4}}{(\Gamma_1^{\,2} + A_i^{\,2})^{l+2}} - \frac{a^{-4}}{(\Gamma_2^{\,2} + A_i^{\,2})^{l+2}}\right).$$

These terms were obtained by differentiating denominators only, and a *weaker* decrease with E_i cannot be obtained with any other allowed differentiating scheme. For example, the operator for $m_l = 0$ could have been applied to the numerators $l/2$ or $(l+1)/2$ times, but this produces terms that decrease more strongly with E_i. Thus it is clear that these parts of the amplitudes vary as $E_i^{-(l+5)}$ for $m_l = \pm l$, and this variation also is correct for the calculated part of the contribution from $m_l = 0$.

Next we calculate the contribution to G of Eq. A3.3*a* obtained by peaking at $\mathbf{S}_2 = 0$ in the last two factors of G and use the associated integrals $I_1(\Gamma)$ and $I_2(\Gamma)$ for the rest of the constituent amplitude. $I_1(\Gamma)$ of Eq. A3.2*a* (see

Eq. A3.2g also) is differentiated first, and the derivatives needed are recorded in Eq. A3.3d:

$$\mathbf{V} = \frac{\mathbf{A}}{a} \to \mathbf{K}_f, \qquad R = \Gamma_2^2$$

$$S \to K_f^2, \qquad \mathbf{Q} \to \mathbf{K}_f, \qquad D \to (R++)(R+-),$$

$$\frac{\partial S}{\partial A_m} = \frac{2xA_m}{a^2} \to \frac{2xK_{fm}}{a},$$

$$\frac{\partial Q^2}{\partial A_m} = \frac{2x^2A_m}{a^2} + \frac{2x(1-x)K_{fm}}{a} \to \frac{2xK_{fm}}{a},$$

$$\frac{\partial D}{\partial A_m} = \frac{4(Rx + K^2 + S)A_m}{a^2} - \frac{8xK^2}{a} \qquad \text{(A3.3}d\text{)}$$

$$\times \left[\frac{xA_m}{a} + (1-x)K_{fm}\right] \to \frac{4xK_{fm}}{a}(Rx + K_f^2 - K^2),$$

$$\left(\frac{\partial}{\partial A_m}\right)^2 D = \frac{8x^2A_m^2}{a^4} + 4x\frac{(Rx + K^2 + S)}{a^2}\delta_{mo} - \frac{8x^2K^2}{a^2}\delta_{mo} \to \frac{8x^2K_{fm}^2}{a^2}$$

$$+ \left[4x\frac{(Rx + K^2 + K_f^2)}{a^2} - \frac{8x^2K^2}{a^2}\right]\delta_{mo},$$

$$\left(\frac{\partial}{\partial A_m}\right)^3 D = \frac{16x^2A_m}{a^4}\delta_{mo} + \frac{8x^2A_m}{a^4}\delta_{mo} \to \frac{24x^2K_{fm}}{a^3}\delta_{mo},$$

$$\left(\frac{\partial}{\partial A_m}\right)^4 D = \frac{24x^2}{a^4}\delta_{mo}, \qquad \delta_{mo} = \begin{pmatrix} 1, \ m = 0 \\ 0, \ m \neq 0 \end{pmatrix}.$$

It is observed that $I(\Gamma)$ is independent of $\mathbf{B}$, this parameter occurring only in one of the external factors, but the parameter $\mathbf{A}$ appears both in $I(\Gamma)$ and externally. Differentiation with respect to A_m in the external factor causes the associated asymptotic amplitude to decrease with E_i by $E_i^{-1/2}$ for each differentiating operation; consequently, we investigate the corresponding operations on $I_1(\Gamma)$ to determine how its variation with E_i is altered.

The case $m_l = 0$ is studied first. Since the dominant contribution is sought, only the integrals are studied since the variations of the external factors containing β are easily obtained. If l is an even integer $2n$, the weakest decrease with E_i is obtained from the term

$$\left[\frac{(-1)^n(n+1)!}{D^{n+1}}\right]\left[\left(\frac{\partial}{\partial A_m}\right)^2 D\right]^n$$

of the terms constituting

$$\left[\left(\frac{\partial}{\partial A_m}\right)^{2n} D^{-1}\right].$$

This part of the relevant integral is

$$2\pi^2 iK \frac{\partial}{\partial R}\int_0^1 dx \frac{(-1)^n(n+1)!}{D^{n+1}}\left[\left(\frac{\partial}{\partial A_m}\right)^2 D\right]^n$$

$$= \frac{2\pi^2 iK(n+1)!}{(K_f+K)^{2n+2}}\frac{\partial}{\partial R}\int_0^1 dx \frac{[(-4/a^2)x]^n(K_f^2+K^2)^n}{(R+-)^{n+1}}$$

$$= \frac{2\pi^2 iK(K_f^2+K^2)^n}{(K_f+K)^{2n+2}}\left(-\frac{4}{a^2}\right)^n(n+1)!\frac{\partial}{\partial R}\frac{1}{R^n}\int_0^1 \frac{dx}{(R+-)} \qquad \text{(A3.3}e\text{)}$$

$$= i2\pi^2(n+1)!\frac{(-4/a^2)^n K(K_f^2+K^2)^n}{(K_f+K)^{2n+2}}\frac{\partial}{\partial R}\frac{1}{R^{n+1}}$$

$$\times \log\frac{R+(K_f-K)^2}{(K_f-K)^2}, \qquad m=0.$$

In Eq. A3.3e the equal signs signify equality to the dominant part of the integral; then, corresponding to a given power of E_i in the numerator, the part that contains the lowest power of x is selected since this choice yields the weakest decrease with E_i for a given power, $(1+n)$ in $D^{-(1+n)}$, and a glance at a table of integrals corroborates this fact. The approximation

$$\frac{x^n}{D^{n+1}} = \frac{1}{(K_f+K)^{2n+2}R^n(R+-)}$$

provides the dominant term since R^{-n} is the first term of Taylor's expansion of $x^n(R+-)^{-n}$ about $R^{-1}(K_f-K)^2=0$. Thus $(\partial/\partial A_m)^l I_1(\Gamma)$ varies as $E_i^{-1/2}\log E_i$ for the part of the state $m_l=0$ originating from $(\cos\theta)^l$ if l is even, and the associated constituent amplitude varies as

$$E_i^{-(l+7)/2}\log E_i.$$

[The external peaked factors of G yield the variation $E_i^{-(l+6)/2}$, and it is obvious that operating on $I(\Gamma)$ instead of the external factor produces a weaker decrease with E_i.]

Next let l be the odd integer $2n+1$, and we find that the dominant part stems from

$$\left[\frac{(-1)^n(n+1)!}{D^n}\right]\left[\left(\frac{\partial}{\partial A_m}\right)^2 D\right]^n\frac{\partial}{\partial A_m}D^{-1}$$

with dominant part proportional to

$$\left[\frac{4x(K^2+K_f^2)}{Da^2}\right]^n\frac{4x(K_f^2-K^2)K_{fm}}{aD^2},$$

and this term leads to the variation

$$E_i^{-1}\log E_i.$$

Therefore the associated constituent amplitude for odd l varies as

$$E_i^{-(l+8)/2} \log E_i$$

for the part of the state $m_l = 0$ originating from $(\cos\theta)^l$. The amplitude for $m_l = l$ is calculated next since the case $m_l = -l$ can be inferred from this result. According to Eq. A3.3*d* for $l = 2n$, the term

$$\frac{(-1)^n(n+1)!}{D^{n+1}}\left[\left(\frac{\partial}{\partial A_m}\right)^2 D\right]^n$$

again provides the dominant term, which in this case is

$$\begin{aligned} i2\pi^2(n+1)!\left(\frac{-8}{a^2}\right)^n \frac{KK_{fm}{}^{2n}}{(K_f+K)^{2n+2}}\frac{\partial}{\partial R}\int_0^1 \frac{x^{2n}\,dx}{D^{n+1}} \\ = i2\pi^2(n+1)!\left(\frac{-8}{a^2}\right)^n \frac{KK_{fm}{}^{2n}}{(K_f+K)^{2n+2}}\frac{\partial}{\partial R}\int_0^1 \frac{x^{n-1}\,dx}{R^{n+1}}. \end{aligned} \qquad \text{(A3.3}f\text{)}$$

This term varies as $E_i^{-1/2}$, and the corresponding amplitude for $m_l = l$ varies as $E_i^{[-(l+7)/2]}$. A similar calculation for l equal to an odd integer $2n+1$ obtains the dominant variation from

$$\left[\frac{(-1)^n n!}{D^n}\right]\left[\left(\frac{\partial}{\partial A_m}\right)^2 D\right]^n \frac{\partial D^{-1}}{\partial A_m},$$

which produces an E_i^{-1}-dependence, and the associated amplitude for $m_l = l$ varies as $E_i^{-(l+8)/2}$.

Next we investigate the effect of the differentiating operations on $I_2(\Gamma_2)$ which is defined by Eq. A3.2*a* and A3.2*g*. Attention is first directed to the fact that the only nonvanishing derivative of $(Rx + S - Q^2)$ at $\mathbf{A} = a\mathbf{K}_f$ is

$$\left(\frac{\partial}{\partial A_m}\right)^2 (Rx + S - Q^2) = \frac{2x(1-x)}{a^2}\,\delta_{mo}$$

and the corresponding value of $(Rx + S - Q^2)$ is Rx. Attention is next confined to the operator $(\partial/\partial A_m)^l$ occurring in the state $m_l = 0$. Thus only even-ordered derivatives, including zero, of $[Rx + S - Q^2]^{-1/2}$ give a nonvanishing result; this derivative is

$$\frac{\partial^{2n}}{\partial A_m{}^{2n}}[Rx + S - Q^2]^{-1/2} = \frac{[2x(1-x)/a^2]^n}{(Rx)^{n+1/2}} = \frac{[2(1-x)/a^2]^n}{R^{n+1/2}x^{1/2}}, \qquad \text{(A3.3}g\text{)}$$

and it is evident that the x-dependence near $x = 0$ is the same for all n. As noted previously, this is the dependence that determines the dominant terms since this procedure is equivalent to a Taylor's expansion of the integrand about $E_i^{-1} = 0$. According to Eq. A3.2*h*, $I_2(\Gamma_2)$ varies as $E_i^{-1/2}$, and we must determine whether there are combinations of derivatives of $I_2(\Gamma_2)$ that dominate this variation. However, the derivatives in Eq. A3.3*g* do not alter this variation, so we determine the maximum growth with E_i for l odd, as we did for $I_1(\Gamma)$; this result is

$$\left(\frac{\partial}{\partial A_m}\right)^l I_2 = \pi^2 \frac{\partial}{\partial R}\int_0^1 \frac{dx[(\partial/\partial A_m)^2 D]^n(\partial/\partial A_m)(Rx + S - K^2)(n+1)!\,(-1)^n}{R^{1/2}x^{1/2}D^{n+1}}$$

$$= \frac{\pi^2(n+1)!}{(K_f + K)^{2n+2}}\frac{\partial}{\partial R}\int_0^1 dx\,\frac{x^n(-4[K^2 + K_f^{\,2}]/a^2)^n\cdot(2xK_{fm}/a)}{R^{1/2}x^{1/2}(R+-)^{n+1}}$$

$$= \frac{(2/a)(-4/a^2)^n(n+1)!\,K_{fm}(K^2 + K_f^{\,2})^n}{(K_f + K)^{2n+2}}\frac{\partial}{\partial R}\frac{2}{R^{n+3/2}}, \qquad l = 2n+1. \tag{A3.3h}$$

This equation is interpreted the same as Eq. A3.3*e*, and it is obvious that the odd derivatives vary as $E_i^{-1/2}$. If l is even, moreover, the variation is unaltered, as the following calculation confirms:

$$\left(\frac{\partial}{\partial A_m}\right)^l I_2 = \pi^2 \frac{\partial}{\partial R}\int_0^1 dx\,\frac{[(\partial/\partial A_m)^2 D]^n(-1)^n(n+1)!\,(K_f^{\,2} - K^2)}{R^{1/2}x^{1/2}D^{n+1}}$$

$$= \pi^2(K_f^{\,2} - K^2)\frac{(n+1)!\,(-4/a^2)^n(K_f + K^2)^n}{(K_f + K)^{2n+2}}\frac{\partial}{\partial R} \tag{A3.3i}$$

$$\times \frac{2}{R^{(n+1)}(K_f - K)}\tan^{-1}\left(\frac{R^{1/2}}{K_f - K}\right).$$

Since the leading term of the $\tan^{-1}$ function is $\pi/2$, we find, with the aid of Eq. A3.2*a*, that the variation is $E_i^{-1/2}$ again. For this part of the state, $m_l = 0$, the dominant contribution originates from I_2 if l is odd, but it stems from I_1 if l is even.

In the case of the states $m_l = \pm l$ we first recall that all derivatives of $(Rx + S - Q^2)$ vanish since $\delta_{mo} = 0$ here. Next, suppose that $l(=2n+1)$ is odd. The dominant term is easily obtained from the results leading to Eq. A3.3*f* and originates from

$$\frac{(-1)^n(n+1)!}{D^{n+1}R^{1/2}x^{1/2}}\left[\left(\frac{\partial}{\partial A_m}\right)^2 D\right]^n \frac{\partial}{\partial A_m}(Rx + S - K^2),$$

a result which integrates to produce an $E_i^{-1/2}$-variation. If l is an even integer $2n$, the preceding term with $(\partial/\partial A_m)(Rx + S - K^2)$ replaced by $(K_f^{\,2} - K^2)$ (see Eq. A3.2*a*) yields

$$\left(\frac{\partial}{\partial A_m}\right)^l I_2 = \frac{\pi^2(n+1)!\,(-8/a^2)^n K_{fm}^{\,2n}(K_f^{\,2} - K^2)}{(K_f + K)^{2n+2}}\frac{\partial}{\partial R}\int_0^1 \frac{x^{2n}\,dx}{R^{1/2}x^{1/2}(R+-)^{n+1}}. \tag{A3.3j}$$

Since $n \geq 1$, the leading term of the preceding integral is independent of E_i, and this term accordingly varies as E_i^{-1} for $m_l = l$ and l even. The results are summarized in Eq. A3.3*k*:

$$m_l = \pm l, \begin{bmatrix} l = 2n + 1, & E_i^{-(l+7)/2}(I_2) \\ l = 2n, & E_i^{-(l+7)/2}(I_1); \end{bmatrix}$$

$$\underline{(\cos\theta)^l \text{ term}} \tag{A3.3k}$$

$$m_l = 0, \begin{bmatrix} l = 2n + 1, & E_i^{-(l+7)/2}(I_2) \\ l = 2n, & E_i^{-(l+7)/2}\log E_i(I_1). \end{bmatrix}$$

The parenthetical enclosures denote the contributing integral. Particular mention is made of the *np*-states ($l = 1$) since there is only one cosine term in the state $m_l = 0$, and for these states, defined by $l = 1$, the asymptotic variation is the same as for the *ns*-states. If the same calculations are repeated for the integrals containing S^2, in Eq. A3.3*a*, the same variations are obtained; however, the signs are reversed for the odd-ordered derivatives. Consequently, the structure of Eqs. A3.3*a* and A3.3*h* and the corresponding term for $m_l = \pm l$ collectively reveal that cancellation of the dominant terms of

$$(1 - a)^{-4}\frac{\partial I}{\partial A_m}(\Gamma_1) - a^{-4}\frac{\partial I}{\partial A_m}(\Gamma_2)$$

would occur for $l = 1$ if there were no reversal of sign. This concludes the analysis of contributions from G whose variations with E_i do not change with scattering angle θ.

In the case of the function F of Eq. A3.3*b*, however, the variation with E_i changes with θ for exact resonance, $b = a$. Only this feature is reported since the contributions for θ not near π are the same as the one just obtained for G. First, I_V of Eq. A3.3*b* is evaluated at its peaks in **V**-space:

$$I_V = \frac{\lambda'}{|a\mathbf{U} - \mathbf{B}|^2[\lambda'^2 + |(1 - a)\mathbf{U} + \mathbf{B} - a\mathbf{K}_i|^2]^2} + \frac{\beta'}{|\mathbf{U} - a\mathbf{K}_i|^2[(\beta')^2 + |(1 - a)\mathbf{U} + \mathbf{B} - a\mathbf{K}_i|^2]^2} \tag{A3.3l}$$

Only the states corresponding to $m_l = l$ are calculated since this case exemplifies the variations of *nl*-states near $\theta = \pi$ for exact resonance. For this purpose the peaks of Eq. A3.3*l* are used in $F(S_1, S_2)$ of Eq. A3.3*b* to get the following two results:

$$F_1 = \frac{256\pi\mu}{m} N^2(nl)(\lambda + \beta)\lambda' \int \frac{d\mathbf{U}}{(K^2 + i\delta - U^2)|\mathbf{U} - \mathbf{K}_f|^2}$$

$$\times \frac{(1-a)^{-4}}{[\Gamma_1^2(\lambda') + |\mathbf{U} + (\mathbf{B} - a\mathbf{K}_i)/(1-a)|^2]^2} \frac{1}{|\mathbf{B} + a(\mathbf{B} - a\mathbf{K}_i)/(1-a)|^2}$$

$$\times \left\{ \frac{(1-a)^{-4}}{[\Gamma_1^2 + |(a\mathbf{K}_i - \mathbf{B} + \mathbf{A} - \mathbf{K}_f)/(1-a)|^2]^2} - \frac{a^{-4}}{[\Gamma_2^2 + |(\mathbf{B} - a\mathbf{K}_i)/(1-a) + \mathbf{A}/a|^2]^2} \right\};$$

$$F_2 = \frac{256\pi\mu}{m} N^2(nl)(\lambda + \beta)\beta' \int \frac{d\mathbf{U}}{(K^2 + i\delta - U^2)|\mathbf{U} - \mathbf{K}_f|^2} \tag{A3.3m}$$

$$\times \frac{(1-a)^{-4}}{[\Gamma^2(\beta') + |\mathbf{U} + (\mathbf{B} - a\mathbf{K}_i)/(1-a)|^2]^2} \frac{1}{|a\mathbf{K}_i + (\mathbf{B} - a\mathbf{K}_i)/(1-a)|^2}$$

$$\times \left\{ \frac{(1-a)^{-4}}{[\Gamma_1^2 + |(a\mathbf{K}_i - \mathbf{B} + \mathbf{A} - \mathbf{K}_f)/(1-a)|^2]^2} - \frac{a^{-4}}{[\Gamma_2^2 + |(\mathbf{B} - a\mathbf{K}_i)/(1-a) + \mathbf{A}/a|^2]^2} \right\},$$

$$\Gamma(\lambda') = \frac{\lambda'}{1-a}, \qquad \Gamma(\beta') = \frac{\beta'}{1-a}.$$

Next several derivatives are studied:

$$\frac{\partial}{\partial A_m} \left| \frac{a\mathbf{K}_i - \mathbf{B} + \mathbf{A} - \mathbf{K}_f}{1-a} \right|^2 = \frac{2}{(1-a)^2} [A_m - K_{fm} - B_m + aK_{im}],$$

$$\frac{\partial}{\partial A_m} \left| \frac{\mathbf{B} - a\mathbf{K}_i}{1-a} + \frac{\mathbf{A}}{a} \right|^2 = 2\left[\frac{A_m}{a^2} + \frac{B_m - aK_{im}}{a(1-a)} \right].$$

At $\mathbf{B} = \mathbf{K}_i$, $\mathbf{A} = a\mathbf{K}_f$, each of these derivatives vanishes at $\theta = \pi$; however,

$$\frac{\partial^2}{\partial B_{-m}\partial A_m} \left| \frac{a\mathbf{K}_i - \mathbf{B} + \mathbf{A} - \mathbf{K}_f}{1-a} \right|^2 = \frac{-4}{(1-a)^2},$$

$$\frac{\partial^2}{\partial B_{-m}\partial A_m} \left| \frac{\mathbf{B} - a\mathbf{K}_i}{1-a} + \frac{\mathbf{A}}{a} \right|^2 = \frac{4}{a(1-a)}.$$

If the operators $(\partial/\partial A_m \, \partial/\partial B_{-m})^l$ are applied only to the function $F(S_1, S_2)$ evaluated at

$$\mathbf{U} = -\frac{\mathbf{B} - a\mathbf{K}_i}{1-a}, \qquad \mathbf{A} = a\mathbf{K}_f, \qquad \mathbf{B} = \mathbf{K}_i,$$

the m_l-states vary as $E_i^{-1.5} \log E_i$ since the integrals occurring in Eq. A3.3*m* vary as $E_i^{-0.5} \log E_i$ according to Eq. A3.2*g*. All the dependence on E_i originates from the $I_1(\Gamma)$ and the peaked terms $|a\mathbf{U} - \mathbf{B}|^{-2}$ and $|\mathbf{U} - a\mathbf{K}_i|^{-2}$ of I_V; $F(S_1, S_2)$ contributes no power of E_i since its argument $(\mathbf{K}_i + \mathbf{K}_f)$ vanishes at $\theta = \pi$. Although this approximating technique should not be used if two of the peaks merge—as they do here at $\mathbf{U} = \mathbf{K}_f$, $\mathbf{U} = -\mathbf{K}_i(\mathbf{A} = a\mathbf{K}_f,\ \mathbf{B} = \mathbf{K}_i)$—this calculational procedure usually gives the correct variation with E_i notwithstanding the associated incorrect numerical factors.[9]

REFERENCES

1. G. Racah, *Phys. Rev.*, **62**, 438 (1942).
2. G. Racah, *Phys. Rev.*, **63**, 367 (1943).
3. A. de-Shalit and I. Talmi, *Nuclear Shell Theory*, Academic Press, New York (1963).
4. E. U. Condon and G. H. Shortley, *Theory of Atomic Spectra* (1935); reprint corr. ed., Cambridge University Press, New York (1953), Chapter 3.
5. *Ibid.*, p. 78, Eq. (7).
6. de-Shalit and Talmi, *op. cit.*, p. 514.
7. R. A. Mapleton, *Proc. Phys. Soc.* (*London*), **85**, 1109 (1965).
8. de-Shalit and Talmi, *op. cit.*, p. 519.
9. R. A. Mapleton, *Proc. Phys. Soc.* (*London*), **83**, 895 (1964).
10. R. A. Mapleton, "Classical and Quantal Calculations on Electron Capture," Air Force Cambridge Research Laboratories Report AFCRL 6-0351, Physical Research Papers, No. 328, 1967, App. F, Pt. I.
11. R. A. Mapleton, *Phys. Rev.*, **126**, 1477 (1962).
12. L. Pauling and E. B. Wilson, Jr., *Introduction to Quantum Mechanics*, McGraw-Hill, New York (1935), p. 131, Eq. (20.9).

AUTHOR INDEX

SUBJECT INDEX